Good Manufacturing Practices for Pharmaceuticals

Good Manufacturing Practices for Pharmaceuticals

Sixth Edition

edited by

Joseph D. Nally
Nallianco LLC
New Vernon, New Jersey, U.S.A.

informa
healthcare

New York   London

Informa Healthcare USA, Inc.
270 Madison Avenue
New York, NY 10016

No claim to original U.S. Government works
Printed in the United States of America on acid-free paper
10 9 8 7 6 5 4 3 2 1

International Standard Book Number-10: 0-8493-3972-3 (Hardcover)
International Standard Book Number-13: 978-0-8493-3972-1 (Hardcover)

Visit the Informa Web site at
www.informa.com

and the Informa Healthcare Web site at
www.informahealthcare.com

Preface

The world of Pharmaceutical CGMP compliance has certainly evolved since the 5th edition of this book was published about six years ago. We are now living in the post 9-11 era; national and global pharmaceutical business consolidations and expansions are continuing at a brisk pace; the FDA is changing its approach to regulating manufacturers; global harmonization of regulatory requirement and quality standards has been accelerated; availability and access to internet sources of information has exploded and counterfeiting and bioterrorism are part of the daily dialogue.

From a GMP impact perspective, the FDA changes are potentially significant. They have initiated a new framework for regulating manufacturing that now supports risk based regulation, quality-by-design, a quality systems approach, process analytical technology (PAT) application and similar initiates. It was recognized that historical regulatory approaches and initiatives were constraining manufacturing progress and the agency needed to change. FDA is showing signs of being committed to these new approaches. They have restructured the Office of Regulatory Affairs (ORA) to support risk based inspections. The offices of Compliance (OC) and Pharmaceutical Science (OPS) within CDER (Center for Drug Evaluation & Research) have been reorganized to handle quality-by-design initiatives. A quality-by-design pilot initiative for CMC (Chemistry, Manufacturing and Controls) submissions has resulted in a product approval. They have withdrawn guidance documents that conflict with its 21st Century Initiatives, adopted a number of the International Conference on Harmonization (ICH) standards and the FDA Quality Systems Guidance was just finalized and ahead of the ICH Q10 document.

These are all signs of progress and any efforts to make regulations and compliance more flexible by allowing for a variety of modern approaches and more science based evaluations is appreciated and should be embraced by a typically conservative pharmaceutical industry.

These changes do not come without controversy. The reality is that FDA has less financial resources to deal with all these new technologies and harmonization efforts. This has resulted in fewer inspections (total numbers have decreased since 2003) and some of this decrease is by-design and coincides with their 21st century approach with risk based inspections. Recent criticisms of FDA have come from a number of areas. For example Representative Henry A.Waxman questioned FDA's enforcement record stating concern over an agency that is in a state of "decline" over the past five years and that enforcement actions represented by Warning letters have declined by 50% from 2000 to 2005. Seizures also declined by 44%. Whether the reductions are a sign of enforcement weakness or the industry is getting better is yet to be seen. The truth is most likely somewhere in the middle.

An industry association recently filed a citizen's petition to the FDA requesting that the agency increase inspections of drug manufacturing facilities located outside the U.S.A. They asserted that nearly half of all drugs marketed in the United States are produced or manufactured in foreign facilities but the vast majority of FDA inspections are of domestic facilities. And with the FDA facing budget cuts, most foreign facilities are not likely to be inspected. It would appear that asking for more FDA inspections would not be an industry position but when it comes to competition, fair play and a level playing field for U.S. drug requirements—everyone has a vested interest.

The current state of affairs can best be summed up by a discussion I had with a colleague that recently returned from a FDA/Industry Conference. When I asked about the overall theme coming from the speakers and participants he confirmed what many of us had suspected.

There is more dialogue and science based regulations/approaches appear to be the future. The chances of being inspected are less but if non-compliance is found during an inspection, the penalties are going to be more severe. This is especially true for firms having a history of repeat observations or recidivist behaviors.

This brings us back to this book. The 5th and previous editions have been successful in providing a wide audience of professionals and educators with the fundamental knowledge of CGMP regulations and current industry application. This text is invaluable for anyone directly involved in pharmaceutical manufacturing and control operations. It is also essential for GMP and quality auditors (private industry and government), quality assurance and control professionals, contractors and suppliers to the industry, industry consultants, pharmacy and pharmaceutical science educators.

The 6th edition continues in that tradition. Building on the excellent foundation a team of industry experts was assembled to update the text by applying their knowledge and experience in specific subject matter areas. The contributing authors collectively, have hundreds of year's experience with most in industry and some in academic settings. We updated each chapter focusing on current practices, the priorities in the regulatory climate and included explanations or links to the relevant regulatory guidance documents that are readily available on the internet. In order to provide this value added new information, we re-focused on the basic GMPs and consolidated or eliminated some 5th edition chapters and appendices.

The basic information and structure of chapters one through twelve that cover the GMP regulations - Subparts A to K have been kept and expanded upon. While many of the new initiatives may change the approach, complying with the fundamental GMPs outlined in the first twelve chapters is the basis of GMP compliance. The remaining chapters provide valuable supplementary information in understanding and applying the regulations.

In addition to being updated with the latest information Chapters 1 through 12 have some expanded sections. Chapter 3 includes more on quality professionals' role/responsibility and also on GMP training and training programs. Chapter 5 on Equipment has an extensive maintenance & calibration section. Production & Process Controls in Chapter 7 has updates on SOP's and validation practices and PAT initiatives. New labeling regulations and applications are covered in Chapter 8 and Chapter 11 on records/reports (documentation) has a current section on failure investigations.

There are a number of new and relevant chapters. Chapter 14 deals with the quality systems approach and risk management processes with a focus on understanding and application. Chapter 15 is quite extensive and specific about GMPs for Clinical Trial Materials. Contracting and Outsourcing continues to be a popular activity in this day of consolidation and contracting so Chapter 16 has been added. Chapter 18 covers the world of Bulk Pharmaceutical Excipient GMPs and compliance.

Chapter 17 on API's has been updated according to the most recent ICH Q7A Guideline. Updated enforcement alternatives (Recalls, Warning Letters, Seizures and Injunctions) are included in Chapter 19 and updates on inspection procedures, the pre-approval inspection process and the Food & Drug Modernization Act are included in Chapters 21 and 22.

We conclude the 6th edition with Chapters 22 and 23 that provide updates on worldwide GMP regulation and other quality approaches namely ISO 9000, Malcolm Baldrige and six sigma methodologies.

No book is complete without appendices and we have included the current and best list of guidance documents: CDER List of Guidance Documents, ICH Quality Guidance Document List and ORA Compliance Policy Guides.

In the many years I have been involved with applying GMP regulations to operations, it has always surprised me that since 1978 the discussions, arguments and debates continue over mainly the intent and "how to" of GMP compliance. This all goes back to the original 1978 regulations and the intent of Congress. They intended that the agency (FDA) determine what constitutes current or the "C" in CGMP, based on their experience. The Congress also interpreted current as not necessarily widely prevalent. They did not require that a majority of manufacturers had to be following a practice before it was accepted as current. If a practice was shown to be feasible and valuable in assuring drug quality it could be considered current. This is what makes life in Pharmaceutical operations and CGMP compliance interesting.

For those professionals starting in pharmaceuticals, I would recommend a thorough read of the preamble to the GMPs, the GMP regulations, the related FDA guidance documents that provide further information/background and then look at industry practice and application.

Finally, I would like to thank and extend my gratitude to all the contributing authors who spent their personal time to research, collect and complete the chapters. Their expertise and experience in applied GMP is much appreciated.

Joseph D. Nally

Contents

Contributors

Graham Bunn   GB Consulting LLC, Berwyn, Pennsylvania, U.S.A.

Joseph T. Busfield   Pharmaceutical Technical Services, Warrington, Pennsylvania, U.S.A.

P. Denis Celentano   Percels Consulting Inc., Orlando, Florida, U.S.A.

Joanne W. Cochran   JWC Training Associates, Jamison, Pennsylvania, U.S.A.

Robert Del Ciello   Northshire Associates, Martinsville, New Jersey, U.S.A.

Wayne J. DeWitte   W. J. DeWitte Consulting, Suffern, New York, U.S.A.

Alex M. Hoinowski   Quantic Group Ltd., Livingston, New Jersey, U.S.A.

Michael D. Karaim   Brewster, Massachusetts, U.S.A.

Joseph D. Nally   Nallianco LLC, New Vernon, New Jersey, U.S.A.

Laura L. Nally   Nallianco LLC, New Vernon, New Jersey, U.S.A.

Steven Ostrove   Ostrove Associates Inc., Elizabeth, New Jersey, U.S.A.

Arlyn R. Sibille   Consultant, Harmony, Pennsylvania, U.S.A.

Irwin Silverstein   IBS Consulting in Quality LLC, Piscataway, New Jersey, U.S.A.

1 Status and Applicability of U.S. Regulations
Current Good Manufacturing Practices in Manufacturing, Processing, Packaging, and Holding of Drugs

Joseph D. Nally
Nallianco LLC, New Vernon, New Jersey, U.S.A.

This chapter addresses Food and Drug Administration (FDA) and some other federal regulations that have been promulgated for statutory effectuation and implementation in the main. The major statute underlying such regulations is the Federal Food, Drug, and Cosmetic (FFDC) Act as amended, which may be found in the U.S. Code at 21 USC 321 through 392.

The regulations discussed in the main are found at Title 21 of the Code of Federal Regulations. The latter is composed of nine volumes. The parts in these volumes are arranged in the following order: Parts 1–99, 100–169, 170–199, 200–299 [containing the bulk of the current good manufacturing practices (CGMPs)], 300–499 [containing the bulk of the investigational new drug (IND) application, new drug application (NDA), and abbreviated new drug application (ANDA) materials], 500–599, 600–799, 800–1299, and 1300–end. This last volume addresses matters subject to Drug Enforcement Administration (DEA), the Department of Justice (DOJ), and the Office of National Drug Control Policy.

The text also addresses guidelines, recommendations, and agreements that for the most part are governmental in derivation.

The CGMP regulations (21 CFR 210–226) are promulgated by the Commissioner of the FDA under Section 701 (a) of the FFDC Act [21 USC 371 (a)] in furtherance of the requirement of Section 501(a)(2)(B) of the Act [21 USC 351(a)(2)(B)], which specifies that a drug is deemed adulterated "if the methods used in, or the facilities or the controls used for, its manufacture, processing, packing, or holding do not conform to or are not operated or administered in conformity with current good manufacturing practice." The purpose of Section 501(a)(2)(B) is to assure that such drug meets the requirements of the act as to safety and has the identity and strength and meets the quality and purity characteristics that it purports or is represented to possess. The FDA is, of course, committed to various programs and systems designed to assure the quality of all drug products by careful monitoring of drug manufacturer's compliance with CGMP regulations. In order to identify their regulatees, Section 510(b) and (c) of the FFDC Act requires the registration of all producers of drugs and devices. Congressional language that accompanied this amendment stated it was "necessary to provide for the registration and inspection of all establishments in which drugs were manufactured, prepared, propagated, compounded, or processed" since these products were likely to enter interstate commerce. Section 510(h) requires that each registrant be inspected for compliance every two years.

The FDA includes as registrants manufacturers whose final products are homeopathic drugs and requires that such products must be manufactured in conformance with CGMPs, but some requirements of 21 CFR 211 are considered by them to be inapplicable. The authors have included FDA Compliance Guides as to preparation for, and marketing of, homeopathic drugs in Chapter 2.

The approval process for drug marketing applications (original and ANDAs and Antibiotic Forms 5 and 6) includes a review of the manufacturer's compliance with the CGMPs.

In recent years, the FDA has assumed additional roles for assurance to vendees through programs like the Government-Wide Quality Assurance Programs for drug purchase contracts by the Department of Defense and Veterans Affairs and the MAC program (Maximum Allowable Cost), a program that became seminal to the manufacture of generics. Their policy is outlined in the ORA (Office of Regulatory Affairs) Compliance Document Sec. 400.200 Consistent Application of CGMP Determinations (CPG 7132.12) available at: http://www.fda.gov/ora/compliance_ref/cpg/cpgdrg/cpg400-200.html.

Decisions regarding compliance with CGMP regulations are based on inspection of the facilities, sample analysis, and compliance history of the firm. These data are summarized in profiles that represent several years of history of the firms.

The CGMP deficiencies supporting regulatory action by the FDA also support decisions regarding nonapproval of NDA Supplements, as well as the purchasing contracts and candidacy for MAC; hence, some FDA expanded action is likely. Therefore, issuance of a "warning" letter or other regulatory action based on discovery of CGMP deficiencies must be accompanied by disapproval of any pending NDA, ANDA, or Supplement, or any government contract produced under the same deficiencies.

The FFDC Act applies to drugs introduced into interstate commerce in the United States, including drugs exported to or imported from other countries. Manufacturers in other countries who export to the United States are inspected either by the FDA or under reciprocal inspection agreements as part of the NDA approval process and antibiotic drug certification. Currently, such agreements came out of the Food and Drug Administration Modernization Act (FDAMA) program and exist between the United States and Sweden, Switzerland, Canada, United Kingdom, and Australia. "A Plan That Established a Framework For Achieving Mutual Recognition of Good Manufacturing Practices Inspections" is available at: http://www.fda.gov/fdama/fdamagmp.pdf. Individual drug products are subjected to extensive examination, including laboratory testing, before being allowed into the United States.

The FDA has the authority to deny entry to any drug, if there is a question regarding its safety, identity, strength, quality, or purity. This authority is exercised unless factory inspection is permitted or inspection information is available concerning nondomestic firms, in lieu of conducting foreign inspections. Although this authority is exercised more rarely and tempered by Chapter 8 of the Act, the FDA also has the authority to deny exit to questionable drugs.

From the inception of the good manufacturing practices (GMPs), the FDA strived to ensure that the regulated industries comply with a total control of product quality concept through its factory inspection programs and through participation in voluntary CGMP compliance seminars and workshops sponsored jointly with the industries or with educational institutions. More recently, as part of the FDA's Pharmaceutical CGMPs for the 21st Century Initiative, they have introduced quality systems and risk management approaches into existing programs. Regardless of the approaches used, the FDA wants industry to prevent a drug product from being deemed adulterated under Section 501(a)(2)(B) and violative of Section 301(b) of the Food, Drug, and Cosmetic Act as is indicated by 21 CFR 211, Current Good Manufacturing Practice for Finished Pharmaceuticals.

In the following sections, an attempt is made to provide specific guidelines and concepts that can serve as checks for critical operations within the entire organization in order that effective and integrated quality systems and product controls can be achieved. Each requirement that is loosely generalized in GMP regulations will be enlarged and made more specific in order to include measures that the authors believe are necessary for good control systems.

§210.1 STATUS OF CURRENT GOOD MANUFACTURING PRACTICE REGULATIONS

(a) The regulations set forth in this part and in Parts 211 through 229 of this chapter contain the minimum CGMP for methods to be used in, and the facilities or controls to be used for, the manufacture, processing, packing, or holding of a drug to assure that such drug meets the requirements of the act as to safety, and has the identity and strength and meets the quality and purity characteristics that it purports or is represented to possess.

(b) The failure to comply with any regulation set forth in this part and in Parts 211 through 229 of this chapter in the manufacture, processing, packing, or holding of a drug shall render such drug to be adulterated under Section 501(a)(2)(B) of the act and such drug, as well as the person who is responsible for the failure to comply, shall be subject to regulatory action.

§210.2   APPLICABILITY OF CURRENT GOOD MANUFACTURING PRACTICE REGULATIONS

(c)   The regulations in this part and in Parts 211 through 229 of this chapter, as they may pertain to a drug, and in Parts 600 through 680 of this chapter, as they may pertain to a biological product for human use, shall be considered to supplement, not supersede, each other, unless the regulations explicitly provide otherwise. In the event that it is impossible to comply with all applicable regulations in these parts, the regulations specifically applicable to the drug in question shall supersede the more general.

(d)   If a person engages in only some operations subject to the regulations in this part and in Parts 211 through 226 and Parts 600 through 680 of this chapter, and not in others, that person need only comply with those regulations applicable to the operations in which he or she is engaged. (See also 21 CFR 207.)

§210.3   DEFINITIONS

(e)   The definitions and interpretations contained in Section 201 of the act shall be applicable to such terms when used in this part and in Parts 211 through 226 of this chapter.

(f)   The following definitions of terms apply to this part and to Parts 211 through 226 of this chapter.

  (1)   "Act" means the FFDC Act, as amended (21 USC 321, et seq.).

  (2)   "Batch" means a specific quantity of a drug or other material that is intended to have uniform character and quality, within specified limits, and is produced according to a single manufacturing order during the same cycle of manufacture.

  (3)   "Component" means any ingredient intended for use in the manufacture of a drug product, including those that may not appear in such drug product.

  (4)   "Drug product" means a finished dosage form, for example, tablet, capsule, solution, etc., that contains an active drug ingredient generally, but not necessarily, in association with inactive ingredients. The term also includes a finished dosage form that does not contain an active ingredient but is intended to be used as a placebo.

  (5)   "Fiber" means any paniculate contaminant with a length at least three times greater than its width.

  (6)   "Non-fiber-releasing filter" means any filter, which after any appropriate pre-treatment such as washing or flushing, will not release fibers into the component or drug product that is being filtered. All filters composed of asbestos are deemed to be fiber-releasing filters.

  (7)   "Active ingredient" means any component that is intended to furnish pharmacological activity or other direct effect in the diagnosis, cure, mitigation, treatment, or prevention of disease or to affect the structure of any function of the body of man or other animals. The term includes those components that may undergo chemical change in the manufacture of the drug product in a modified form intended to furnish the specified activity or effect.

  (8)   "Inactive ingredient" means any component other than an "active ingredient."

  (9)   "In-process material" means any material fabricated, compounded, blended, or derived by chemical reaction that is produced for, and used in, the preparation of the drug product.

  (10)   "Lot" means a batch, or a specific identified portion of a batch, having uniform character and quality within specified limits; or, in the case of a drug product produced by continuous process, it is a specific identified amount produced in a unit of time or quantity in a manner that assures its having uniform character and quality within specified limits.

(11) "Lot number, control number, or batch number" means any distinctive combination of letters, numbers, or symbols, or any combination of them, from which the complete history of the manufacture, processing, packing, holding, and distribution of a batch or lot of drug product or other material can be determined.

(12) "Manufacture, processing, packing, or holding of a drug product" includes packaging and labeling operations, testing and quality control of drug products.

(13) "Medicated feed" means any "complete feed," "feed supplement," or "feed concentrate" as defined in §558.3 of this chapter and is a feed that contains one or more drugs as defined in Section 201 (g) of the act. Medicated feeds are subject to Part 225 of this chapter.

(14) "Medicated premix" means a substance that meets the definition in §558.3 of this chapter for a "feed premix," except that it contains one or more drugs as defined in Section 201 (g) of the act and is intended for manufacturing use in the production of a medicated feed. Medicated premixes are subject to Part 226 of this chapter.

(15) "Quality control unit" means any person or organizational element designated by the firm to be responsible for the duties relating to quality control.

(16) "Strength" means:
   (i) The concentration of the drug substance (e.g., weight/weight, weight/volume, or unit dose/volume basis), and/or
   (ii) The potency, that is, the therapeutic activity of the drug product as indicated by appropriate laboratory tests or by adequately developed and controlled clinical data (e.g., expressed in terms of units by reference to a standard).

(17) "Theoretical yield" means the quantity that would be produced at any appropriate phase of manufacture, processing, or packing of a particular drug product, based upon the quantity of components to be used, in the absence of any loss or error in actual production.

(18) "Actual yield" means the quantity that is actually produced at any appropriate phase of manufacture, processing, or packing of a particular drug product.

(19) "Percentage of theoretical yield" means the ratio of the actual yield (at any appropriate phase of manufacture, processing, or packing of a particular drug product) to the theoretical yield (at the same phase), stated as a percentage.

(20) "Acceptance criteria" means the product specifications and acceptance/rejection criteria, such as acceptable quality level and unacceptable quality level, with an associated sampling plan, that are necessary for making a decision to accept or reject a lot or batch (or any other convenient subgroups of manufactured units).

(21) "Representative sample" means a sample that consists of a number of units that are drawn on the basis of rational criteria, such as random sampling, and intended to assure that the sample accurately portrays the material being sampled.

As described in the *Federal Register*, this general introduction to what is projected as a series of GMP regulations for all human drug products, as well as specific products or specific processes, is "intended to be general enough to be suitable for essentially all drug products, flexible enough to allow the use of sound judgment and permit innovation, and explicit enough to provide a clear understanding of what is required." The concept places a large burden on the manufacturer of pharmaceuticals. Adherence to the explicit regulations is a required minimum, but it is not adequate to ensure that a manufacturer is in compliance. In addition, the manufacture of a pharmaceutical must be by current methods with current controls, thus setting as a requirement that which is current or generally accepted in the drug industry as appropriate equipment, methodology, controls, and records. Even being "average" in all respects, compared with the industry, does not ensure that a manufacturer is in compliance, because the standard is not only that practices be "current" but that they also be "good." Thus, if a new practice is introduced anywhere in the industry that is better than what is current, then all manufacturers may seem obligated to adopt the better practice.

Therefore, it can be seen that being in compliance with GMP is not a static situation, but requires the manufacturer to be aware not only of what is current in the industry but also to be aware of innovations that may be good. It would seem also that a legitimate inquiry for the evaluation of compliance is the measures taken by a manufacturer to obtain knowledge, on a continuing basis, of both what is current and may be good, and to provide a method for incorporating the necessary changes into the already established system of manufacture and control.

The FDA currently makes available in electronic format various information and data from its centers. The main web page is: http://www.fda.gov. The majority of this 6th edition revisions came from Internet sources.

Those who do analytic work for manufacturers should be certain they are optimizing their knowledge of such accessible data by inquiry of the FDA, their own association resources, and that of the Association of Food and Drug Officials of the United States. Sometimes, these additional resources are also noted at FDA Science Forums on the Regulatory Sciences, and abstracts may be available. For example, information routinely maintained within the Division of Pesticides and Industrial Chemicals can be retrieved. These data are accessible via the Prime Connection electronic bulletin board of the Center for Food Safety and Applied Nutrition (CFSAN), accessible via an 800 number to anyone with a computer and modem, and CFSAN's VAX computer anonymous file transfer protocol (FTP) site, accessible via the Internet. There is a current menu as to other sources of information from and about the FDA at the end of this chapter and in the Appendices.

ROLE AND IMPORTANCE OF THE UNITED STATES PHARMOCOPEIA

The role of the United States Pharmacopeia (USP) is of course not limited to the United States. This was true even prior to the time it became incorporated within the FFDC Act for its ultimate importance in dealing with such portions of that law that recite prohibited acts. There are other pharmacopeias with substantial recognition in the global and local manufacture and distribution of pharmaceuticals, but the USP is a major reference in this regard.

For that reason, and for the fact that in 1999 the USP announced a major structural reorganization for improved focus responsive to the "ongoing transformation of healthcare science and technology," it is perhaps helpful to briefly review its historic and present important role in the manufacture, labeling, and standard setting vital to pharmaceutical manufacture.

Established in 1820 "to ensure that consumers receive medicines of the highest possible quality, strength, and purity in the United States," it was destined to reflect medical and pharmaceutical advances from the major European laboratories and academia from the first. At present, the USP provides standards for more than 3400 drugs and dosage forms for medicines and dietary supplements.

While it has taken on additional duties in the areas of reporting and prevention programs regarding product problems and medication errors, the interest of our readers is better expressed within their traditional roles in standard setting.

These can be summarized (and are currently handled by a newly formed division following the 1999 reorganization) as follows:

Standards—General Policies, Requirements, Nomenclature and Labeling, Veterinary Drugs, Excipients/Pharmaceutical Waters, Biologics and Biotechnology, Dietary Supplements, and Pharmaceuticals, which also includes the Reference Standards Laboratory and the Research and Development Laboratory.

Located nearby the FDA in Rockville, Maryland, U.S.A., they can be reached at 301-998-6821 or at: http://www.usp.org.

Under Uniform State Food and Drug Laws, the FFDC Act, and the recent Food and Drug Administration Modernization Act, USP-National Formulary (NF) standards are legally enforceable.

These standards are published in the USP-NF and legally recognized. And once the FDA approves a new drug product, the USP establishes public standards for some that become similarly enforceable. It is hard to imagine that a manufacturer would not have copies of the current USP29-NF24 compendium available to the staff. All proposed revisions to USP-NF standards are published for review and comment in a USP publication, *Pharmacopeia Forum*. The USP-NF is, of course, also available on CD and in English and Spanish.

The USP provides Reference Standards that are highly characterized specimens of drug substances, major impurities, degradation products, and performance calibrators for use in testing drugs and nutritional supplements. They are used to perform official methods of analysis in pharmaceutical testing. The manufacturer may use other than the official method of analysis, but on FDA inspection and challenge, the substance used and the product manufactured must meet the official specifications contained in the USP-NF, following the official method of analysis. USP Material Safety Data Sheets are available to purchasers of standards. The USP also tests and distributes other authenticated substances not currently included in the USP-NF that are still in sufficient demand; FCC Reference Standards specified in the latest edition of the Food Chemicals Codex; and highly purified samples of chemicals, including drugs of abuse. Readers who seek information as to specific USP Reference Standards, whether or not current, can call 1-800-227-8772 in the United States or 1-301-881-0666 for outside the U.S. and Canada. The USP website http://www.usp.org has links to Reference Standards, USP-NF, Patient Safety, and USP-Verified and other topics.

It goes without saying that constant update acquisition of official and unofficial compendia is necessary for timely attention to the CGMPs as well. The law requires that products meet the requirements of the USP/NF for the monographs applicable to their products as labeled. Since these are not static, good reasoning, good science, and good practical issues arise. For example, where the manufacturer has prepared the product in accord with the pertinent USP monograph and that monograph undergoes a significant change five years later as to its methodology of analytic controls, what is a reasonable approach for the manufacturer? First, the manufacturer should implement the new controls promptly. As to product samples held in reserve, they are examples of product legally placed into Interstate Commerce under analytic controls then in effect, and any obligation to retest them might not be reasonable. The exception might rest on the nature and importance of the change to such a product and on whether the manufacturer might have knowledge that the former method would find the product out of control in the marketplace at this date. In such an instance, the safety and effectiveness of the product would be foremost in any decision. It is typical for many revisions to occur in USP-NF Supplements and Editions. However, it is the responsibility of the drug manufacturer to keep abreast of the proposed changes and potential impact.

As an example, revisions in content uniformity specifications and limits have been known to generate questions on manufacturing process capability and, in some cases, result in the obsolescence of equipment, especially for older products. When is a piece of machinery no longer in compliance? The compliant answer is when it can no longer produce product that meets the specification. The decision will be that of the manufacturer, who must abide by the label claims of the product.

THE MEANING OF "CURRENT"

The most unique and interesting part of the GMP regulations is the concept of what is "current" or the "C" that precedes GMP?

> The Congress intended that the phrase itself (current good manufacturing practice) have a unique meaning ... The agency determines what constitutes "current good manufacturing practice" based upon its experience with the manufacture of drugs through inspectional and compliance activities ... Although the practices must be "current" in the industry, they need not be widely prevalent. Congress did not require that a majority or any other percentage of manufacturers already be following the ... practice ... that ... had been shown to be both feasible and valuable in assuring drug quality. [*Federal Register*]

The FDA also holds that, although it does not manufacture drugs, it has the unique ability to determine CGMPs for drugs, since it alone has access to the facilities and records of every manufacturer of pharmaceuticals in the United States. Given the fact that many processes and controls are considered by individual manufacturers to be trade secrets, competitors, through the various associations of manufacturers or independent third parties, such as the compendial authorities, are not likely to discover what nonpublic practices are current.

Even if current practices were available, the FDA holds that it has special technical and scientific expertise to determine which of the current practices are also good. This expertise is inherent in reviews of production and control techniques in NDAs and ANDAs, supplemental applications, antibiotic certification forms, biological establishment and product licenses, new animal drug applications, and proposed and final compendial standards. Additional experience is based on establishment inspection reports filed by FDA investigators and the monitoring of drug recalls.

A current, although not necessarily predominant, practice is considered "good" if:

1. It is feasible for manufacturers to implement.
2. It contributes to ensuring the safety, quality, or purity of the drug product.
3. The value of the contributions or added assurance exceeds the cost in money or other burdens of implementing or continuing the practice.

LEGAL RESPONSIBILITIES OF THOSE INVOLVED WITH DRUG MANUFACTURING

Also, note that in addition to proceeding against the drug, regulatory action may be taken against the person who is responsible for the failure to comply. Responsibility for failure to comply would seem extensible vertically from senior management to line management, which did not supply adequate supervision or directions, through quality assurance/control and individual production people, who did not follow directions, and horizontally to supplies of raw materials, whose products did not meet purported specifications, as well as to contract laboratories. Since criminal penalties (fines and/or prison sentences) are possible, these regulations impose a standard of responsibility to have knowledge, to train subordinates, and to continually check to ensure compliance with directives.

The legal standard for responsibility for all those engaged in drug manufacture is high. People entering this field of endeavor should be aware of the special burden of complete accountability. Violation of the FFDC Act is handled under unique legal doctrine that does not require proof of criminal intent as a prerequisite for criminal culpability. In order to provide maximum protection of the public health, Congress purposely neither required that actual harm from contamination of a drug product has to be proven for a charge that the product is adulterated, nor that each article in the batch be adulterated before the entire amount is subject to condemnation or other action. The law is aimed not only at removal of the adulterated article from commerce, but for the same offense, and simultaneously, may seek punishment of a person. Note the "person" is defined in Section 201(e) of the act to include corporations and partnerships, as well as individuals. Landmark judicial decisions are (*i*) *United States v. Dotterweich*, 320 U.S. 277 (1943) and (*ii*) *United States v. Park*, 421 U.S. 658 (1975).

Almost all civil and criminal actions initiated by the FDA are derived from violations of statutory definitions of misbranding and adulteration. This fact is magnified by both the expanded areas of definition and prohibitions created by the New Drug Amendments and the prevalent judicial policy of liberal construction of the statute with its prime objective of consumer protection.

Currently, the adulteration statute, aside from the regulations on GMPs, holds that the presence of a foreign substance, even of distinct and contrary appearance from the product itself, could cause it to be adulterated. Previously, distinct substances, such as nails or pieces of a container not commingled in such a manner as to masquerade as a part of the food itself, would not have been considered an ingredient in support of a charge of adulteration.

However, the courts have not been entirely consistent as to this interpretation, and no doubt the subject regulations will be used to strengthen the FDA position of greater inclusion.

In 1952, a landmark decision in the 8th Circuit Court stated that a defendant might enjoy a certain latitude where a "mere possibility" of contamination existed, subject to proof that factory conditions "would with reasonable possibility result in contamination" (*Berger v. The United States*, 200 F.2d 818). Obviously then, today's GMP regulations are viewed as the means for the FDA to present to the court this "reasonable possibility" based on breach of said regulations. To increase the chances of success for enforcement, evidence from examination of samples will frequently be offered to show that the possibility is realized.

RECENT FOOD AND DRUG ADMINISTRATION DRUG-RELATED MILESTONES

A full history summary is available at: http://www.fda.gov/opacom/backgrounders/miles.html.

1997

The Food and Drug Administration Modernization Act reauthorizes the Prescription Drug User Fee Act of 1992 and mandates the most wide-ranging reforms in agency practices since 1938. Provisions include measures to accelerate review of devices, regulate advertising of unapproved uses of approved drugs and devices, and regulate health claims for foods.

1998

The FDA promulgates the Pediatric Rule, a regulation that requires manufacturers of selected new and extant drug and biological products to conduct studies to assess their safety and efficacy in children.

Mammography Quality Standards Reauthorization Act continues 1992 until 2002.

First phase to consolidate FDA laboratories nationwide from 19 facilities to 9 by 2014 includes dedication of the first of five new regional laboratories.

1999

ClinicalTrials.gov is founded to provide the public with updated information on enrollment in federally and privately supported clinical research, thereby expanding patient access to studies of promising therapies.

A final rule mandates that all over-the-counter drug labels must contain data in a standardized format. These drug facts are designed to provide the patient with easy-to-find information, analogous to the nutrition facts label for foods.

2000

The U.S. Supreme Court, upholding an earlier decision in *Food and Drug Administration v. Brown Williamson Tobacco Corp. et al.*, ruled 5-4 that FDA does not have authority to regulate tobacco as a drug. Within weeks of this ruling, FDA revoked its final rule, issued in 1996, that restricted the sale and distribution of cigarettes and smokeless tobacco products to children and adolescents, and that determined that cigarettes and smokeless tobacco products are combination products consisting of a drug (nicotine) and device components intended to deliver nicotine to the body.

Federal agencies are required to issue guidelines to maximize the quality, objectivity, utility, and integrity of the information they generate, and to provide a mechanism whereby those affected can secure correction of information that does not meet these guidelines, under the Data Quality Act.

Publication of a rule on dietary supplements defines the type of statement that can be labeled regarding the effect of supplements on the structure or function of the body.

2002

The Best Pharmaceuticals for Children Act improves safety and efficacy of patented and off-patent medicines for children. It continues the exclusivity provisions for pediatric drugs as mandated under the Food and Drug Administration Modernization Act of 1997, in which market exclusivity of a drug is extended by six months, and in exchange the manufacturer carries out studies of the effects of drugs when taken by children. The provisions both clarify aspects of the exclusivity period and amend procedures for generic drug approval in cases when pediatric guidelines are added to the labeling.

In the wake of the events of September 11, 2001, the Public Health Security and Bioterrorism Preparedness and Response Act of 2002 is designed to improve the country's ability to prevent and respond to public health emergencies, and provisions include a requirement that FDA issue regulations to enhance controls over imported and domestically produced commodities it regulates.

Under the Medical Device User Fee and Modernization Act, *fees are assessed to sponsors of* medical device applications for evaluation, provisions are established for device establishment inspections by accredited third-parties, and new requirements emerge for reprocessed single-use devices.

The Office of Combination Products is formed within the Office of the Commissioner, as mandated under the Medical Device User Fee and Modernization Act, to oversee review of products that fall into multiple jurisdictions within the FDA.

An effort to enhance and update the regulation of manufacturing processes and end-product quality of animal and human drugs and biological medicines is announced: the CGMP initiative. The goals of the initiative are to focus on the greatest risks to public health in manufacturing procedures, to ensure that process and product quality standards do not impede innovation, and to apply a consistent approach to these issues across the FDA.

2003

The Medicare Prescription Drug Improvement and Modernization Act requires, among other elements, that a study be made of how current and emerging technologies can be utilized to make essential information about prescription drugs available to the blind and visually impaired.

The Animal Drug User Fee Act permits the FDA to collect subsidies for the review of certain animal drug applications from sponsors, analogous to laws passed for the evaluation of other products the FDA regulates, ensuring the safety and effectiveness of drugs for animals and the safety of animals used as foodstuffs.

The FDA is given clear authority under the Pediatric Research Equity Act to require that sponsors conduct clinical research into pediatric applications for new drugs and biological products.

2004

Project BioShield Act of 2004 authorizes the FDA to expedite its review procedures to enable rapid distribution of treatments as countermeasures to chemical, biological, and nuclear agents that may be used in a terrorist attack against the United States, among other provisions.

A ban on over-the-counter steroid precursors, increased penalties for making, selling, or possessing illegal steroids precursors, and funds for preventive education to children are features of the Anabolic Steroid Control Act of 2004.

The FDA publishes "Innovation or Stagnation?—Challenge and Opportunity on the Critical Path to New Medical Products," which examines the critical path needed to bring therapeutic products to fruition, and how the FDA can collaborate in the process, from laboratory to production to end use, to make medical breakthroughs available to those in need as quickly as possible.

On the basis of recent results from controlled clinical studies indicating that Cox-2 selective agents may be connected to an elevated risk of serious cardiovascular events, including heart attack and stroke, the FDA issues a public health advisory urging health professionals to limit the use of these drugs.

To provide for the treatment of animal species other than cattle, horses, swine, chickens, turkeys, dogs, and cats, as well as other species that may be added at a later time, the Minor Use and Minor Species Animal Health Act is passed to encourage the development of treatments for species that would otherwise attract little interest in the development of veterinary therapies.

Deeming such products to present an unreasonable risk of harm, the FDA bans dietary supplements containing ephedrine alkaloids based on an increasing number of adverse events linked to these products and the known pharmacology of these alkaloids.

2005

Formation of the Drug Safety Board is announced, consisting of FDA staff and representatives from the National Institutes of Health and the Veterans Administration. The Board will advise the Director, Center for Drug Evaluation and Research, FDA, on drug safety issues and work with the agency in communicating safety information to health professionals and patients.

SUPPLEMENTAL FOOD AND DRUG ADMINISTRATION INFORMATION AVAILABLE TO THE READER

The FDA home page http://www.fda.gov has the following sections and links:
FDA News
- Recent topics: http://fda.gov/bbs/topics/news/
- Recalls, product safety: http://www.fda.gov/opacom/7alerts.html
- Product approvals: http://www.fda.gov/opacom/7approvl.html
- Press releases, testimony, speeches, etc.: http://www.fda.gov/opacom/hpwhats.html

Reference Room
- Laws FDA enforces: http://www.fda.gov/opacom/laws
- CFR: http://www.accessdata.fda.gov/scripts/cdrh/cfdocs/cfcfr/cfrsearch.cfm
- Federal Register: http://www.accessdata.fda.gov/scripts/oc/ohrms/index.cfm
- Guidance documents: http://www.fda.gov/opacom/morechoices/industry/giudedc.htm
- Forms: http://www.fda.gov/opacom/morechoices/fdaforms.html
- Dockets: http://www.fda.gov/ohrms/dockets/default.htm
- Warning letters: http://www.fda.gov/foi/warning.htm
- Manuals and publications: http://www.fda.gov/opacom/7pubs.html.

Let Us Hear from You
- Report a product problem: http://www.fda.gov/opacom/backgrounders/problem.html.
- Comments on proposed regulations: http://www.fda.gov/opacom/backgrounders/voice.html.
- Petition FDA: http://www.fda.gov/opacom/backgrounders/voice.html
- Contact FDA: http://www.fda.gov/comments.html

There are additional information/sections/links for the products FDA regulates, Hot Topics, FDA Activities, and Consumer Information on the home page.

If you have no Internet access, FDA can be reached by mail and telephone:
Food and Drug Administration, 5600 Fishers Lane, Rockville, MA 20857, U.S.A.: 1-888-INFO-FDA (1-888-463-6332)—main FDA phone number (for general inquiries)

HOW TO OBTAIN FOOD AND DRUG ADMINISTRATION REGULATIONS

http://www.cfsan.fda.gov/~lrd/ob-reg.html

The FDA's regulations are printed in Title 21, Code of Federal Regulations (21 CFR). In addition, the FDA and other government agencies publish new regulations and proposals in the Federal Register throughout the year. Readers may purchase the books in 21 CFR from the U.S. Government Printing Office.

Title 21, Code of Federal Regulations (21 CFR) is updated April 1 of each year. The current edition contains nine volumes and is printed in paperback books. Ordering instructions appear at the bottom of this page. The next revision of 21 CFR will be available in late summer. A description of the regulations in each volume of 21 CFR follows:

- 21 CFR 1-99. General regulations for enforcement of the FFDC Act and the Fair Packaging and Labeling Act. Color additives.
- 21 CFR 100-169. General regulations for food labeling (contains the new regulations for nutrition labeling). Infant formula quality control procedures and labeling. Food standards and quality standards for bottled drinking water. The CGMP regulations for food, bottled drinking water, low-acid canned foods, and acidified foods.
- 21 CFR 170-199. Food additives.
- 21 CFR 200-299. General regulations for drugs.
- 21 CFR 300-499. Drugs for human use.
- 21 CFR 500-599. Animal drugs, feeds, and related products.
- 21 CFR 600-799. Biologics and cosmetics.
- 21 CFR 800-1299. Medical devices, radiological health, interstate conveyance sanitation, and control of communicable diseases. Regulations for the Federal Import Milk Act, Federal Tea Importation Act, and Federal Caustic Poison Act.
- 21 CFR 1300-End. Regulations implementing the Controlled Substances Act and the Controlled Substances Import and Export Act.

The *Federal Register*

The *Federal Register* is published Monday through Friday by the U.S. Government Printing Office in paper and microfiche editions, and as a database on Internet.

How to Order
For information on ordering the Federal Register, and for the latest Code of Federal Regulations stock numbers and prices, phone U.S. Fax Watch at (202)-512-1716 from your touch-tone phone. For information on the Federal Register, select option 2, then 3. For information on the CFRs, select option 2, then 1.

Readers may subscribe to the Federal Register and order Title 21, Code of Federal Regulations by making checks or money orders payable to 'Superintendent of Documents' and mail to New Orders, Superintendent of Documents, P.O. Box 371954, Pittsburgh, PA 15250-7954, U.S.A. Purchases may also be made (using Visa, MasterCard, or Superintendent of Documents deposit accounts) by telephoning (202)-512-1800. Back issues of the Federal Register (if available), and the Code of Federal Regulations may be purchased by telephoning regional offices of GPO listed on the back of this page or (202)-512-1800. Readers may subscribe to the Federal Register online database by sending e-mail to the GPO Access User Support Team at the address: gpoaccess@gpo.gov, or by calling (202)-512-1530.

SUGGESTED READINGS

1. Dean E. Snyder. FDA Speak, A Glossary and Agency Guide. 2nd edn. HIS Health Group, 2002.
2. Marie A. Urban. Reinventing the Food and Drug Administration. Food, Drug, Cosmetic, and Medical Device Law Digest. 14(2), 5/97.
3. Gerald F. Meyer. The generic industry: an FDA insiders view. F.D.C.M.L. Digest 1990; 7(2).
4. James L. Vesper. GMP in Practice, Regulatory Expectations for the Pharmaceutical Industry, 2002.

2 | Finished Pharmaceuticals: General Provisions
Subpart A

Joseph D. Nally
Nallianco LLC, New Vernon, New Jersey, U.S.A.

§211.1 SCOPE

(a) The regulations in this part contain thze minimum current good manufacturing practice (CGMP) for preparation of drug products for administration to humans or animals.

(b) The CGMP regulations in this chapter, as they pertain to drug products, and in Parts 600 through 680 of this chapter, as they pertain to biological products for human use, shall be considered to supplement, not supersede, the regulations in this part unless the regulations explicitly provide otherwise. In the event it is impossible to comply with applicable regulations both in this part and in other parts of this chapter or in Parts 600 through 680 of this chapter, the regulation specifically applicable to the drug product in question shall supersede the regulation in this part.

(c) Pending consideration of a proposed exemption, published in the *Federal Register* of September 29, 1978, the requirements in this part shall not be enforced for over-the-counter (OTC) drug products if the products and all their ingredients are ordinarily marketed and consumed as human foods, and which products may also fall within the legal definition of drugs by virtue of their intended use. Therefore, until further notice, regulations under Part 110 of this chapter, and where applicable, Parts 113 to 129 of this chapter, shall be applied in determining whether these OTC drug products that are also foods are manufactured, processed, packed, or held under CGMP.

Since the 1978 publication of the GMPs, the scope of "drug products" and companies producing the same has been fairly well defined. A drug is defined as:

■ Articles recognized in the official USP (*United States Pharmacopeia*), HPUS (*Homeopathic Pharmacopeia of the United States*) or NF (*National Formulary*) or any supplement to any of them.

■ Articles intended for use in the diagnosis, cure, mitigation, treatment, or prevention of disease in man or other animals.

■ Articles (other than food) intended to affect the structure or any function of the body of man or other function of the body of man or other animals . . .

Prescription and OTC drug products and manufacturing have clearly been included in the GMP scope as well as re-packaging and re-labeling operations. For a time, OTC products were under some contention, but Sec. 450.100 CGMP Enforcement Policy—OTC vs Rx Drugs (CPG 7132.10)—makes it very clear.

SECTION 450.100 CGMP ENFORCEMENT POLICY—OTC VS Rx DRUGS (CPG7132.10)

Background

Because of increased visibility and promotion of certain OTC preparations, there are periodic inquiries from district offices regarding whether or not the enforcement policy for CGMP regulations is the same for OTC drug products as it is for prescription (Rx) drug products.

Section 501(a)(2)(B) of the Federal Food, Drug, and Cosmetic Act requires drugs to be manufactured in conformance with CGMP. This section does not differentiate between OTC and Rx products and it was not intended by Congress to do so.

A prescription drug may be toxic or have other potential for harm, which requires that it be administered only under the supervision of a licensed practitioner (Section 503(b)(1) of the Act). For this reason, problems associated with its manufacture are generally more likely to cause serious problems.

Policy

The CGMP regulations apply to all drug products, whether OTC or prescription.

Regulatory Guidance

The selection of an enforcement action to be applied will be based on the seriousness of the deviation, including such factors as potential hazard to the consumer.
Issued: 4/1/82

One area that is still under contention for GMP scope and applicability is pharmacy-compounding operations. Chapter 22 explores some of the recent decisions made, and the applicability is still far from clear.

Dietary Supplements

Unless they meet drug definition, Dietary Supplements are generally regulated as foods and are *not* in the scope of drug GMPs.

> There are special statutory provisions and implementing regulations for dietary supplements that differ in some respects from those covering conventional foods. Moreover, the regulatory requirements for dietary supplements also differ from those that apply to prescription and OTC drug products.
>
> Congress defined the term dietary supplement as a product that, among other things, is ingested, is intended to supplement the diet, is labeled as a dietary supplement, is not represented as a conventional food or as a sole item of a meal or the diet, and that contains at least one dietary ingredient. The dietary ingredients in these products may include vitamins, minerals, herbs or other botanicals, amino acids, and dietary substances such as enzymes. Dietary ingredients also can be metabolites, constituents, extracts, concentrates, or combinations of the preceding types of ingredients. Dietary supplements may be found in many forms, such as tablets, capsules, liquids, or bars. DSHEA (Dietary Supplement Health and Education Act) placed dietary supplements in a special sub-category under the general umbrella of foods, but products that meet the drug definition are subject to regulation as drugs (1).

Homeopathic Drugs

If "finished homeopathic drugs" are offered for the cure, mitigation, prevention, or treatment of disease conditions, they are regarded as drugs within the meaning of Section 210(g)(1) of the Act. This would apply whether or not they are official homeopathic remedies listed in the HPUS. As drugs, they are patently subject to the CGMPs with some reservations. Homeopathic drugs generally must meet the standards for strength, quality, and purity set forth in the Homeopathic Pharmacopeia. Section 501(b) of the Act (21 U.S.C. 351) provides in relevant part:

> Whenever a drug is recognized in both the United States Pharmacopeia and the Homeopathic Pharmacopeia of the United States it shall be subject to the requirements of the United States Pharmacopeia unless it is labeled and offered for sale as a homeopathic drug, in which case it shall be subject to the provisions of the Homeopathic Pharmacopeia of the United States and not to those of the United States Pharmacopeia.

The FDA compliance policy guide Sec. 400.400 Conditions Under Which Homeopathic Drugs May be Marketed (CPG 7132.15) provides supplemental information regarding homeopathic drugs for those readers interested in import, export, and interstate commerce in such products. This guidance is provided in total at the end of this chapter.

§211.2 DEFINITIONS

The definitions set forth in §210.3 of this chapter apply in this part.

In short, this section states that the company's adherence to the requirements of the entire set of regulations determined whether its output will be judged as adulterated or violative. Adherence to the requirements initially necessitates an analysis of all current operations within the company that affect the quality of the finished marketed product. Such an analysis looks at the quality management systems of an organization as well as the effectiveness of decision and information flows between managers, operators, scientists, technicians, and other personnel who regulate product quality. An analysis of current conditions also looks at the flow of materials into discrete, sequential operations from the receipt and sampling of raw materials to final accountability computations during the market distribution, in order that critical procedures can be specified and more closely examined.

The following survey is provided as an example in establishing baseline information about a company involved in drug manufacturing.

Company Survey

1. Name of company

2. Address

3. Telephone

4. Number of years in business

5. How is the company controlled?

    ___ Independent   ___ Subsidiary

    a. Parent company

    b. Address

6. Ownership

    ___ Corporation   ___ Partnership

    ___ Private   ___ Other

7. Field of operation

    ___ Domestic   ___ Foreign

8. Type of operation

    ___ Manufacturer   ___ Repacker

    ___ Packer   ___ Other

9. Extent of operations

    Plant locations _____ *No. of buildings* _____ *No. of employees* ___

10. Current approvals

    ___ DA registration   ___ VA contract

    ___ Defense personnel   ___ Other

11. Membership in trade associations (show professional interest)

    ___ Pharmaceutical Manufacturers Associations (i.e., PhRMA)

    ___ The Proprietary Association

    ___ National Pharmaceutical Council

    ___ Parenteral Drug Association

    ___ Drug and Allied Products Guild

    ___ Other

12. Attendance at pharmaceutical meetings related to manufacturing and quality assurance operations.

    Person ___ Position ___ Meetings attended/Date _____

    Dissemination of proceedings to managers and supervisory personnel.

    Lecturer ___ Subject _____ Personnel in Attendance ____ Date _____

13. To whom does it sell (approximate percentage of sales)?

    ___ Wholesaler _____ Hospital _____ Physician _____ VA

    ___ Direct pharmacy _____ Defense personnel ____ Other

14. How large is the sales force?

15. What consultant services are used (including outside laboratories)?

    Consultant:

    Training:

    Position when not consulting:

    Responsibility:

    Time per month:

16. In order for quality control to function properly, key executives must be appropriately educated, trained, and experienced. They should be approachable and sensitized by training or experience to quality control problems.

    Title Name Education Training Experience:

    President

    Vice-President

    Sales Manager

    Medical Director

    Plant Manager

    Engineering Manager

    Production Manager

    Quality Assurance Manager

    Quality Control Manager

17. Define the functional organization structure, including in detail all functions that contribute to acceptance or rejection decision for a product or its components.

    Who has the authority to:

    a. reject defective material

    b. approve rework of salvageable material

    c. dispose of non-salvageable material

    d. approve operating procedures

The head of quality reports to

The head of manufacturing reports to

It is important that quality control and production be kept separate and equal. The Quality Control function alone should have ultimate responsibility for removing a product at any stage in its processing into or from quarantine or into rejection status.

18. Product Information: In order to initiate quality control procedures, the dimensions of operations should be estimated. This requires the following information for the entire product line of the firm.

Type and Product Name	Quantity Manufactured	Quantity Packaged	
		Own label	Other label
Tablets			
Tablets, coated			
Tablets, multilayer			
Tablets, enteric coated			
Tablets, repeat dosage			
Tablets, sustained release			
Capsules			
Capsules, sustained release			
Liquids, external			
Liquids, oral			
Liquids, oral, sustained release			
Ophthalmic solutions			
Parenteral, sterile fill			
Parenteral, sterilized			
Syringe, prefilled			
Suppositories			
Granules, oral			
Powders			
Aerosols			
Aerosols, metered dose			
Sterile dressings			

How are the products promoted? (Obtain samples and package inserts.)
___ Professional journal ___ Lay journal ___ Internet ___ Newspaper ___ Other

The same procedures should be followed in assessing the operations of all outside contractors who contribute to the production of the finished pharmaceutical.

Attention should be focused on the critical concepts of a quality management system. The production cycle for each drug must be controlled so that optimum quality levels can be attained for each manufacturing sequence. The efforts of all personnel making product integrity decisions during processing must be coordinated and standardized to attain these desired levels. The materials and accompanying information flow through production must demonstrate that control, engineering, and production management have determined potential sources of error and have introduced control procedures to minimize the possibility.

A model of material and information flows for operations should show complete quality assurance and control surveillance of all operations involved with drug production, adequate information exchange to monitor and control this surveillance, and records that document all activity. The flow chart in Figure 1 depicts one model for an analysis of

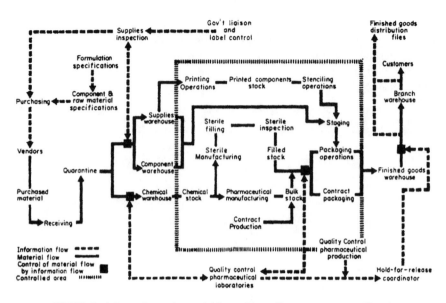

FIGURE 1   Information and material flow with quality assurance surveillance.

current operations that incorporates these considerations. More specific documentation and information requirements necessary to achieve control will be suggested in relevant CGMP chapters.

REFERENCE

1. Background on Regulation of Dietary Supplements, http://www.fda.gov/ola/2003/dietarysupplements1028.html, accessed July 2, 2006.

SEC. 400.400 CONDITIONS UNDER WHICH HOMEOPATHIC DRUGS MAY BE MARKETED (CPG 7132.15)

Background

The term "homeopathy" is derived from the Greek words *homeo* (similar) and *pathos* (suffering or disease). The first basic principles of homeopathy were formulated by Samuel Hahnemann in the late 1700s. The practice of homeopathy is based on the belief that disease symptoms can be cured by small doses of substances that produce similar symptoms in healthy people.

The Federal Food, Drug, and Cosmetic Act (the Act) recognizes as official the drugs and standards in the *Homeopathic Pharmacopeia of the United States* (HPUS) and its supplements [Sections 201 (g)(1) and 501 (b), respectively]. Until recently, homeopathic drugs have been marketed on a limited scale by a few manufacturers who have been in business for many years and have predominantly served the needs of a limited number of licensed practitioners. In conjunction with this, homeopathic drug products historically have borne little or no labeling for the consumer.

Today, the homeopathic drug market has grown to become a multimillion dollar industry in the United States, with a significant increase shown in the importation and domestic marketing of homeopathic drug products. Those products that are offered for treatment of serious disease conditions must be dispensed under the care of a licensed practitioner. Other products, offered for use in self-limiting conditions recognizable by consumers, may be marketed over-the-counter (OTC).

This document provides guidance on the regulation of OTC and prescription homeopathic drugs and delineates those conditions under which homeopathic drugs may ordinarily be marketed in the United States. Agency compliance personnel should particularly consider whether a homeopathic drug is being offered for use (or promoted) significantly beyond recognized or customary practice of homeopathy. If so, priorities and procedures concerning the agency's policy on health fraud would apply. (See CPG 7150.10 "Health Fraud-Factors in Considering Regulatory Action" 6/5/87.)

Definitions

The following terms are used in this document and are defined as follows.

- *Homeopathy*: The practice of treating the syndromes and conditions that constitute disease with remedies that have produced similar syndromes and conditions in healthy subjects.
- *Homeopathic Drug*: Any drug labeled as being homeopathic which is listed in the HPUS, an addendum to it, or its supplements. The potencies of homeopathic drugs are specified in terms of dilution, that is, 1× (1/10 dilution), 2× (1/100 dilution), etc. Homeopathic drug products must contain diluents commonly used in homeopathic pharmaceutics. Drug products containing homeopathic ingredients in combination with non-homeopathic active ingredients are not homeopathic drug products.
- *Homeotherapeutics*: Involves therapy which utilizes drugs that are selected and administered in accordance with the tenets of homeopathy.
- *Homeopathic Pharmacopeia of the United States (HPUS)*: A compilation of standards for source, composition, and preparation of homeopathic drugs. The HPUS contains monographs of drug ingredients used in homeopathic treatment. It is recognized as an official compendium under Section 201(j) of the Act.
- *Compendium of Homeotherapeutics*: An addendum to the HPUS that contains basic premises and concepts of homeopathy and homeotherapeutics; specifications and standards of preparation, content, and dosage of homeopathic drugs; a description of the proving[a] process used to determine the eligibility of drugs for inclusion in HPUS; the technique of prescribing the therapeutic application of homeopathic drugs; and a partial list of drugs which meet the criteria of the proving process and are eligible for inclusion in HPUS and other homeopathic texts.
- *Extemporaneously Compounded OTC Products*: Those homeopathic drug products which are often prepared by dilution to many variations of potency from stock preparations, and which: (i) have at least one OTC indication; (ii) are prepared pursuant to consumers' oral or written requests; and (iii) are not generally sold from retail shelves. Those products which are prescription drugs only cannot be provided to consumers as extemporaneously compounded OTC products, but may only be prepared pursuant to a prescription order.
- *Health Fraud*: The deceptive promotion, advertisement, distribution or sale of articles, intended for human or animal use, that are represented as being effective to diagnose, prevent, cure, treat, or mitigate disease (or other conditions), or provide a beneficial effect on health, but which have not been scientifically proven safe and effective for such purposes. Such practices may be deliberate, or done without adequate knowledge or understanding of the article.

Discussion

Section 201(g)(1) of the Act defines the term "drug" to mean articles recognized in the official United States Pharmacopeia (USP), the official HPUS, or official National Formulary (NF) or any supplement to them; and articles intended for use in the diagnosis, cure, mitigation, treatment, or the prevention of disease in man or other animals; articles (other than food) intended to affect the structure or any function of the body of man or other animals; and articles intended for use as a component of any articles specified in the above. Whether or not they are official homeopathic remedies, those products offered for the cure, mitigation, prevention, or treatment of disease conditions are regarded as drugs within the meaning of Section 201(g)(l) of the Act.

---

[a]A "proving" is synonymous with the homeopathic procedure (identified in HPUS as a Research Procedure) that is employed in healthy individuals to determine the dose of a drug sufficient to produce symptoms.

Homeopathic drugs generally must meet the standards for strength, quality, and purity set forth in the Homeopathic Pharmacopeia. Section 501(b) of the Act (21 U.S.C. 351) provides in relevant part:

Whenever a drug is recognized in both the USP and the HPUS, it shall be subject to the requirements of the USP unless it is labeled and offered for sale as a homeopathic drug, in which case it shall be subject to the provisions of the HPUS and not to those of the United States Pharmacopeia.

A product's compliance with requirements of the HPUS, USP, or NF does not establish that it has been shown by appropriate means to be safe, effective, and not misbranded for its intended use.

A guide to the use of homeopathic drugs (including potencies, dosing, and other parameters) may be found by referring to the following texts: *A Dictionary of Practical Materia Medica* by John Henry Clarke, M.D. (3 volumes; Health Science Press) and *A Clinical Repertory to the Dictionary of Materia Medica* by John Henry Clarke, M.D. (Health Science Press). These references must be reviewed in conjunction with other available literature on these drug substances.

Policy

Labeling

Homeopathic drug product labeling must comply with the labeling provisions of Sections 502 and 503 of the Act and Part 201 Title 21 of the Code of Federal Regulations (CFR), as discussed subsequently, with certain provisions applicable to extemporaneously compounded OTC products. Those drugs in bulk packages intended for manufacture or preparation of products, including those subsequently diluted to various potencies, must also comply with the provisions of Section 502 of the Act and Part 201 (21 CFR 201).

General Labeling Provisions

- *Name and Place of Business*: Each product must bear the name and place of business of the manufacturer, packer, or distributor in conformance with Section 502(b) of the Act and 21 CFR 201.1.
- *Directions for Use*: Each drug product offered for retail sale must bear adequate directions for use in conformance with Section 502(f) of the Act and 21 CFR 201.5. An exemption from adequate directions for use under Section 503 is applicable only to prescription drugs.
- *Statement of Ingredients*: Ingredient information shall appear in accord with Section 502(e) of the Act and 21 CFR 201.10. Labeling must bear a statement of the quantity and amount of ingredient(s) in the product in conformance with Section 502(b) of the Act, as well as 21 CFR 201.10, expressed in homeopathic terms, for example, $1\times$, $2\times$.

  Documentation must be provided to support that those products or ingredients which are not recognized officially in the HPUS, an addendum to it, or its supplements are generally recognized as homeopathic products or ingredients.
- *Established Name*: The product must be in conformance with Section 502(e)(1) of the Act and must bear an established name in accord with Section 502(e)(3) of the Act and 21 CFR 201.10. Many homeopathic products bear Latin names which correspond to listings in the HPUS. Since Section 502(c) of the Act and 21 CFR 201.15(c)(1) require that all labeling be in English, the industry is required to translate these names from Latin to their common English names as current labeling stocks are depleted, or by June 11, 1990, whichever occurs first. It is permissible for industry to include in the labeling both English and Latin names.
- *Container Size—Labeling Exemption*: For those products packaged in containers too small to accommodate a label bearing the required information, the labeling requirements provided under Section 502 of the Act and 21 CFR 201 may be met by placing information on the carton or outer container, or in a leaflet with the package, as designated in 21 CFR 201.10(i) for OTC drugs and in 21 CFR 201.100(b)(7) for prescription drugs. However, as

a minimum, each product must also bear a label containing a statement of identity and potency, and the name and place of business of the manufacturer, packer, or distributor.

■ *Language*: The label and labeling must be in the English language as described and provided for under 21 CFR 201.15(c)(1), although it is permissible for industry to include foreign language in the labeling as well.

Prescription Drugs

The products must comply with the General Labeling Provisions mentioned earlier, as well as the provisions for prescription drugs mentioned subsequently.

- *Prescription Drug Legend*: All prescription homeopathic drug products must bear the prescription legend, "Caution: Federal law prohibits dispensing without prescription," in conformance with Section 503(b)(1) of the Act.
- *Statement of Identity*: The label shall bear a statement of identity as provided for under 21 CFR 201.50.
- *Declaration of Net Quantity of Contents and Statement of Dosage*: The label shall bear a declaration of net quantity of contents as provided in 21 CFR 201.51 and a statement of the recommended or usual dosage as described under 21 CFR 201.55.
- *General Labeling Requirements*: The labeling shall contain the information described under 21 CFR 201.56 and 21 CFR 201.57. For all prescription homeopathic products, a package insert bearing complete labeling information for the homeopathic practitioner must accompany the product.

OTC Drugs

Product labeling must comply with the General Labeling Provisions mentioned earlier and the provisions for OTC drugs mentioned subsequently, as current labeling stocks are depleted or by June 11, 1990, whichever occurs first.

- *Principal Display Panel*: The labeling must comply with the principal display panel provision under 21 CFR 201.62.
- *Statement of Identity*: The label shall contain a statement of identity as described in 21 CFR 201.61.
- *Declaration of Net Quantity of Contents*: The label shall conform to the provisions for declaring net quantity of contents under 21 CFR 201.62.
- *Indications for Use*: The labeling for those products offered for OTC retail sale must bear at least one major OTC indication for use, stated in terms likely to be understood by lay persons. For extemporaneously compounded OTC products, the labeling must bear at least one major OTC indication for use, stated in terms likely to be understood by lay persons. For combination products, the labeling must bear appropriate indications(s) common to the respective ingredients. Industry must comply with the provisions concerning indications for use as current labeling stocks are depleted, or by June 11, 1990, whichever occurs first.
- *Directions for Use*: See the General Labeling Provisions above.
- *Warnings*: OTC homeopathic drugs intended for systemic absorption, unless specifically exempted, must bear a warning statement in conformance with 21 CFR 201.63(a). Other warnings, such as those for indications conforming to those in OTC drug final regulations, are required as appropriate.

Prescription/OTC Status

The criteria specified in Section 503(b) of the Act apply to the determination of prescription status for all drug products, including homeopathic drug products. If the HPUS specifies a distinction between nonprescription (OTC) and prescription status of products, which is based on strength (e.g., 30×)—and which is more restrictive than Section 503(b) of the Act— the more stringent criteria will apply. Homeopathic products intended solely for self-limiting disease conditions amenable to self-diagnosis (of symptoms) and treatment may be marketed OTC. Homeopathic products offered for conditions not amenable to OTC use must be marketed as prescription products.

- *Home Remedy Kits*: Homeopathic home remedy kits may contain several products used for a wide range of conditions amenable to OTC use. When limited space does not allow for a list of those conditions on the labels of the products, the required labeling must appear in a pamphlet or similar informational piece which is enclosed in the kits. However, as a minimum, each product must also bear a label containing a statement of identity and potency.

Other Requirements

All firms which manufacture, prepare, propagate, compound, or otherwise process homeopathic drugs must register as drug establishments in conformance with Section 510 of the Act and 21 CFR 207. Further, homeopathic drug products must be listed in conformance with the sections above. (*Note*: For a given product, variations in package size and potency are not required to be listed on separate forms 2657, but instead, may be listed on the same form.) Homeopathic drug products must be packaged in accordance with Section 502(g) of the Act. Homeopathic drug products must be manufactured in conformance with CGMP, Section 501(a)(2)(B) of the Act and 21 CFR 211. However, due to the unique nature of these drug products, some requirements of 21 CFR 211 are not applicable, as follows:

1. Section 211.137 (Expiration dating) specifically exempts homeopathic drug products from expiration dating requirements.
2. Section 211.165 (Testing and release for distribution): In the Federal Register of April 1, 1983 (48 FR 14003), the Agency proposed to amend 21 CFR 211.165 to exempt homeopathic drug products from the requirement for laboratory determination of identity and strength of each active ingredient prior to release for distribution.

Pending a final rule on this exemption, this testing requirement will not be enforced for homeopathic drug products.

Regulatory Action Guidance

Those firms marketing homeopathic drugs which are not in compliance with the conditions described above will be considered for regulatory follow-up. ⟨ ⟩ The Office of Compliance, HFD-304, Center for Drug Evaluation and Research, should be consulted before *warning* letters are issued.

Recommendations for the issuance of *warning* letters or other regulatory sanctions must be submitted in conformity with the Regulatory Procedures Manual and other Agency guidance concerning the review of regulatory actions.

Material between asterisks is new or revised
⟨ ⟩ Indicates material has been deleted
Issued: 5/31/88
Revised: 3/95

The following section was included in the Fifth Edition of this book and is retained in this edition for historical purposes. Chapter 2 deals with the scope of drug CGMPs. An important part of CGMP history includes the device CGMPs which used the same "umbrella" approach to CGMP regulation that was the underpinning of the existing CGMP regulation.

SUPPLEMENTARY INFORMATION AS TO CGMPS FOR MEDICAL DEVICES AS PROVIDED BY FOOD AND DRUG ADMINISTRATION (HFZ)341) 2098 GAITHER RD., ROCKVILLE, MD 20850 (301-594-4648)

Manufacturers establish and follow quality systems to help ensure that their products consistently meet applicable requirements and specifications. The quality systems for Food and Drug Administration (FDA)-regulated products (food, drugs, biologics, and devices) are known as CGMPs. The CGMP requirements for devices [part 820 (21 CFR part 820)] were first authorized by section 520(f) of the Federal Food, Drug, and Cosmetic Act (the Act) (21 U.S.C. 360j(f)), which was among the authorities added to the Act by the Medical Device Amendments of 1976 (Pub. L. 94-295). The Safe Medical Devices Act (SMDA) of 1990 (Pub. L. 101-629), enacted on November 28, 1990, amended Section 520(f) of the Act, providing FDA with the explicit authority to add preproduction design validation controls to the CGMP regulation. The SMDA also added a new section 803 to the Act (21 U.S.C. 383), which, among other things, encourages FDA to work with foreign countries toward mutual recognition of CGMP requirements.

The FDA undertook the revision of the CGMP regulation in part to add the design controls authorized by the SMDA to the CGMP regulation, and in part because the agency believes that it would be beneficial to the public, as well as the medical device industry, for the CGMP regulation to be consistent, to the extent possible, with the requirements for quality systems contained in applicable international standards, namely, the International Organization for Standards (ISO) 9001:1994 "Quality Systems—Model for Quality Assurance in Design, Development, Production, Installation, and Servicing" (Ref. 1), and the ISO working draft revision of ISO/DIS 13485 "Quality Systems—Medical Devices—Supplementary Requirements to ISO 9001," among others. The preamble to the November 23, 1993 proposal contained a detailed discussion of the history of the device CGMP regulation, from the agency's initial issuance of the regulation through the FDA's decision to propose revising the regulation.

The agency's working draft embraces the same "umbrella" approach to CGMP regulation that is the underpinning of the existing CGMP regulation. Thus, because this regulation must apply to so many different types of devices, the regulation does not prescribe in detail how a manufacturer must produce a specific device. Rather, the regulation lays the framework that all manufacturers must follow, requiring that the manufacturer develop and follow procedures, and fill in the details, that are appropriate to a given device according to the current state-of-the-art manufacturing for that specific device. The FDA has made further changes to the proposed regulation, as the working draft evidences, to provide manufacturers with even greater flexibility in achieving the quality requirements.

The FDA met with the Global Harmonization Task Force (GHTF) Study Group in early March 1994, in Brussels, to compare the provisions of the proposal with the provisions of ISO 9001:1994 and European National (EN) standard EN 46001 "Quality Systems—Medical Devices—Particular Requirements for the Application of EN 29001." The GHTF includes: Representatives of the Canadian Ministry of Health and Welfare; the Japanese Ministry of Health and Welfare; the FDA; and industry members from the European Union, Australia, Canada, Japan, and the United States. The participants at the GHTF meeting favorably regarded FDA's effort toward harmonization with international standards. The GHTF submitted comments, however, noting where the FDA could more closely harmonize to achieve consistency with quality system requirements worldwide. Since the proposal was published, the FDA has also attended numerous industry and professional association seminars and workshops, including ISO Technical Committee 210 "Quality Management and Corresponding General Aspects for Medical Devices" meetings, where the proposed revisions were discussed. See also ISO 9000 and Medical Device Regulation in the European Union-James W. Kolka, Vol 11, no. 3 (Oct 1990), in the F.D.C.M.D. Law Digest.

The original period for comment on the proposal closed on February 22, 1994, and was extended until April 4, 1994. Because of the heavy volume of comments and the desire to increase public participation in the development of the quality system regulation, the FDA decided to publish this notice of availability in the *Federal Register* to allow comment on the working draft, to be followed by two public meetings, as described subsequently, before issuing a final regulation.

Having addressed the many comments received, the agency has framed a final rule that achieves the public health goals to be gained from implementation of quality systems in the most efficient manner.

3 | Organization and Personnel
Subpart B

Joanne W. Cochran
JWC Training Associates, Jamison, Pennsylvania, U.S.A.

Joseph D. Nally
Nallianco LLC, New Vernon, New Jersey, U.S.A.

§211.22 RESPONSIBILITIES OF QUALITY CONTROL UNIT

(a) There shall be a quality control unit (QCU) that shall have the responsibility and authority to approve or reject all components, drug product containers, closures, in-process materials, packaging materials, labeling, and drug products and the authority to review production records to assure that no errors have occurred, or, if errors have occurred, that they have been fully investigated. The QCU shall be responsible for approving or rejecting drug products manufactured, processed, packed, or held under contract by another company.

(b) Adequate laboratory facilities for the testing and approval (or rejection) of components, drug product containers, closures, packaging materials, in-process materials, and drug products shall be available to the QCU.

(c) The QCU shall have the responsibility for approving or rejecting all procedures or specifications impacting on the identity, strength, quality, and purity of the drug product.

(d) The responsibilities and procedures applicable to the QCU shall be in writing; such written procedures shall be followed.

The regulations clearly assign to the QCU responsibility for approval or rejection of components, in-process materials, and products. At one time, the attainment of quality standards relied heavily on testing and inspection by quality control (QC)—quality was inspected into components and products. As QC testing is usually after the event and also relies on the evaluation of a relatively small sample, this approach was seen by some as both ineffective and inefficient. It also tended to separate accountability for production and quality.

The current good manufacturing practices (CGMPs) continue to focus most responsibility for quality onto the QCU despite the fact that since 1978, there have been more progressive regulations and standards such as the Device GMPs (CFR 820), EC GMPs, ISO, and ICH standards. There is no mention of Quality Assurance (QA) or Quality Management Systems in §211.22; however, it is implied that the QCU is responsible for all QA aspects. The responsibility for quality should clearly be a collaborative and functional responsibility between the quality groups and the functional areas (manufacturing, engineering, logistics, etc.). The new Food and Drug Administration (FDA) inspectional approach by quality systems may result in more recognition of the functional area quality responsibility.

There are no defined responsibilities for production management, unlike the European and World Health Organization guidelines that define both separate and joint responsibilities for these functions. This latter approach more clearly emphasizes that the consistent achievement of quality standards requires a team effort.

A more effective approach has been to design quality into products during the development phase and then to build in additional assurance during production. The regulations support this approach. New products are usually developed by a research and development unit, which will draft appropriate specifications. These must then be reviewed and approved by the QCU, which serves as an independent double check on this important parameter.

Within the production operations, all quality-impacting procedures and systems are to be approved by the QCU. Typically, these include manufacturing and control documents such as master and production batch records, standard operating procedures (SOPs), process validation protocols/reports, supplier certification protocols, complaint handling, and even buildings/equipment design protocols and work orders. As some of these systems are "owned" by other functions, it is essential that the QCU has effective procedures to ensure that such systems are reviewed in a timely manner and that changes cannot be introduced without approval.

The QCU is also responsible for approving or rejecting labeling. This responsibility lies in two areas. First, new or modified labeling should be reviewed to ensure that it complies with the Abbreviated New Drug Application (ANDA), New Drug Application (NDA), Over the Counter

Business Knowledge/Understanding
- Policies/Standards/Regulations
- Manufactured Products
- Pharm/Bio Industry Knowledge (Rx, OTC) – local
- Global
- General Business (Mfg., Logistics, TS, Eng.)
- HR, Finance, Marketing/Sales, IS

Leadership & Management Skills
- Facilitiation and Training
- Positive Regard and Motivation
- Performance Assessment/Feedback/Coaching
- Team Building
- Networking

Communication Skills
- Oral
- Written
- Presentation
- Influencing
- Negotiation and Conflict Management
- Language

Process Skills
- Time Management
- Quality Planning
- Proposal Preparation
- Project Management & Planning
- Process Management
- Change Management
- Problem Solving/Decision Making
- Management Tools
- Risk Analysis & Management

Quality Design & Prevention
- Basic Quality Tools
- Process Capability and Statistical QC
- Process/System Design
- Design of Experiments (DOE)
- Failure Mode Effect Analysis (FMEA)
- Value Engineering/Analysis & Re Engineering
- Benchmarking

Validation
- Facilities/Critical Plant Systems (IQ/OQ/PQ)
- Production Equipment (IQ/OQ/PQ)
- Manufacturing Process Validation
- Retrospective Process Validation (data review)
- Equipment Cleaning Validation
- Computer System Validation

Manufacturing
- Formulations & Manufacturing Procedures
- Statistical Process Controls
- Process Capability
- Equipment & Processing Parameters
- Environmental Monitoring

Chemistry/Microbiology
- Good Laboratory Practices
- Latest Instrumentation & Automation
- LIMS or Lab Management systems
- Methods Development
- Methods Validation

Audit/Assessment (Auditor) Skills
- Pharmaceutical/Biological Operations
- Packaging Materials Operations
- Bulk Pharmaceutical Chemical (BPC) Op.
- ISO 9000
- Process/Systems Approach

Suppliers/Contractors/Third Parties
- Quality Assurance Of Suppliers
- Quality Assurance Of Third Parties
- Partnership Management

Quality Systems
- Policies/Procedures
- Annual Product Review (APR)
- Complaints
- Failure Investigations/Materials Decisions
- Product Release
- Change Control
- Components, Materials/Warehousing/Dist.
- Calibration & Maintenance
- Management Notification
- Training
- Recall
- Technology Transfer

Customer Awareness/Understanding
- Complaint Handling
- Product Audits/Competitive Comparisons
- Customer Visits/Interviews
- Market/Customer Surveys
- Next Operation as Customer (NOAC)
- Six Sigma Approach

FIGURE 1 Knowledge and skill requirements for today's quality professional.

(drugs) (OTC) monograph, or other official requirements that are applicable. This checking may be delegated to other functions, but the QCU must assure that the checkers are qualified to perform their function and that they have done so. Secondly, incoming labeling supplies are to be evaluated to assure that they are correct. These responsibilities do not apply to promotional literature. The approval/rejection responsibility also applies to operations contracted out to other companies. This does not necessarily require any additional or duplicate testing. Provided the contractor has adequate procedures and is in full compliance with CGMPs, it should only be necessary to compare the test data with specification and with data from previous batches to identify trends.

Confirmation of the adequacy of the contractor will normally involve an audit and testing in parallel for a period of time; periodic re-evaluation should also occur.

The regulations require adequate laboratory facilities to be available to the QCU. This clearly allows the use of outside laboratories where necessary, but these should be comprehensively evaluated before use.

The FDA emphasis for the QCU is on release and/or rejection authority. Although this is important, the regulations ignore the ever-increasing importance of other activities by QC/QA that provide positive impact on quality. These include creation of quality awareness, involvement in product design and development, design and provision of quality training, facilitation for quality improvement, analysis of quality trend data to identify improvement needs and opportunities, identification of quality metrics, and collection and dissemination of quality benchmarking data. These additional activities all enhance the awareness and involvement of senior management, thereby assuring greater emphasis and attention to quality by all functions.

In order to effectively monitor and control virtually all GMP documents/activities in a facility, the quality professional should have a very high level of knowledge, skills, and experience. Figure 1 represents a comprehensive list of the knowledge and skills needed for high-level quality professional in the 21st century (1).

QCU Organization

No guidance is provided in the regulations about the actual organization of the QCU; a wide range of viable alternatives are in effect and often depend on the size of the company. The more typical organizational structure used in the industry involves three quality groups:

Group	Quality System Responsibility or Involvement
QC	Laboratory Systems (QC and Stability): Sample Management Reference Standards SOPs and Test Methods Method Validation and Transfer Instrument Qualification/Calibration and Maintenance Data Analysis, Records, and Document Control Change Control Contract Laboratory Management
QA	Site Quality Systems: SOPs and Document Control Master and Production Batch Records (MBR, PBR) Batch Record Review and Product Release Failure Investigations and CAPA Training Site Change Control Validation (Facilities, Equipment, and Computer) Supplier Management and Control Complaints Annual Product Review Management Notification
Compliance	Policies and Standards Audit (Internal and External) Regulatory Commitments and Documents Recall

In the above model, the three groups typically report to one quality head or leader. The activities and responsibilities in QA and Compliance can vary, and in smaller organizations, there may be only a QA and QC organization.

The regulations essentially expect the QCU or QC/QA and compliance functions to provide an independent policy-type role and to monitor the entire production process from purchasing of materials to product distribution and use. While organizational structures can vary, what is important is that the QCU is actively involved in all of those aspects of the operations. The quality function should also be proactive by evaluating data on processes, materials, and suppliers and by recommending changes that will improve efficiency and consistency. From a business perspective, the QCU should be a resource that plays a positive role in improving profitability.

The responsibility of the QCU with respect to acceptance/rejection has led to extensive discussion on organizational reporting lines. Obviously, the QCU function cannot report to the person who is held directly accountable for production. This could result in undue pressure being brought to bear to release marginal materials or products. In an enlightened company where everyone is fully aware of the importance of quality and committed to its achievement, this should not be an issue. However, some companies have gone a step further by insisting that the QCU should report outside of the plant operations to a group quality function or other scientific or technical function. This arrangement certainly provides an added level of independence and appears to be favored by the FDA. However, as previously expressed, quality standards cannot be assured by a police approach only. With a totally independent QCU, there is likely to be a chance of divided responsibility—production to produce and QC to confirm quality. It is preferable that the entire plant operate as a quality-aware team, every individual being expected to perform their job in such a manner as to achieve the quality standards. The QCU then becomes a supporting resource. This is more likely to occur if the QCU reports directly to the leader of the plant team, the plant manager. An adequate degree of independence can be incorporated into the organization by having a clearly defined functional reporting ("dotted line") relationship to a suitable scientific professional in the organization. This approach encourages a team spirit, which will result in a higher and more consistent achievement of quality standards. Even the review of potential accept/reject decisions should be handled by the management team so that everyone is involved in understanding the cause of the problem, the implications of the ultimate decision, and the need for appropriate corrective action. In the event that the team is in favor of acceptance when the QCU considers rejection to be correct, the final decision resides with QCU; the plant manager cannot override.

Some companies have taken the team approach a stage further by the introduction of self-managed work teams. The various functions or disciplines are incorporated into the team, which can be responsible for all its operational requirements. This can include, in extreme cases, hiring of new team members, discipline, allocation of wage increases or bonuses, work scheduling, product testing, and release/rejection decisions. In these instances, there still needs to be an independent QCU evaluation for final release/rejection to satisfy the regulations.

§211.25 PERSONNEL QUALIFICATIONS

(a) Each person engaged in the manufacturing, processing, packing, or holding of a drug product shall have education, training, and experience, or any combination thereof, to enable that person to perform the assigned functions. Training shall be in the particular operations that the individual performs and in CGMP (including the CGMP regulations in this chapter and written procedures required by these requirements), as they relate to the individual's functions. Training in CGMP shall be conducted by qualified individuals on a continuing basis and with sufficient frequency to assure that individuals remain familiar with CGMP requirements applicable to them.

(b) Each person responsible for supervising the manufacture, processing, packing, or holding of a drug product shall have the education, training, and experience, or any combination thereof, to perform assigned functions in such a manner as to provide assurance that drug product has the [safety, identity, strength, quality and purity] that it purports or is represented to possess.

(c) There shall be an adequate number of qualified personnel to perform and supervise the manufacture, processing, packing, or holding of each drug product.

Because the individual is the keystone to the manufacturing systems in the plant, it is important that individuals receive standardized training so that a high level of proficiency and competency can be maintained in the work area. To ensure that training produces a qualified individual, a defined training system must be established, in place, and in use.

TRAINING SYSTEM

Because the quality of the product is directly affected by actions that personnel take in their jobs, there must be assurance that they are properly trained. This assurance is built by having a training system that is robust, compliant, and sustainable and is able to produce individuals who are qualified.

Elements that are needed in a strong training system include the following:

- an accurate description of the job or role;
- specific training requirements for each job or role;
- training plan to accomplish the training;
- training materials that are applicable to each type of training;
- qualified trainers to perform the training;
- evaluations to measure the effectiveness of the training;
- a documentation and record keeping system for storage and retrieval of training records and materials.

Before a training system can be established, the basic question—what is the individual's job in the organization—has to be answered.

JOB DESCRIPTION

A job description should define the job and role of the individual. It should be fairly high level and include major job functions, not tasks the individual performs. The function may be divided into duties and responsibilities, competencies that an individual may have, and prerequisites needed (e.g., must be a college graduate, must have five years of experience, etc.). The manager is expected "to define appropriate qualifications for each position to help ensure individuals are assigned appropriate responsibilities (2)."

TRAINING REQUIREMENTS

When a job description has been created and approved by both Human Resources (HR) and the functional area, training requirements can then be defined. The knowledge and skills the individual needs to successfully perform the job should be identified. The desired skills and knowledge are compared against the individual's skills and knowledge when entering the position; gaps are identified. The training requirements will be derived from the identified gaps. Training requirements must be updated on a periodic basis. The training requirements must be reviewed when new processes and new equipment are introduced, regulations are changed, and job responsibilities and duties are changed. Training requirements should be established for all levels within the organization.

TRAINING PLAN

To ensure that the individual receives the "right" training at the "right" time, an individual training plan should be created and executed for each individual. The training plan should include the following (Fig. 2):

- an individual's curriculum or training topics or courses
- how the training will be performed
- the sequence of the training
- approximate training time
- a clear indication of when the individual will be fully qualified

The individual's curriculum should include both procedural (knowledge) training, usually SOPs, and competency-based skills training (on-the-job training or OJT). Both these types of training should be standardized as much as possible with the same training material used for all trainees.

The individual's curriculum will include different levels of training. These can be divided into three levels.

Levels of Training

The first level is an overview or general training conducted by the site HR or corporate training group as part of a new hire or induction training. The second level is held within the functional area. The third level, most specific to the employee, is one-on-one training.

The training plan should also indicate how personnel will receive the training. Will it be self-study by reading, classroom training, computer-based training, or one-on-one OJT?

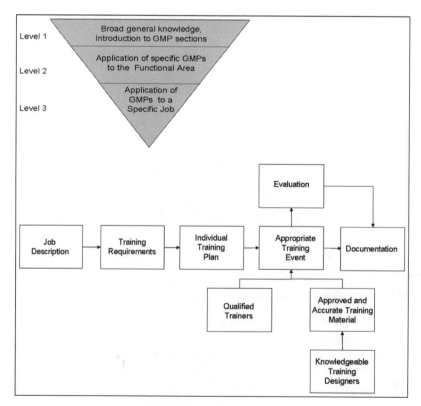

FIGURE 2 Training system.

training? This should be determined. The sequence of training should be clearly defined. An example would be learning about equipment. A logical sequence would be the assembly, operation, disassembly, and cleaning of equipment. The functional area may have specific equipment personnel are trained on first. By having a sequence defined, it is easier for both the trainee and the supervisor to determine where the trainee is in the training plan and to estimate a completion or qualification date.

The training plan should include an approximate training time. Training can be monitored to see whether it is going according to plan. This can help to identify problems early in the training process and make adjustments. There should be an agreement within the company and functional area as to what constitutes qualification. It can be one document reflecting a structured training program that is entered into the record keeping system, or a series of courses that the individual has completed.

The site may also have an annual or semi-annual training plan that defines what GMP training should be given to functional areas at the site and when.

QUALIFIED TRAINERS

Since 211.25(a) requires GMP training to be given by qualified personnel, the company should have a procedure and process for qualifying trainers. Minimum requirements for trainers may include some formal education (e.g., Train-the-Trainer course) or experience in presenting training, subject matter expertise in the subject they will be training, understanding of GMPs in terms of how it impacts the specific training they are responsible for, and the knowledge about how to train adults. Trainers should be selected for their ability to help individuals learn. If they are an OJT trainer, they should be able to demonstrate the skills and also clearly explain how to perform the skill. Often the person who has the best technical knowledge is not necessarily the best trainer.

TRAINING MATERIALS

Training materials should be designed and developed for most training. Whether the training is given once or many times, the information should be the same. The training material should be clear and well organized. Training materials should contain stated objectives. In addition to the content being trained, the reason behind the training should be explained and stressed. If training GMPs, the impact of the particular training on the production of the product should be explained.

The design of the training materials should be as important as the content. The design goal should be to optimize maximum retention by the employee. The designer should also understand that the retention rates vary based on the delivery method. To illustrate this, a Learning Pyramid, as defined by the National Training Laboratories in Bethel, Maine based on Edgar Dale's "Cone of Learning," found the following average retention rates for different training and teaching methods (3):

- 5% lecturing;
- 10% reading;
- 20% audiovisual;
- 30% demonstration;
- 50% discussion;
- 75% practice by doing;
- 90% teaching others.

GMP training should be designed by someone with a knowledge of adult learning theory. Adults learn differently from children because of their experience and knowledge. Once the training method is chosen, a template may be developed for the design of training materials. This helps to standardize the training material and provide consistency.

Training materials for GMP training should be approved by Quality. It should review the material for compliance with GMPs, company policies, and guidelines and procedures.

Training materials should be version-controlled, and if developed in conjunction with an SOP, reviewed, and updated, if required, when the SOP is updated or reviewed.

EVALUATIONS

Job skills and GMP training should be evaluated. For SOPs and other GMP training, evaluation tools such as questionnaires, case studies, discussions, and other tools may be used. If using questionnaires and seeking a percentage of correct answers, there should be a process in place if the percentage is not achieved. For job skills, a performance-based evaluation is used. The usual method is to have an observer watch the individual in training and complete a checklist. The individual should be able to demonstrate the correct practice without the coaching or help from another individual. The performance checklist is usually signed by the trainee and by a member of management. This document is retained. Evaluations should be performed after the training.

DOCUMENTATION/RECORD KEEPING

The training system and training processes should be documented, possibly in an SOP, describing how the training system works and the type of training included in the training system. Training records should be retained in a documentation system. There should also be a method to ensure that training curricula and training requirements are up-to-date in the event that an individual transfers to another job within the company. Any changes made to curricula should be documented and approved.

Training documentation should be readily retrievable. Often the FDA or other regulators request an individual's training record. The record should be made available to the regulators within a short period of time after their request. Performance evaluations should also be retained.

CONTINUING GOOD MANUFACTURING PRACTICES TRAINING

For personnel to be current with GMPs, GMP training must be a continuous process. It should begin when the individual first enters the company. At this time, they will attend induction/new hire training. Topics that typically may be included are information about the pharmaceutical industry, background and history of the GMPs, company policies, guidelines and general procedures required for all individuals, safety information, sometimes correct documentation practices, and possibly site-wide SOPs. The training may also include information about the individual's impact on the quality of the product and their responsibility as an individual of the company.

Once the individual enters their functional area, they will receive additional training.

The first will be an orientation to the area, which may include a tour, work rules for the area, technical training on the processes and equipment used in the functional area, information about specific products the functional area produces, and SOPs that personnel in the functional area use. At this point, specific GMP regulations that impact their functional area should be emphasized.

The individuals will begin their job skills training or on the job training. Even though this is not regulatory training, it is still considered GMP training. This training should be structured to include a consistent content using standardized training materials, qualified trainers, and a performance evaluation. There should be an agreement within the company and functional area as to what constitutes qualification.

Throughout the year, the individual should also attend ongoing GMP training. This may consist of awareness about specific GMP topics, current practices at the company, new industry practices, and new or amended GMP regulations. It may also be a refresher training to ensure that individual's skills and knowledge are up-to-date on specific GMP topics.

MANAGEMENT'S ROLE IN ASSURING QUALIFICATION

Within the Quality System, as defined by the FDA, functional area management has the responsibility to ensure that all of their individuals are qualified. If individuals are not fully qualified, they should be restricted from performing work for which they are not qualified. Once individuals have received basic information about GMPs, they then may begin training within their functional areas. During this time, they must be closely supervised. Management should also be accountable for assuring that their individuals attend the scheduled training.

Supervisors should encourage the transfer of skills gained from training into the day-to-day performance of the individual.

SUPERVISOR TRAINING

Supervisors should also have a training curriculum and be fully knowledgeable about GMPs. On the basis of their knowledge of GMPs, supervisors should be able to apply the GMP concepts to determine whether the processes for which they are responsible are in control.

Because of their role in the organization, they should also have strong process and technical knowledge. Training for this may be external to the company. This training may need to be updated periodically because of technological changes or new processes.

§211.28 PERSONNEL RESPONSIBLITES

(a) Personnel engaged in the manufacture, processing, packing, or holding of a drug product shall wear clean clothing appropriate for the duties they perform. Protective apparel, such as head, face, hand, and arm coverings, shall be worn as necessary to protect drug products from contamination.

(b) Personnel shall practice good sanitization and health habits.

(c) Only personnel authorized by supervisory personnel shall enter those areas of the buildings and facilities designated as limited access areas.

(d) Any person shown at any time (either by medical examination or supervisory observation) to have an apparent illness or open lesions that may adversely affect the safety or quality of drug products shall be excluded from direct contact with components, drug product containers, closures, in-process materials, and drug products until the condition is corrected or determined by competent medical personnel not to jeopardize the safety or quality of drug products. All personnel shall be instructed to report to supervisory personnel any health conditions that may have an adverse effect on drug products.

The purpose of this section of the regulations is to prevent harm to the drug product.

Individuals entering production areas or areas designated as GMP areas should dress appropriately to prevent contamination to the product or themselves if exposed to potent materials. Appropriate uniforms or protective clothing should be provided by the company, as needed. If protective clothing is needed, personnel should be trained in donning or removing the protective clothing.

Individuals should be physically fit to work. A pre-employment physical, based on the company's requirements, is usually a condition of employment. An employee who has been absent due to illness should report to the company's medical facility before reporting to work in their functional area.

Guidelines for personnel responsibilities are suggested in the following:

1. If aseptic processing is being conducted, personnel should cover skin and hair. The types of components recommended are:
    a. face masks;
    b. hoods;

    c.   beard/moustache covers;
    d.   protective goggles;
    e.   elastic gloves;
    f.   sterilized and nonshedding gowns that cover the whole body.

2. Only authorized personnel who have received the appropriate and relevant training should enter limited access areas. If it is necessary for a contractor, service technician, or personnel not employed by the company to enter these areas, they should receive minimum training necessary to perform their job and be accompanied by company personnel.
3. Eating, drinking, chewing, and smoking, and storing food, alcohol, cigarettes, or personal medication in the production and storage areas should be prohibited (4).
4. The wearing of jewelry in the production or quality control area should be evaluated because it can be a safety hazard and a contamination issue.
5. The wearing of make-up should be evaluated because it can provide a source of contamination.
6. If there is an equipment malfunction or process deviation, personnel should immediately report it to the supervisor.
7. If an individual is feeling ill, has a medical problem, or open lesions, the individual should report immediately to a supervisor.

§211.34 CONSULTANTS

Consultants advising on the manufacture, processing, packing, or holding of drug products shall have sufficient education, training, and experience, or any combination thereof, to advise on the subject for which they are retained. Records shall be maintained stating the name, address, and qualifications of any consultants and the type of service they provide.

Consultants are retained on the basis of their education and experience. Consultants' résumés should be kept on file for inspection by regulatory agencies. It is advisable for the company that hires the consultant to require that the consultants attend some type of training that includes GMPs before entering the site. This will help the consultant to understand the local procedures of the company and GMPs as it relates to the company's business.

In recent years, the FDA has looked seriously at training programs, personnel responsibilities, and the QCU. The following observations from warning letters emphasize their intent.

CITATIONS FROM WARNING LETTERS

1. "Failure to have a quality control unit that shall have the responsibility and authority to approve or reject all components, drug product containers, closures, in-process materials, packaging material, labeling, and drug products, and failure to follow written responsibilities and procedures applicable to the quality control unit." WL FLA-05-13, 12/04.
2. "Failure of the quality control unit to review production records to assure that no errors have occurred or, if errors have occurred, that they have been fully investigated." WL CBER-04-015, 8/04.
3. "Your firm failed to establish written procedures applicable to the function of the quality control unit." WL CBER-05-023, 6/05.
4. "Failure to assure that individuals responsible for supervising the manufacture and processing of a drug product have the education, training and experience to perform assigned functions in such a manner as to provide assurance that the drug product has the safety, identity, strength, quality, and purity it purports or is represented to possess." WL #6-NWJ-12, 5/06.
5. "Failure to conduct GMP training on a continuous basis." WL KAN 2006-15, 3/06.
6. "Failure to train employees in the particular operations performed or in current good manufacturing practice regulations as they relate to the employee's functions." Reference WL SEA 05-28, 8/05.

REFERENCES

1. Nally J, Kieffer RG. GMP Compliance, Productivity and Quality. Ch. 13. Interpharm, 1998:465–466.
2. Guidance for Industry, Quality Systems Approach to Pharmaceutical Current Good Manufacturing Practice Regulations. Draft Guidance, September 2004.
3. NTL Institute from Retention Rates from Different Ways of Learning.
4. 4 EUDRALEX Volume 4—Medicinal Products for Human and Veterinary Use: Good Manufacturing Practice.

SUGGESTED READINGS

1. Litterer J. Organizations: Structure and Behavior. New York: Wiley, 1968.
2. Murray E. Motivation and Emotion. Englewood Cliffs, NJ: Prentice-Hall, 1965.
3. Schein E. Organizational Psychology. Englewood Cliffs, NJ: Prentice-Hall, 1965.
4. Delmore F. Industry associations and self-regulation. Food Drug Cos Law J 1969; 24(11):557–564.
5. Saengen J. The key to quality programs. Bull Parent Drug Assoc 1969; 23:197–285.
6. Willig SH. Drug Abuse in Industry and Business 1971, Symposium Enterprises.
7. Donato BJ, Olsen KA. Food Drug Cosmetic "Legal Liability For Regulatory Affairs Professionals. . ." and Medical Device Law Digest. Vol. 11(2), July 1994.
8. Clark RH, Dimond KL. "Compliance Plan: A Drug Company's first line of defense . . ." Food, Drug, Cosmetic and Medical Device Law Digest. Vol. 13(2), May 1996.
9. Drug Free Work Place Requirements, Fed. Register, Jan 31, 1989.
10. Employee Benefits Bulletin. A review and analysis of recent developments. Alan M. Koral. Free Copies Available, Sep 1999.
11. Gallup D, Beauchemin K, Gillis M. A comprehensive approach to compliance training in a pharmaceutical manufacturing facility. PDA J Pharm Sci Technol 1999.
12. Vesper JL. Defining your GMP training program with a training procedure. BioPharm 2000; 13(11): 28–32.
13. Guidance for Industry, Sterile Drug Products Produced by Aseptic Processing—Current Good Manufacturing Practice. U.S. Department of Health and Human Services, Food and Drug Administration, Sep 2004.

4 | Buildings and Facilities
Subpart C

Robert Del Ciello
Northshire Associates, Martinsville, New Jersey, U.S.A.

§211.42 DESIGN AND CONSTRUCTION FEATURES

(a) Any building(s) used in the manufacture, processing, packing, or holding of a drug product shall be of suitable size, construction, and location to facilitate cleaning, maintenance, and proper operations.

Regarding buildings and facilities, there are two major areas of concern: the external environment and the internal environment. The external environment must be amenable to the location of well-designed and constructed buildings. It is insufficient that the buildings in which the production operations are to occur are clean and orderly and are of suitable size and construction. If the land, air, or water resources that surround the plant offer the potential for water damage, infestation, or contamination of any type, the facilities are in jeopardy of being judged unsuitable.

Several professional resources and functions will be involved in site selection. These are likely to include legal, real estate, and state and local government agencies, utility companies, engineers, and architects. These functions not only provide professional expertise but also are able to identify possible sources of financial incentives for building in specific geographical locations.

Pertinent consideration prior to purchase, construction, or alteration of existing facilities includes the following:

1. adequate space for future expansion;
2. zoning laws to allow anticipated development while restricting undesirable developments in the vicinity;
3. availability of water (quality and quantity), power, fuel, sewage, and waste-stream removal;
4. accessibility for employees (availability of public transportation), materials, and visitors (customers and suppliers);
5. environmental issues such as site history; soil, water, and air quality; and geological and topological issues (potential for flooding, earthquakes, and foundation instability);
6. proximity of undesirable activities—such as other industries, disposal sites, or open mining—that are likely to pollute or act as a source of vermin, insects, odor, or microorganisms;
7. availability of a suitable labor force (people, skills, wage expectations, labor relations and attitudes, and access to further education sources);
8. ability to provide adequate security arrangements;
9. proximity or accessibility to interrelated operations of the company—research and development (R&D), marketing, and internally produced intermediates or components;
10. political situation—government stability, trade policies and taxation (for foreign-based operations), and financial incentives.

Having identified a suitable location for the facility, the site-development plan is prepared and will include:

1. compliance with appropriate laws and regulations and any additional company standards;

2. site resources and infrastructure such as amenities, green spaces, parking for employees and visitors' vehicles and delivery and distribution vehicles, road and rail access, recreation areas, site utilities, tank farms and other external storage, and protection of wetlands and other restricted environments;
3. stormwater and waste management;
4. site security and access—fences, guard posts, and cameras;
5. buildings—siting, layout, usage, function interrelationships for efficiency, possible expansion, and surface finishes;
6. utilities—design, layout, and backup (especially for critical utilities as electricity and nitrogen for some chemical operations);
7. equipment—design, layout, spares, and capacity;
8. traffic flow—pedestrian arid vehicular (internal and external);
9. safety—for personnel and equipment, containment for hazardous materials, sprinkler system, emergency egress, and emergency services access;
10. external architecture to take into account local environmental conditions (wind, snow, and humidity) and aesthetic appearance blending local atmosphere, comparative image, and functionality;
11. ease of maintenance—accessibility to services (service ducts), ease of cleaning, and access for equipment;
12. selection and use of experienced contractors;
13. identification of project management responsibility;
14. validation plans and an effective change control procedure. Provision of design and "as-built" drawings;
15. construction materials.

The choices of materials of construction for manufacturing facilities are numerous. Some examples are presented subsequently. It is important, when choosing a material, to keep in mind the characteristics of the manufacturing process. Parenteral manufacturing operations have different requirements than oral dosage manufacturing operations do. Similarly, biotech and vaccine manufacturing operation requirements can differ. It is imperative that these requirements be considered when specifying wall, floor and ceiling construction, and finishes. The ISPE Guides provide guidance on these choices (refer to Suggested Readings).

a. *Walls.* The position of walls should provide an orderly movement of materials and personnel and should also take into account noise levels to provide acceptable working conditions. The interrelationship of different operations should minimize the potential for cross-contamination and component mix-up during storage and interdepartmental shipping.

    Walls in manufacturing areas, corridors, and packaging areas should be of plaster finish on high-quality concrete blocks or gypsum board. The finish should be smooth, usually with enamel or epoxy paint.

    Prefabricated partitions may be used in packaging areas where flexibility of layout is important. Prefabricated units have also been used in other areas, including sterile suites where panel joints must be given particular attention. Where possible, walls should be flush and projections should be avoided.

b. *Floors.* Floor covering should be selected for durability as well as for cleanability and resistance to the chemicals with which it is likely to come into contact.
    i.   Terrazzo provides a hard-wearing finish; both tiles and poured-in-place finishes are available. The latter is preferable for manufacturing areas; if tiles are used, care must be taken to ensure effective sealing between the tiles, which, otherwise, could become a harboring area of dirt and microorganisms.
    ii. Usually, ceramic and vinyl tiles are not recommended for production areas. However, if used, the between-tile sealing should be flush and complete.
    iii. Welded vinyl sheeting provides an even, easy to clean surface. This is not practical for heavy traffic areas, but can be of value in production areas, especially for parenteral and biotech products. Here, the lack of joints improves the ease of cleaning and sanitation.

    iv.   Epoxy flooring provides a durable and readily cleanable surface. However, the subsurface finish is extremely important.

c.   *Ceilings.* Suspended ceilings may be provided in office areas, laboratories, toilets, and cafeterias. They usually consist of lay-in acoustical panels of nonbrittle, nonfriable, non-asbestos, and noncombustible material.

       Manufacturing areas require a smooth finish, often of seamless plaster or gypsum board. All ceiling fixtures such as light fittings, air outlets and returns, PA system, and sprinkler heads should be designed to assure ease of cleaning and to minimize the potential for accumulation of dust.

d.   *Services.* In the building design, provisions must be made for drains, water, steam, electricity, and other services to allow for ease of maintenance. Access should, ideally, be possible without disruption of activity within the actual rooms provided with the services.

> (b)   Any such building shall have adequate space for the orderly placement of equipment and materials to prevent mix-ups between different components, drug product containers, closures, labeling, in-process materials, or drug products and to prevent contamination. The flow of components, drug product containers, closures, labeling, in-process materials, and drug products through the building(s) shall be designed to prevent contamination.

The requirements of this section involve the design and layout of the facility, which must minimize the possibility of mix-ups or contamination. Sufficient space must be provided to allow adequate separation of adjacent equipment and operations. An example of this includes the spatial separation of packaging lines so that packaging components, bulk product, and finished product cannot intermix between lines and that dust or spillage from one line cannot result in the contamination of adjacent equipment. For example, a common practice is to introduce a physical barrier between the packaging lines. This need not be a permanent wall; a moveable partition serves the purpose.

The layout of the manufacturing and support operations must account for efficient material, personnel, and equipment flow patterns. Adequate access control is required to restrict entrance to manufacturing areas. The most efficient and compliant flow pattern is the one that provides for unidirectional flow. This methodology minimizes backtracking and thus the potential for cross-contamination and mix-ups during the manufacturing operation, which includes cleaning operations. The separation of visitor and employee entrances to the manufacturing building should be considered. Visitor entrances to the manufacturing and support areas should be minimized and restricted to a single point.

> (c)   Operations shall be performed within specifically defined areas of adequate size. There shall be separate or defined areas or other such control systems for the firm's operations as are necessary to prevent contamination or mix-ups during the course of the following procedures:
>
> (1)   receipt, identification, storage, and withholding from the use of components, drug product containers, closures, and labeling, pending the appropriate sampling, testing, or examination by the quality control unit before release for manufacturing or packaging;
>
> (2)   holding rejected components, drug product containers, closures, and labeling before disposition;
>
> (3)   storage of released components, drug product containers, closures, and labeling;
>
> (4)   storage of in-process materials;
>
> (5)   manufacturing and processing operations;
>
> (6)   packaging and labeling operations;
>
> (7)   quarantine storage before release of drug products;
>
> (8)   storage of drug products after release;
>
> (9)   control and laboratory operations.

This subsection has on occasion been interpreted to mean that separate discrete areas must be provided for each of the listed operations. Although there is no dispute with respect to (5), (6), and (9), the other areas are more controversial. However, in the preamble to the regulations, it is specifically stated that "separate or defined is not intended necessarily to mean a separate room or partitioned area, if other controls are adequate to prevent mix-ups and contamination." The Federal Register of February 12, 1991 (56 FR 5671) proposed the inclusion of "as necessary" to qualify the requirement for separate or defined areas. This was to clearly indicate that separate rooms or partitioned areas are not necessary if other controls exist to prevent mix-ups or contamination. This clarification is also present in the preamble to the 1978 final rule. The intent has now been confirmed in the revision (effective February 21, 1995) that added the words "or other control systems." Facilities and equipment should be designed and operated to minimize the potential for mix-ups or contamination. Where there is reliance on systems, paper, or computer, it must be demonstrated that such systems are effective and followed. As with all key systems, employees must be fully trained in their use and routine audits should be performed. Systems control of storage, and flow of materials and product, can be more effective than physical separation and is certainly more efficient with respect to space utilization and materials handling. Physical movement of materials into and out of quarantine, for example, not only adds cost but, by adding another action, actually increases the potential for error. The further sophistication of bar-coding materials throughout the various plant operations and linking this into a computer materials handling procedure greatly minimize the chance of unreleased or substandard materials being used inadvertently.

Some companies have found segregation using flexible physical areas to be a satisfactory alternative. For example, in a warehouse, a quarantine area can be designated around the goods simply by roping off the quarantine goods or by placing floor markings. This arrangement allows easy expansion or contraction of the area to meet changing volumes. It should be noted, however, that even physical separation will be ineffective in preventing mix-ups and contamination, unless accompanied by adequate support procedures.

Although segregation of materials by systems is acceptable, this is obviously inadequate for any materials requiring specific storage conditions such as low temperature or controlled humidity. In these instances, the required conditions would need to be provided. Receiving areas, where materials are unloaded from delivery transportation, are an access point for airborne contamination such as dirt, dust, insects, vermin, birds, and even engine fumes from the delivery vehicles themselves. Where possible, these access points should be protected by flexible curtains to minimize the gap to the outside when vehicles are unloading; air curtains between the receiving bays and the warehouse proper may also be used to provide additional protection to the warehouse environment. Insect and rodent traps are usually required.

Sampling, particularly of chemical components, requires separate comment. When containers of components are opened for sampling purposes, the contents are exposed, albeit for short periods, to ambient conditions. It should be demonstrated that normal warehouse conditions do not expose the materials to unacceptable contamination from other components, particulate matter, or microorganisms. Otherwise, separate facilities will need to be provided for sampling. This is addressed in Section 211.80(b).

Traditionally, most warehouses for components and finished products have been operated under ambient conditions. Generally, this has been adequate since most pharmaceutical products are sufficiently stable under such conditions and stability data are available to support defined shelf lives. The prevailing conditions in a warehouse must be monitored and any particularly sensitive products or components should be provided appropriate environments. For relatively stable products, it has been a common practice to omit any specific storage conditions on the labeling; it was then assumed that the United States Pharmacopeia (USP) conditions of "room temperature" applied. Recently, some sections of the Food and Drug Administration (FDA) have been insisting that in order to obtain approval for new products, defined storage conditions must be stated on the labeling of all products. This has been further complicated by the revision to the USP definition of Controlled Room Temperature, from 15°C to 30°C to "A temperature maintained thermostatically that encompasses the usual and customary working environment of 20° to 25°C (68° to 77°F); that results in a mean kinetic temperature calculated to be not more than 25°C and that allows for excursions between 15°C and 30°C (59° and 86°F) that are experienced in pharmacies, hospitals and warehouses. Articles may be labeled for storage at 'controlled room temperature' or at 'up to 25°C' or other wording based on the same mean kinetic temperature." Although it may seem relatively easy to calculate the mean kinetic temperature, in practice, this is not so simple. Products remain in facilities for differing lengths of time and it is possible, for example, that a batch of product stored in a warehouse for three summer months may be exposed to a higher mean kinetic temperature than if it were stored for three summer and three fall months. The impracticality of evaluating mean kinetic temperatures for the storage of individual batches is obvious, and it is hoped that a general calculation possibly based on average storage periods will be acceptable. This revised definition, which does not allow excursions above 30°C, is also resulting in many warehouses having to install air conditioning, at considerable expense, to control relatively brief exposure to higher temperatures in summer months.

The issue of labeling is still unresolved. It is considered that the use of the term "controlled room temperature" would be meaningless for certain areas of trade that are not aware of the USP. The more extensive wording of USP, Controlled Room Temperature, is too verbose for most labels and incomprehensible to many areas of the trade. Many companies have retained the old USP definition (15–30°C) on their labeling until this is resolved by the FDA.

The subject of stability studies is addressed in more detail in Section 211.66, and since ICH agreement, future stability studies will be performed at $25 \pm 2°C$.

The storage period for rejected materials awaiting destruction should be kept as short as possible. These materials take up valuable space and there is always a risk that they may be inadvertently used. Even with the use of a validated storage procedure, it may be advisable to maintain physical segregation. FDA investigators have found reference to stored reject materials a useful way to identify production deviations.

Many FDA investigators consider that the presence of reject materials and products demonstrates failure with procedures and, consequently, is evidence of good manufacturing practice (GMP) violations. Rejections of materials could be due to inadequate definition/agreement of specifications, different test methodologies used by the supplier and the customer, nonvalidated analytical methods, nonvalidated production procedures at the supplier resulting in variable quality, use of untrained analysts, or expected data variability around a specification limit. For products, many of the same potential causes apply plus noncompliance with procedures. Frequently an investigator will consider product rejection evidence of an inadequately validated process. Obviously, the cause of any rejection does need to be thoroughly investigated and appropriate corrective action taken and documented.

The degree of separation of individual manufacturing and processing operations will be dependent on the nature of these operations. Raw materials are usually dispensed in an area specifically designed to minimize the potential for mix-ups and cross-contamination. Scales are separated by partitions and supplied with dust extraction and sometimes laminar airflow. Where a manufacturing process requires several different pieces of equipment (e.g., blender, granulator, and dryer), these may all be contained in one room or suite of rooms. Processes for different products should use completely segregated facilities. Where this is not possible, adequate physical separation should be maintained along with documented evidence to demonstrate the adequacy of the arrangement. This evidence could include data from the analysis of air samples, which confirms that the potential for cross-contamination is negligible. However, where such arrangements are necessary, it would be advisable to provide separate facilities for any particularly potent or sensitive products or manufacture in campaigns.

Packaging and labeling operations are usually kept separate from manufacturing. Even when a highly automated process is used, and packaging immediately follows manufacturing, the packaging is usually performed in an adjacent area.

The need to provide physical barriers between packaging lines has already been mentioned. Particular attention needs to be paid to the on-line storage of bulk product, labeling, and filled but unlabeled containers. During packaging operations, it is not uncommon for individual pieces of line equipment to break down. Under these circumstances, it may be economically viable to continue the operation and to accumulate part-packaged product until the effective unit is repaired. When the labeling unit breaks down, special care must be taken to ensure that unlabeled containers do not get onto another line, or even intermixed with a different batch of the same product. Where possible, accumulation tables should be an integral part of a packaging line, thereby enabling short down-times on equipment to be handled without the need to remove part-packaged product from the line. Protracted breakdown of labeling equipment may, on occasion, result in amounts of unlabeled product in excess of the capacity of accumulation tables. Also, some processes are designed to produce filled unlabeled product. This includes sterile products such as ampoules and vials, which are labeled outside of the sterile suite. Obviously, in such situations, great care must be taken to prevent mix-ups. When labeling is to be performed later, security can be enhanced by holding the unlabeled product in sealed or locked containers (see also chapter 8).

The requirement of separate areas for control and laboratory operations does not preclude the use of in-process testing within the manufacturing and packaging areas. However, the environmental conditions in these areas must be suitable for the proper operation of the equipment and performance of the testing. In some instances, it may be necessary to site in-process test equipment in designated areas or rooms within the manufacturing or packaging facilities.

(c) (10) Aseptic processing, which includes as appropriate:
   (i)   floors, walls, and ceilings of smooth, hard surfaces that are easily cleanable;
   (ii) temperature and humidity controls;
   (iii) an air supply filtered through high-efficiency particulate air (HEPA) filters under positive pressure, regardless of whether flow is laminar or nonlaminar;
   (iv) a system for monitoring environmental conditions;
   (v)   a system for cleaning and disinfecting the room and equipment to produce aseptic conditions;
   (vi) a system for maintaining any equipment used to control the aseptic conditions.

This subsection emphasizes the special requirements associated with aseptic processing. Some companies also apply aseptic processing techniques during the production of terminally sterilized products. In these cases, compliance with the regulations is not mandatory, although it does make good sense. The absence of a terminal sterilization process and the relative ineffectiveness of end product sterility testing place a critical reliance on the environmental conditions associated with aseptic processing. Recognizing the importance of aseptic processing in the production of injections, the FDA issued a "Guideline on Sterile Drug Products Produced by Aseptic Processing" in 1987. However, with respect to facilities, the guide provides guidance only on air quality, airflow, and pressure differentials. There is no information on surface finishes.

Floors, walls, and ceilings in sterile suites are subject to intensive and frequent cleaning and sanitization; they must be composed of smooth, hard surfaces with a minimum of joints. Additionally, they should be resistant to abrasion, not shed particles, free from holes, crevices, and cracks, sufficiently flexible to accommodate building strains, and impervious to water and cleaning and sanitization solutions. Regular examinations should be performed to identify and repair any cracks in the surfaces or around service fittings and windows. Critical rooms, such as those for filling of final containers, should preferably have windows to allow supervision without the necessity for access. All service fittings should be flush with surrounding surfaces for ease of cleaning and sanitization.

Temperature and humidity need to be controlled primarily for the comfort of operators. The gowning requirements to minimize the potential for microbial contamination from operators are rather stringent and can easily cause personal discomfort, which could, in turn, adversely impact on the aseptic processing. Conditions in the order of 68°F and 45% relative humidity have been found to be suitable.

The most critical factor in aseptic processing is the microbial and nonviable particulate condition of the air. This air is provided by way of high-HEPA filters and the quality of the air is adjusted to meet the varying needs of the different processing areas. The requirements of the environment in these areas vary slightly between the FDA and European Union (EU) regulations. It is important to understand these requirements and the differences between these regulations. It is recommended that one refers to the ISPE Guide on Sterile Manufacturing Facilities.

Such heavy emphasis on air quality necessitates appropriate systems and procedures for monitoring. This will include evaluation of pressure differentials between rooms, particulate levels (viable and nonviable), and also temperature and humidity.

Air-pressure differentials should be monitored automatically and audible or visual warning alarms are an added advantage. The number of rooms interlinked in a sterile suite makes the balancing of air-pressure differentials very difficult. Movement of people and materials, involving opening and closing of doors, adds to the complexity. Computer control can provide a more rapid response to these changing conditions. HEPA filters must be tested at regular intervals for the presence of leaks; such leaks would also be likely to affect pressure differentials. Particulate levels are usually monitored during each work shift or part shift when operators leave and return; air-sampling devices are most commonly used since they do provide a quantitative measure of the volume of air sampled. However, for microbial evaluation, settle plates can also be of value since they provide a measure of the microbial impact over a more protracted period.

Cleaning and disinfection of aseptic facilities and equipment are of obvious importance, especially in the critical areas. Procedures must be validated with respect to both removal of previous product and to demonstrate effective disinfection. Residual amounts of any cleaning or disinfectant agents should be at an acceptably low level. In order to minimize the possibility of microbial resistance, the disinfectant should be changed periodically. After cleaning and disinfecting, rooms and equipment must be maintained in such a manner that these conditions are not impaired.

For certain pieces of equipment, a "clean-in-place" procedure is most effective. This is particularly valuable with tanks and pipelines where access or dismantling may be difficult. The procedure basically consists of applying sequential wash, sanitization, and flush cycles to the assembled equipment. Sanitization is often accomplished with high-pressure steam.

Having established suitable conditions for aseptic processing, it is necessary to have a defined maintenance program for equipment and facilities. In addition to the servicing of HVAC equipment and checking of ducts, filters, and service ports for leaks, the physical condition of walls, floors, and ceilings should be monitored. Slight shifts in building position, which are not uncommon, can result in cracks, which then need to be repaired.

The environmental conditions are essentially established by flushing the area with high-quality air. Any disruption in this flushing process will affect pressure differentials and possibly adversely affect the conditions. Consequently, the provision of auxiliary generating capacity to maintain essential air-handling equipment can be a valuable asset. This equipment should switch on automatically in the event of a power failure and will allow completion of ongoing sensitive operations and maintain the environment until normal power is restored. Where such auxiliary power is not available, aseptic operations should cease immediately if there is a power failure and restarting will not be possible until the re-establishment of the defined condition has been confirmed.

The industry is utilizing barrier technology systems as a means of enhancing levels of sterility assurance and, for new facilities, reducing costs. Two main variations exist: barrier isolation systems, which protect the product from the operators and the external environment, and barrier containment, which additionally protects the operator from the product. Barrier containment is frequently used when handling high hazard or cytotoxic agents. Both systems contain the aseptic operating environment within a closed system with no direct access to operators. Access is via glove ports with sterilized components being fed directly from a sterilizing/depyrogenizing tunnel or after batch processing via a rapid transfer port. Clean and sterilize in place procedures are required.

The space required to be maintained at a Class 100 (Class A) level is significantly less than that for traditional aseptic rooms and this can have important cost implications. The overall facility can be smaller, the Class 100 (Class A) space costs less than one-quarter the cost of clean space and gowning areas can be eliminated. Additional benefits include more consistent assurance of sterility since the microbial and nonviable particulate content of the processing environment is constant—no operator involvement and more comfortable working conditions for employees. The actual level of sterility assurance should be greatly enhanced, possibly from 1×10^{-3} to 10^{-5} or 10^{-6}.

(d)   Operations relating to the manufacture, processing, and packing of penicillin shall be performed in facilities separate from those used for other drug products for human use.

This is the first of several portions of the regulations that pertain specifically to the production of products containing penicillin (see also Sections 211.46(d) and 211.176).

The industry has taken the steps to have completely separate manufacturing facilities, and usually manufacturing sites, for these types of products rather than attempt to separate these operations in the same building. This is due to the fact that it has been almost impossible to prevent the migration of penicillin products throughout the building.

§211.44 LIGHTING

Adequate lighting shall be provided in all areas.

In order to meet lighting requirements, it is necessary for the manufacturer to define the term "adequate." It is defined as the amount of light (lux or foot-candles) reaching the working surface for each area involved in the production of pharmaceuticals. Public standards exist for some types of work. Normally, a range of 30 to 50 foot-candles ensure worker comfort and ability to perform efficiently and effectively; however, 100 foot-candles may be needed in some areas, as well as special lighting for some operations, such as inspection of filled vials. Once the light levels have been defined, it is necessary that they be measured periodically and the results recorded. The specifications should call either for replacement of light sources when some level above the established minimum has been reached or, alternatively, routine replacements of light sources on some schedule that has been shown adequate to ensure that light levels do not drop below the established minimum.

§211.46 VENTILATION, AIR FILTRATION, AIR HEATING, AND COOLING

(a)   Adequate ventilation shall be provided.
(b)   Equipment for adequate control over air pressure, microorganisms, dust, humidity, and temperature shall be provided when appropriate for the manufacture, processing, packing, or holding of a drug product.
(c)   Air filtration systems, including prefilters and particulate matter air filters, shall be used when appropriate on air supplies to production areas. If air is recirculated to production areas, measures shall be taken to control recirculation of dust from production. In areas where air contamination occurs during production, there shall be adequate exhaust systems or other systems adequate to control contaminants.
(d)   Air-handling systems for the manufacture, processing, and packing of penicillin shall be completely separate from those for other drug products for human use.

The regulations provide minimal guidance by stating that ventilation should be adequate. It is then up to the producer to demonstrate adequacy with respect to the operations being performed. The conditions necessary for aseptic processing have already been described [Section 211.42(c)(10)].

Air-handling systems should consider the following factors.

1.   Placement of air inlet and outlet ports. These should be sited to minimize the entry of air-borne particulates or odors from the surrounding areas. Outlets should not be sited near inlets.
2.   Where recirculation of air is acceptable, adequate precautions must be taken to ensure that particulates from a processing area are removed. This will usually require an alarm system or an automatic cutoff in the event that a filter develops a hole. Dust extraction systems should be provided, where appropriate, to further minimize this potential problem.

3. The degree of filtration and the air volumes should be matched to the operations involved.
4. Temperature and humidity conditions should provide personnel comfort, which will enhance employee performance.
5. Where differential pressures are required between adjacent areas, suitable monitoring equipment must be provided. For example, solids manufacturing areas are usually maintained at a negative pressure in relation to adjacent rooms and corridors in order to minimize the possibility of dust migration to these other areas.
6. The siting of final air filters close to each room being serviced eliminates concerns regarding the possibility of small leaks in the air duct system. Air usually enters rooms near the ceiling and leaves from the opposite side near the floor.

As with all systems, operating requirements should be defined and monitored at appropriate frequency to ensure compliance. If conditions are shown to have fallen below the required standards, it may be necessary to more thoroughly evaluate any products that were produced during the period in question.

It is important to monitor filters to ensure proper operation. After initial mounting and testing with a smoke generator of defined particle size range, the use of a differential manometer to monitor pressure drop across the filter gives warning of both breaks in the filter and buildup of retained particulates necessitating filter replacements. A specification of maximum permissible pressure drop before replacement should be defined.

As indicated previously for sterile areas, computer control of HVAC systems is more likely to allow the delicate balancing of the various air pressures, airflows, temperatures, and humidity. When this is expanded to the entire plant systems, the computer control can additionally optimize energy utilization, thereby reducing costs.

The regulations also make specific reference to the handling of penicillin products [211.46(d)]. Not only shall such operations be performed in separate facilities from those for other drug products for human use [211.42(d)], they must also have separate air-handling systems.

§211.48 PLUMBING

(a) Potable water shall be supplied under continuous positive pressure in a plumbing system free of defects that could contribute contamination to any drug product. Potable water shall meet the standards prescribed in the Environmental Protection Agency's (EPA's) Primary Drinking Water Regulations set forth in 40 CFR Part 141. Water not meeting such standards shall not be permitted in the potable water system.
(b) Drains shall be of adequate size and, where connected directly to a sewer, shall be provided with an air break or other mechanical device to prevent back-siphonage.

The Public Health Service Drinking Water Standards are now administered by the EPA. The standard is somewhat variable in that the frequency of examination of the water is dependent on the size of the population served. This leads to some uncertainty in water quality if potable water becomes part of a pharmaceutical product. This problem does not arise for products of the official compendia (USP and National Formulary), for which purified water is always required. The text of the drinking water standards is found in 40 CFR 141.

The FDA usually will not inquire into whether the potable water does meet the standard, if the manufacturer connects the potable waterline to a public supply that meets the standard. A quality control problem arises, however, in that the public supply ensures the quality only to the edge of the manufacturer's property, and even then tests, usually, only at the central reservoir. The water can lose quality in transmission through the public piping system and, of course, through the manufacturer's system. The prudent manufacturer will test potable water periodically. If potable water is obtained from wells under the control of the manufacturer, periodic testing is mandatory.

Drains, particularly those in production areas, can be a potential source of microbial hazard. The requirement to include an air break between drain and sewer is an attempt to minimize this by eliminating the chance of back-siphonage. Drains should also be regularly disinfected.

§211.50 SEWAGE AND REFUSE

Sewage, trash, and other refuse in and from the building and immediate premises shall be disposed of in a safe and sanitary manner.

A pharmaceutical plant may consider disposal in several different ways.

1. *Product Disposal.* Any product requiring disposal should initially be separated from its packaging if appropriate. For example, any product to be disposed of in an approved landfill site should not be left in impermeable glass, plastic, or other containers, which would significantly delay destruction. There are risks associated with the destruction of products—potential for the product to get diverted, legitimately or otherwise, during the disposal sequence and contamination of groundwater. Disposal procedures should involve agents with a proven record of dealing with such sensitive materials or the use of company personnel to accompany the material from plant to disposal. Ideally, incineration procedures have preference over landfill. Where incineration is used, product in plastic or other flammable packaging may not need to be returned to bulk.
2. *Printed Packaging Disposal.* The disposal of printed packaging components including labels, inserts, and cartons poses no health risk. However, ineffective disposal, such as into public landfill, can give rise to public concern that product may be associated with the packaging. Such materials should preferably be incinerated.
3. *General Trash and Sewage.* Normal local services will usually be adequate for trash and sewage. However, internal procedures should be sufficiently rigorous and monitored, to ensure that product and packaging waste do not get intermixed. Containers used within the plant to accumulate waste materials should be clearly marked to denote their designated use.

§211.52 WASHING AND TOILET FACILITIES

Adequate washing facilities shall be provided, including hot and cold water, soap or detergent, air driers or single-service towels, and clean toilet facilities easily accessible to working areas.

In addition to GMP regulations, Occupational Safety and Health Administration regulations impact on washing and toilet facilities (see 29 CFR 1019.141). These require toilet rooms to be separate for each sex except where individual locked toilet rooms are available and also define the minimum number of water closets based on the number of users. The legal requirements of GMPs specify minimum facilities for personnel. Management concern with employee morale and extra measures to ensure minimum probability of contamination suggest additional emphasis and activities.

1. Eating facilities:
   a. Eating and drinking are permitted only in separate eating facilities, well segregated from all production areas [see also 29 CFR 1910.141(s)(g)(2)]. Smoking is now usually prohibited in manufacturing building.
   b. Prominent signs indicating these rules are posted at entrances to production areas.
   c. Enforcement procedures against violators are taken by management.
   d. Permanent facilities for breaks and people bringing lunches are required. Cafeterias serving hot meals are ideal to reduce the amount of food, a potential contamination source, being brought into the plant.
2. For production and materials processing areas:
   a. Drinking, eating, smoking, tobacco chewing, and expectoration are prohibited.
   b. Tissues and closed disposal containers are readily available.

3. Lavatories and lockers:
    a. Adequate in number for the number of personnel employed.
    b. Conveniently located at all areas.
    c. Hot shower facilities are provided [see also 29 CFR 1910.141-(s)(d)(d)].
    d. Disinfectant soaps are utilized.
    e. Adequate ash and waste receptacles are provided.
    f. Periodic cleaning of the area during each shift with logging of times and conditions is mandatory.
    g. Complete cleaning with cleansing and disinfectant agents daily. Follow-up inspection by supervisory personnel is logged.
    h. Specific rest areas for female employees are provided.
    i. Eating and drinking are not permitted. Foods and beverages for meals and breaks may be stored only in lockers and then removed to a separate eating area.
    j. Areas separated from all aseptic spaces by an air lock.

§211.56 SANITATION

(a) Any building used in the manufacture, processing, packing, or holding of a drug product shall be maintained in a clean and sanitary condition. Any such building shall be free of infestation by rodents, birds, insects, and other vermin (other than laboratory animals). Trash and organic waste matter shall be held and disposed of in a timely and sanitary manner.
(b) There shall be written procedures assigning responsibility for sanitation and describing in sufficient detail the cleaning schedules, methods, equipment, and materials to be used in cleaning the buildings and facilities; such written procedures shall be followed.
(c) There shall be written procedures for the use of suitable rodenticides, insecticides, fungicides, fumigating agents, and cleaning and sanitizing agents. Such written procedures shall be designed to prevent the contamination of equipment, components, drug product containers, closures, packaging, labeling materials, or drug products and shall be followed. Rodenticides, insecticides, and fungicides shall not be used unless registered and used in accordance with the Federal Insecticide, Fungicide, and Rodenticide Act (7 U.S.C. 135).
(d) Sanitation procedures shall apply to work performed by contractors or temporary employees as well as to work performed by full-time employees during the ordinary course of operations.

This requirement relates to the availability of effective cleaning and sanitation programs and confirmation that they have been followed. No details are given, nor should they be, on how to achieve the desired conditions. Cleaning and sanitation programs should be adjusted to meet the specific needs of each facility. In addition to the cleaning of floors, walls, and ceilings, there should be attention to dust extraction and air input systems. Ductwork, especially for dust extraction systems, can become a potential explosion hazard if dust is allowed to accumulate.

Cleaning procedures should be written in sufficient detail, with respect to materials, equipment, process, and frequency, such that they are unambiguous. Where appropriate, data should be accumulated to confirm the adequacy of the cleaning procedure.

The total elimination of rodents, birds, and insects is virtually impossible and the regulations do refer to freedom from "infestation." The use of rodenticides, fungicides, fumigating agents, and other techniques should be combined with good hygienic practices. Spilled materials, such as sugar, that might attract creatures should immediately be eliminated. Holes in buildings that could provide additional means of access should be blocked.

Where traps and other lethal techniques are used, there should be frequent examination and removal of "corpses," which could in time become a source of further contamination. If these traps consistently yield results, attempts should be made to identify and eliminate the source of the problem.

Frequently, rodenticides and other treatments are contracted out. As with any contracted service, the company must assure that the procedures used are viable, achieve the desired results, and that they are followed.

§211.58 MAINTENANCE

Any building used in the manufacture, processing, packing, or holding of a drug product shall be maintained in a good state of repair.

Deterioration of buildings not only presents a poor image of the facility, but can also impact on product quality. Cracks and holes in walls, floors, or ceilings can provide access for insects, rodents, birds, dirt, or microorganisms. They can also hinder cleaning and sanitation, thereby increasing the potential for cross-contamination or microbial multiplication. Floor cracks can also become a safety hazard for people or even dislodge materials from trucks.

The ingress of water from roof leaks can cause significant damage to materials and equipment, give rise to electrical failures and fires, and result in damage to the basic structure of the building. Additionally, holes in the roof or near the tops of buildings provide ready access to birds, which may then be encouraged to nest within the building.

BUILDINGS AND FACILITIES

Damage to insulation or pipes and ductwork will detract from the basic purpose of such insulation. It may also result in freezing and eventual leakage of pipes and in the shedding of insulation material into product and equipment.

Light fittings need regular cleaning to remove any accumulated dust, which can act both as a potential source of contamination and reduce light intensity.

Where the proper correction of building deficiencies requires shutdown of the area, it may be necessary to resort to temporary repair until adequate time can be made to enact a permanent repair. Building inspection and maintenance programs should be defined in writing and a record kept confirming compliance and referencing any repairs performed.

This regulation specifically refers to buildings, whereas Section 211.67 relates to equipment. This appears to ignore the maintenance of services, some of which are included in other sections of Subpart C, but without reference to maintenance. Clearly, services can impact directly on processing and product quality and they must undergo routine maintenance. Essential services will include HVAC, water (all types), steam, vacuum, compressed air and other gases, electricity, dust extraction, product/material pipelines, drainage, and sprinkler system.

GENERAL OBSERVATIONS

Building new or renovating present pharmaceutical or related manufacturing facilities represents special problems and the need for special expertise. A number of architectural and engineering firms take on the majority of such projects. They should be selected with great care and ongoing review of the project, preferably by outside experts as well as the internal team, should be the rule.

Because the author's experience amounts mainly to some situations in which some ultimate dissatisfaction has led to legal review of the potential for recovery, we are presenting additionally a series of observations that were developed in such retrospective.

First, bear in mind that the usual contract between the owner and the architect is a version of a "Standard Form of Agreement Between the Owner and the Architect" prepared by the American Institute of Architects. It contemplates arbitration, rather than recourse to the court system, in the case of a disagreement as to its terms or its performance. Therefore, competent and experienced legal advice prior to entry into the contract is required. Not all of the boilerplate of the standard agreement need be accepted, and riders can be added to afford greater protection to the owner. In larger projects, a contractual relationship to the production manager is likewise important to define. There are agreements with contractors and subcontractors that must contemplate a schedule process that is reasonably enforceable by the production managers and the owner's team. It is a major undertaking that in the last instance needs to meet the requirements of the federal and state and local agencies involved in a timely manner. People involved in such a project should, I recommend, be absolved from other responsibilities, but they should have the quality control "savvy" that the average architect and engineer may not possess.

The project manager will usually be supported by a project architect, project mechanical services engineer, project electrical engineer, project structural engineer, project civil engineer, and project landscape architect on a large project.

When we recall the nature of later complaints, we urge a careful and extensive reading of the foregoing material in this chapter. Legal recourse is limited by the expertise of those who would provide it. It is limited by time as well, since a claim based on negligence of those contracted with must usually be brought within two years from actual or constructive discovery of the harm. A Breach of Contract Claim has a four-year statute of limitations from the time of breach.

Some of the typical claims that we recall include "unbalanced" air systems, undersized exhaust fans, problems with heat exchangers, insufficient light levels, pressure monitoring problems, seal dampers not meeting specifications, inadequate roof or room drains, condensate problems, undersized electrical systems, improper laminar flow hoods, flawed water supply, inadequate geologic surveys, etc.

Obviously, close supervision must be accomplished by knowledgeable employees and agents of the owner. Otherwise the requisite expansion may become a large drain swollen by lost time and financial resources. In addition, of course, failures in planning, in equipment, and personnel handicapped by these result in the FDA citations following.

EXAMPLES OF OBSERVATIONS FROM FOOD AND DRUG ADMINISTRATION 483 CITATIONS

1. "The sterility test room was not designed and constructed to facilitate cleaning and disinfection."
2. "The HVAC and dust collection systems are not validated."
3. "The direction of airflow is not monitored in the manufacturing rooms."
4. "There are no approved procedures for maintaining the HVAC and dust control systems throughout the plant."
5. "The WFI system is not designed in a manner to minimize microbial contamination and endotoxin load. For example, in the past year, there have been 10 incidents of WFI samples that exceeded specifications for microbial contamination."
6. "The written procedures covering pest control within the buildings are not signed or dated by the personnel who prepared and authorized them."
7. "Inspection of the reverse osmosis water system revealed dead legs, which are potential sites for microorganisms to lodge, multiply, and enter the effluent."
8. "There are no temperature or humidity specifications for the area."
9. "Sensors for monitoring warehouse temperature have not been calibrated since their installation three years ago."
10. "Air recirculated in the compressing area has never been tested for particulate matter. Validation of the air-handling system is inadequate—samples for cross-contamination were collected from only ... cubicles in the compressing area."

SUGGESTED READINGS

1. Clean Room and Work Station Requirements. Federal Standard 209E, Washington, DC.
2. Mead WJ. Maintenance: its interrelationship with drug quality. Pharm Eng 1987; 7:29–33.
3. Fornalsaro T. Design and operation of a new sterile manufacturing facility. Bull Parent Drug Assoc 1970; 24:110.
4. Goddard K. Designing a parenteral manufacturing facility. Bull Parent Drug Assoc 1969; 23:69.
5. Loughhead H. Parenteral production under vertical laminar flow. Bull Parent Drug Assoc 1969; 23:17.
6. FDA. Guideline on Sterile Drug Products Produced by Aseptic Processing. June 1987.
7. Haas PJ. Engineering design considerations for barrier isolation technology. Pharm Technol 1995; 19(2):26.

8. Wintner B, Divelbiss JD. Isolator evaluation using computer modeling. Part I. Airflow within the minienvironment. Pharm Eng 1994; 14(6):8.
9. Divelbiss J, Wintner B. Isolator evaluation using computer modeling. Part II. Hydrogen peroxide sterilization with two diluent fluids. Pharm Eng 1995; 15(2):84.
10. Melgaard HL. Barrier isolation design issues. Pharm Eng 1995; 14(6):24.
11. Blanchard JA, Signore AA. Cost effective cGMP facilities. Pharm Eng 1995; 15(2):44.
12. Brader WR, Hsu PT, Lorenz BJ. Impact of implementing barrier technology on existing aseptic fill facilities. Pharm Eng 1995; 15(2):30.
13. Collentro WV. USP purified water and water for injection storage systems and accessories, part I. Pharm Technol 1995; 19(3):78.
14. Stark S, Vichl S. Laboratory facility renovations: 15 considerations that can't be ignored. Pharm Eng 1995; 15(1):28.
15. Collentro WV. USP purified water systems: discussion of ion exchange. part II. Pharm Technol 1994; 18(10):56.
16. Wood JP. Containment in the Pharmaceutical Industry. Marcel Dekker, 2001.
17. Avis KA, Liebermann HA, Lachman L. Pharmaceutical Dosage Forms, Vols 1, 2, 3. Marcel Dekker, 1993.
18. deSpautz JF. Automation and Validation of Information in Pharmaceutical Processing. Marcel Dekker, 1998.
19. Signore AA, Jacobs T. Good Design Practices for GMP Pharmaceutical Facilities. Marcel Dekker, 2005.
20. ISPE Baseline Guides. Bulk Pharmaceutical Chemicals. Vol. 1, ISPE.
21. ISPE Baseline Guides. Oral Solid Dosage Forms. Vol. 2, ISPE.
22. ISPE Baseline Guides. Sterile Manufacturing Facilities. Vol. 3, ISPE.
23. ISPE Baseline Guides. Water and Steam Systems. Vol. 4, ISPE.
24. ISPE Baseline Guides. Commissioning and Qualification. Vol. 5, ISPE.
25. ISPE Baseline Guides. Biopharmaceutical Manufacturing Facilities. Vol. 6, ISPE.

5 | Equipment
Subpart D

Robert Del Ciello
Northshire Associates, Martinsville, New Jersey, U.S.A.

Joseph T. Busfield
Pharmaceutical Technical Services, Warrington, Pennsylvania, U.S.A.

Steven Ostrove
Ostrove Associates Inc., Elizabeth, New Jersey, U.S.A.

INTRODUCTION

The specification, design, qualification, and use of equipment in the various manufacturing processes have taken on more technology-driven characteristics recently. With the advent of a systems approach to compliance, the attributes of equipment that need integration with other quality systems have been emphasized. This has occurred in the areas of specification through a stronger integration with process validation, instrumentation having a stronger integration with calibration/maintenance quality system, and embedded controls that enable continuous improvement programs. These are results of significant initiatives that occurred in the industry in the past 10 years. These programs have sprung from different points. The Food and Drug Administration (FDA) had initiated the Process Analytical Technology (PAT) program. This program's objectives encompass a thorough knowledge of the manufacturing process coupled with the appropriate instrumentation and controls that address the critical processing parameters in such a manner so as to allow improvements in the manufacturing process without significant regulatory filings. The advent of an integrated commissioning and qualification approach to the start-up and qualification of new equipment has streamlined this entire process. The focus of this program is to test the equipment system according to procurement specifications during commissioning and then verify whether the equipment systems operate according to manufacturing process requirements for qualification. The development of Good Automation Manufacturing Practices (GAMP) guidelines and calibration program guidelines has substantially transformed the process used, not only for the procurement and qualification of computer systems, but also is affecting the entire system for the acquisition of equipment and systems.

§211.63 EQUIPMENT DESIGN, SIZE, AND LOCATION

> Equipment used in the manufacture, processing, packing, or holding of a drug product shall be of appropriate design, adequate size, and suitably located to facilitate operations for its intended use and also for its cleaning and maintenance.

In order to properly specify equipment and systems, the pharmaceutical manufacturer needs to identify the manufacturing process. This requires that during the development of the product, the critical quality attributes (CQA) of the product, and the associated critical process parameters (CPP) that directly affect the attributes, be identified and the effect of the parameters on the attributes be understood. These requirements are to be included in a user requirements specification (URS) of the equipment/system. This specification is to focus only on the manufacturing requirements, for example, range of critical processing parameters, sensitivity of the control system to maintain these parameters, and so on. The purpose of this focus is to ensure that the qualification testing is limited only to the equipment capabilities that are required to manufacture an effective product.

In the past it was common to have qualification testing address all equipment/system capabilities. However, this type of testing is normally addressed during the start-up and commissioning of the equipment/system. Only capabilities that affect the manufacturing characteristics are tested during qualification. Therefore, commissioning tests the equipment/system against the requirements of the purchased specification, whereas the qualification testing addresses those requirements included in the URS.

An important tool used during the selection of equipment is to perform an impact assessment. The impact assessment provides a scientific rationale to the effect the equipment system has on the quality attributes of the product being manufactured. The assessment segregates equipment systems into three categories: direct impact, indirect impact, and no impact. This categorization enables the appropriate attention to be directed towards the specification, installation, and qualification of equipment systems.

Once the use of equipment system has been properly identified, as indicated in the previous edition, several parameters are to be considered when evaluating the equipment:

1. Availability of *spares and servicing*.
2. The frequency and ease of *maintenance* will significantly impact on productivity and even quality. Equipment breakdown during processing could adversely affect quality. Included in the maintenance evaluation should be the cleanability of the equipment. This will involve accessibility to the parts to be cleaned and the relative ease of disassembly and reassembly.
3. *Environmental issues* are important constructions. Is the design of the equipment conducive to the application? Such attributes as the ability to contain toxic products, the ability to contain dust, the ability to maintain aseptic conditions, etc. need to be reviewed.
4. *Construction materials and design* (see §211.65 below).
5. The type of *process controls* such as automatic weight adjustment on tablet presses and temperature recorders on ovens. The use of these controls has become a routine and is expected in today's manufacturing environment. The PAT initiative depends upon these controls to demonstrate that the manufacturing process is under control and to facilitate a continuous improvement program.

New equipment should not be used for commercial production until it has been qualified and the process in which it is to be used has been validated; this applies equally to laboratory and other test equipment. All equipment should be appropriately identified with a unique number, to allow reference in maintenance programs and in batch records [see also §211.105(a)].

The location of the equipment in the facility must enable an efficient flow of the manufacturing process. Manufacturing trains should be uni-directional whenever feasible. Backflow or crossflow within the process are to be minimized, as these incidences inherently have a high capability to cause mistakes.

The equipment system is to be placed in such a manner so as to enable all parts requiring maintenance, instrumentation, and calibration to be easily accessible. The key is easily accessible. Locating equipment in areas that are inaccessible usually means the maintenance and cleaning operations are not performed adequately, and thus leads to errors during the manufacturing of products.

§211.65 EQUIPMENT CONSTRUCTION

(a) Equipment shall be constructed such that the surfaces that come into contact with components, in-process materials, or drug products shall not be reactive, additive, or absorptive so as to alter the safety, identity, strength, quality, or purity of the drug product beyond the official or other established requirements.

The pharmaceutical manufacturer is required to fully understand the manufacturing process and all its constituents. This sub-part requires that all product contact surfaces be inert with regards to the active ingredients, incipients, and critical utilities, that is, water (USP or WFI), compressed gases, and so on. Traditionally, product contact surfaces are constructed of stainless steel (usually a 316 grade to facilitate cleaning), Teflon®, viton (or other inert elastomer), or other inert material. The choice of the material depends upon the unit operation within which the equipment will be placed. For example, for bulk chemical API operations, reaction/crystallization vessels can be constructed of stainless steel, hastelloy C, or glass for the chemical reactions to manipulate molecules and develop the active ingredient. For bulk biotech API operations, the equipment is usually constructed of 316L stainless steel for cleaning of proteins due to the inert nature of the stainless steel with respect to the biological processes. Formulation and filling operations, whether for parenteral, biotech, or oral dosage forms, have been and still are constructed from 316L stainless steel. Again, stainless steel is used primarily because of its inert nature as well as its ability to be cleaned and sterilized.

> (b) Any substances required for operation, such as lubricants or coolants, shall not come into contact with components, drug product containers, closures, in-process materials, or drug products so as to alter the safety, identity, strength, quality, or purity of the drug product beyond the official or other established requirements.

This requirement affects the design, construction, and placement of the manufacturing equipment. Motors, drive belts, gears, and other potential sources of lubricant contamination should be located away from equipment, vessel, or package openings that could result in product contamination. For the equipment where this is not possible, such as some mixers, tablet, and encapsulating machines, lubrication needs to be controlled and monitored. Lubricants are to be of good grade to minimize the impact of a contamination.

Gaskets and other connecting surfaces should be monitored to ensure that they do not break down, thereby allowing environmental contamination or gasket particles into the product.

§211.67 EQUIPMENT CLEANING AND MAINTENANCE

> (a) Equipment and utensils shall be cleaned, maintained, and sanitized at appropriate intervals to prevent malfunctions or contamination that would alter the safety, identity, strength, quality, or purity of the drug product beyond the official or other established requirements.

This subsection requires the establishment of a cleaning/sanitization program and a maintenance program, two separate and significant elements of the overall quality system. It is of interest to note that the regulation provides two rationales for the establishment of these programs—proper functioning of the equipment and the potential for cross contamination. Therefore, these concepts need to be at the core of each program.

Sections (b) and (c) go on to provide additional details of the content of the two programs.

> (b) Written procedures shall be established and followed for cleaning and maintenance of equipment, including utensils used in the manufacture, processing, packing, or holding of a drug product. These procedures shall include, but are not necessarily limited to, the following:
>   (1) Assignment of responsibility for cleaning and maintaining equipment;
>   (2) Maintenance and cleaning schedules, including appropriate sanitizing schedules wherever required;
>   (3) A description in sufficient detail of the methods, equipment, and materials used in cleaning and maintenance operations, and the methods of disassembling and reassembling equipment as it is necessary to assure proper cleaning and maintenance;
>   (4) Removal or obliteration of previous batch identification;
>   (5) Protection of clean equipment from contamination prior to use;
>   (6) Inspection of equipment for cleanliness immediately before use.

(c) Records shall be kept for maintenance, cleaning, sanitizing, and inspection as specified in §211.180 and §211.182.

There are various quality system element processes that can be utilized in each of these programs. The details of each depend upon the products being manufactured, business approach utilized, and the preferences of the operating unit. No matter what the process is, each program needs to have distinct attributes.

CLEANING PROGRAM

The cleaning program also needs to be documented in SOPs. In addition, the cleaning processes are required to be validated. The cleaning program identified in this subpart refers to equipment and utensils used in the manufacturing process. It is of note that this subpart identifies utensils. Any utensil, such as scoop, used in a weighing operation or other processing step must also be part of the cleaning program. Facility cleaning and sanitization is covered in subpart 211.56.

The cleaning program consists of two parts—the validation of the processes and detergents/sanitization agents and the day-to-day use of the validated processes and qualified detergents and sanitizers. Many cleaning processes are automated. These fall into two classes: clean in place (CIP) and clean out of place (COP). The CIP systems have their equipment provided with hard-piped services with cleaning solutions. The cleaning process is automated and usually is documented through a printout of the automation system. The COP systems consist of bringing the equipment to a cleaning station or placing the equipment in an automated washer. Again, the cleaning process is automatic with the appropriate documentation of the cleaning batch. CIP and COP cleaning processes are validated.

While the best cleaning systems involve automated process and/or washers, many cleaning processes are manual. The manual systems are executed by operators on a daily basis. Because of the variability of having operators execute the cleaning processes, these cleaning process cannot be truly validated in the same manner as the CIP or COP systems. Manual processes are usually qualified along with the qualification of the operators.

The following attributes are to be addressed in setting up a cleaning program:

1. An equipment and utensil list containing the product contact surfaces, their materials of construction.
2. Justification of sampling site locations based on which criteria—worst case, most difficult to clean, solubility of active, and so on.
3. Identification of operating cleaning procedures.
4. Rationale or justification for the cleaning agents selected. A list of approved cleaning agents, including the active ingredient or residual component tested, test method used, cleaning agent concentration, or effectiveness testing performed (to verify temperature, pressure, and time of working concentration, etc.) is to be identified.
5. Critical parameters to be monitored and controlled (i.e., wash and rinse temperatures, pressures and times) for automated processes.
6. Validated analytical and microbial test method identified.
7. Scientific rationale rinse and swab sampling acceptance limits, including calculations, pre-established acceptance limits, with references.
8. For finished pharmaceutical dosage forms, the cleaning limit is based on allowing not more than a fraction of a therapeutic dose to be present in subsequent products. A scientific rationale or justification is to be developed for residual values.
9. Interferences are to be determined for the cleaning agent and product(s) test methods.
10. A scientific rationale that describes the worst-case challenge selection process is to be developed.
11. Scientific rationale describing the product/ingredient matrix.
12. Definition of hardest-to-clean areas is to be provided.
13. Swab selection criteria and rationale. Criteria used for swab selection including:

a) extractables, minimal interferences with analytical method sensitivity-low carbon extracting swabs
b) surface cleaning ability-soluble and insoluble challenges
c) abrasion resistance-ease of the sorptive material to be removed by surfaces
d) solvent resistance
e) sorption and retention properties
14. Cleaning agent selection criteria and rationale including:
   a) cleaning effectiveness-ability to remove product residue
   b) cleaning capability-statistical limits based on historical data from acceptable study runs (pre-validation activities)
   c) safety assessment-relative to the equipment, does cleaning agent cause pitting or microscopic deterioration?
   d) solubility studies-ease of cleaning agent removal after final rinse
   e) process clearance capability-limits based on removal of residues by the process itself, is the cleaning agent the hardest to remove residue?
   f) validated analytical method-limit of detection and quantitation
   g) toxicity of material based on LD_{50} value
15. Parameters affecting cleaning effectiveness or attributes of the cleaning process
   a) coverage of cleaning agent
   b) wash and rinse temperatures, pressures, and times
   c) storage conditions and testing of the cleaning agent working solution
16. Criteria for acceptance after cleaning, including supporting documentation.
   a) Rationale for maximum limit set for product and cleaning agent residues.
   b) Approved method of sampling (based on swab and cleaning agent profiles and recovery study data).
   c) Validation of the analytical and microbial test methods, including LOD and LOQ.
17. Recovery criteria for sampling methods.
18. The cleaning procedures used during protocol execution are to be listed, and an evaluation of the procedures' effectiveness provided.

Overview—Maintenance/Calibration Programs:

1. All equipment and instruments are to be provided with unique identification. This can be of any format, but usually is an alphanumerical identifier, which can be easily utilized by a computer planning system. Although not mandatory, a computer management system is usually used for the schedule of preventive maintenance (PM) and calibration tasks and the record keeping for calibration data. There are numerous systems available varying widely in cost. If such a system is used it will require validation.
2. All preventive, maintenance, and calibration tasks and regimes are required to be documented as standard operating procedures (SOPs). These SOPs are to be managed according to the site documentation quality system element program.
3. The maintenance and calibration programs are to address both PM activities as well as corrective maintenance activities.
4. The PM and calibration activities are to be documented in SOPs. The frequencies of conducting these activities are to be placed in a scheduling system (whether manual/computerized). A method of documenting the PM and calibration task that has been completed is required.
5. Systems are required to identify how corrective maintenance and calibration activities are conducted. These systems are to be documented in SOPs. Many firms use a work order (WO) system to document corrective activities. Appropriate approvals of the WOs are required.
6. There is to be a process of entering new equipment and instrumentation items into the maintenance or calibration system. These processes are to be documented in SOPs. The processes are to identify what information is required before an equipment item can be entered. Information such as, but not limited to:

  a. unique identifier
  b. equipment assessment or instrument classification
  c. URS
  d. procurement specification
  e. recommended spare parts
  f. recommended lubricants
  g. vendor manuals and drawings

7. A history of maintenance and calibration work is to be kept for each equipment and instrument item.

8. Process to report abnormal situations, that is, out of tolerance for instruments, unapproved parts for maintenance. The system procedures are to identify the conditions that are considered abnormal.

9. Examples of a compliant system, with its attributes are as follows.

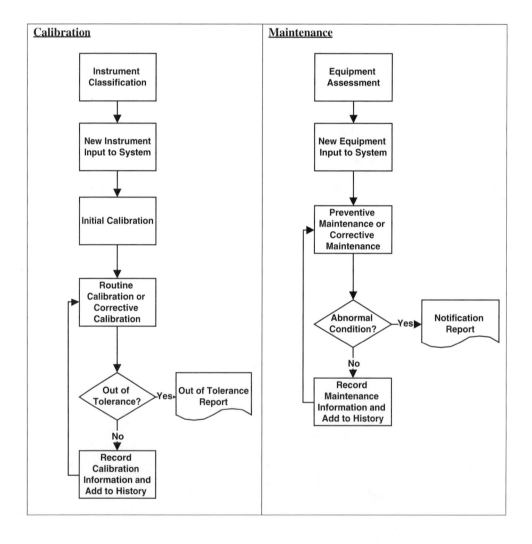

10. Supporting these systems is a spare parts program. This program is to address the specification, acquisition, storage, and issuance of spare parts. How spare parts are approved and how "like-for-like" replacements that are made need to be included in the system SOP.

In the editor's experience, the important aspects of a well-run and compliant maintenance and calibration program is poorly understood in this industry. The following additional sections go into some needed detail regarding compliant programs.

MAINTENANCE AND CALIBRATION PROGRAMS

Joseph T. Busfield

Maintenance and its specialized subset, calibration, are integral parts of all manufacturing industries. The need to maintain equipment and calibrate process controls is not based on any law or regulation of man, rather it is a natural result of the second law of thermodynamics. Without getting too technical, the second law simply states that all things wear out. Everything goes from a state of order to a state of greater disorder. In this context, it is readily seen why regulatory agencies, which have an interest in equipment operation correction, require maintenance and calibration programs to counter the effects of the second law and (in the pharmaceutical and related industries) to maintain the validated state of the equipment as required in 21 CFR 211.67 on cleaning and maintenance, and 21 CFR 211.68a on automatic, mechanical, and electronic equipment.

One of the more interesting quirks of the pharmaceutical and related industry is that while the CFR is often open to interpretation in many areas, allowing for the development of the "C" (current) in CGMPs, there are more specific directions in the maintenance and calibration areas. Yet, these areas are very often overlooked when developing compliance programs. Too often these areas are dismissed as "cost centers" and not considered as part of the compliance program. This ought not to be. A manufacturing site can run 24 hours a day, 365 days a year, and not allow time for maintenance or calibration. It will make lots of money but be completely out of compliance. On the other hand, a site can perform maintenance and calibration according to written programs and schedules but not have any production, and the site will be in compliance but out of business shortly. There needs to be a balance that incorporates maintenance and calibration efforts into the compliance program as much as any other quality system element, but works within the needs for production. The program elements described in this section are designed to do just that. Certainly, there may be other ways to maintain compliance in maintenance and calibration programs, but this section will provide considerations for compliant maintenance and calibration programs that have worked at many sites.

Common Considerations

In general, compliance programs need at least three components: procedures, practices, and paperwork. It needs *procedures* to provide instructions on what to do and how. It must have *practices* that follow the approved procedures. And there must be *paperwork*, which demonstrates that the practices were completed as per the procedures. The need for these three components is just as true in the maintenance and calibration areas. There is the need for procedures (SOPs) to define the administration of the program as well as the actions required to maintain/calibrate specific equipment. Certainly, the procedures must be followed by the practices in the field and office to maintain the equipment and program. And finally, these activities or practices must be documented (paperwork), to maintain a history or evidence of performing the required tasks.

Beyond the requirements common to compliance programs in general, there are several areas of additional requirements or requirements specific to maintenance and calibration that they share.

First, is the need for the program to be codified in procedures. This need is for a written program that outlines such areas as:

- Responsibilities of equipment owners, technical personnel, and quality unit
- Procedures, forms and approval for adding, deleting, modifying equipment in the maintenance or calibration program
- A scheme to identify equipment (if this scheme is not included in the validation program)

- Directions of how to classify equipment and instruments
- How to add, change, or remove spare parts from inventory, including equivalent parts evaluation
- Directions on how to establish and modify schedules for PM and calibration
- How to perform the work, for example, calibration points, corrective WOs, emergency WOs, closing notes for WOs, and so on.
- What reports and metrics need to be regularly generated and reviewed
- How to handle abnormal events, for example, out-of-tolerance calibrations, equipment malfunctions that could impact product, and so on.
- How are technicians qualified
- The training requirements for technicians
- How to qualify contracted companies and their employees

Both maintenance and calibration programs need procedures that follow the life cycle of the equipment (of facility) from addition to the program of retirement and scrapping.

In addition to the program defining procedures or administrative procedures, there is the requirement for specific procedures governing how to maintain or calibrate equipment. What is the PM procedure for a tablet press? A lyophilizer? How is a pressure indicator calibrated? Or a conductivity meter? The CFR is specific in requiring procedures for the actual maintenance activities performed on equipment, and it is not only compliant to have such procedures established and approved, but also just good business.

Another area that the maintenance and calibration programs have in common is the need for the quality unit input into the approval process for specific maintenance and calibration procedures. There is often a cry that quality does not have the background to evaluate maintenance or calibration procedures and this area should be left to the engineering folks to evaluate apart from quality. Of course, the same could be said for a procedure to operate a lyophilizer or purified water system. Quality personnel are not the subject matter experts (SME) on most of the procedures they are called upon to evaluate and approve. But the CFR requires procedures for maintenance and calibration. And the CFR requires the quality unit to approve all procedures that could impact product (21 CFR 211.22c), and certainly maintenance and calibration procedures can impact product. Rather than avoiding the requirement for the quality unit to approve maintenance and calibration procedures, the programs should build such approvals with definitions based on what each approver specifically approves.

Qualifications: Maintenance and calibration technicians must be qualified by education, training, or experience to perform their functions. This includes contracted technicians. Consider the qualifications of house technicians.

Each technician should have a file indicating the qualifications of that person to perform the assigned functions. This file can include any education, training, and experience records that demonstrate the qualification of the technician. Likewise, outside technicians require a review of their qualifications to do the work. Even manufacturer's representatives need to be qualified to some extent. The author once interviewed a "factory" technician who, as it turned out, had only worked for the manufacturer two weeks and came to the manufacturer from an unrelated industry. He was not qualified to do the work, regardless of the business card that had the manufacturer's logo imprinted. A company does not fulfill its compliance requirements by signing a purchase order. A purchase order is not a regulatory document. Suppliers do not get 483s, which is a privilege reserved for the drug or related product manufacturer. That is the entity responsible for the quality of the product, not the Ajax Maintenance, Calibration, and Chimney Sweep Company!

Although maintenance and calibration share many activities and have common requirements, there are also areas where they have specific requirements. Some of those specifics are identified in the following sections.

Maintenance Programs

Maintenance programs in the pharmaceutical or related industry have some specific requirements for approvals and documentation that should be incorporated in the program to ensure compliance.

Preventive Maintenance: One of the first requirements is (as has been mentioned) the need for a procedure for PM of specific equipment. It is most common that the preventive (and corrective) maintenance procedures are incorporated in the computerized maintenance management system (CMMS) of the site. Thus, the PM procedure is developed as a series of tasks and frequencies. For example, an air handler could have a PM such as:

Frequency	Task
3 months	Check belts for wear, tension and alignment
3 months	Check pre-filters for blockage
6 months	Replace pre-filters
6 months	Lubricate bearings
12 months	Clean coils

There needs to be a defined method to add (or modify) equipment and the tasks to be performed into the system. This method must be a documented system, provide the history and have appropriate approvals. Generally, the approvals are technical management, operations (or owner) management, and quality (assurance or control). Once equipment has been identified and its "boilerplate" documented—model and serial number, manufacturer, location, and so on—the requirements for maintenance have to be defined. These maintenance requirements are based on the manufacturer's recommendations and supplemented by the site's experience, conditions, and operating load for the equipment. These maintenance requirements must be developed by someone familiar with PM and the type of equipment. In practice, most PM can be reduced to common activities across many types of equipment. As examples:

- Hard-to-hard (metal-to-metal) moving contact surfaces require lubrication (e.g., bearings)
- Hard-to-soft (metal-to-rubber) moving contact requires inspection and replacement of the softer surfaces (e.g., belts)
- Filters require inspection/testing for leakage and/or blockage
- Hydraulics require inspection for leakage, level, and/or fouling, and so on

Using such basic building blocks, one can construct a simplified maintenance regimen that recognizes the commonality among seemingly different types of equipment. One can also develop basic instructions to use as a pick list to develop a PM procedure. Using such a pick list would ensure common terms and prevent variation based on individual language styles and ensure comprehensive statements of maintenance activities.

Once the PM regimen has been developed, it needs to be reviewed and approved. It should be reviewed and approved by a technical representative for appropriate content and schedules. It should also be reviewed and approved by an operations person familiar with the operation to ensure any concerns around equipment availability and cleaning are considered. It should also be reviewed and approved by a quality representative to ensure the requirements of the program and overall compliance issues are being met.

Often the quality approval is a sticking point. But think about this: Quality is required to approve all procedures that can impact product, and maintenance activities are required to be performed by a written procedure. Suppose Company A did not have a CMMS but performed PM through the use of SOPs. Such procedures would be typed into a word processing program, printed, and approved as an SOP. Now, Company B has a validated state-of-the-art CMMS and the procedures are typed into it instead of a word processor. Too often companies think that using a CMMS somehow enables bypassing the approval process that a regular SOP would require. The approval requirements do not depend on the software on the screen when preparing a PM regimen. There is an appropriate quality review required of PM activities, not as an SME but as a compliance gatekeeper. And this role needs to be outlined in the administrative procedures of the maintenance program.

Note that the same rigor applied to adding equipment to the maintenance program must be applied to changes in PM regimens and removing equipment from the program.

Once a PM regimen has been established (the procedure), it must be executed (the practice). PMs must be performed by technicians qualified to do so, completed in a timely manner, done with regards to operations considerations, and documented (the paperwork). There are several caveats to be aware of when performing PM. First, that which is stated on the procedure is that which is done. No more, no less. If a PM instruction is wrong, it should be noted and the process to correct or modify the regimen followed before the next PM is due. Second, if additional work is necessary, a corrective WO should be generated to cover that work. The problem may be corrected at the time of the PM, but should be done via the corrective WO, else the history of that equipment is muddied and difficult to follow if one has to sort through PM notes to identify corrective work. Third, the technician must be indoctrinated to report any findings that could potentially have impacted product prior to maintenance for example, a seal-leaking lubricant into the product. Finally, the PM must be recorded in an equipment logbook. Not in detail— leave that to the PM sheet—but referenced by date and PM WO number.

PMs are often checklist-type operations with a space for comments. These completed PMs must be reviewed by a supervisor to ensure completion of all the activities and to initiate or verify any remediation required—either to the PM regimen or the equipment.

A short comment on the handling of overdue PM procedures. These should never be overdue. PMs are procedures with due dates. Every effort should be made to complete them on time to maintain compliance. Not doing them as scheduled is a deviation from the compliance program. Some organizations have gone to taking equipment out of service if the PM is overdue, the same way instruments are generally taken out of service if the calibration is overdue. Certainly, this approach cannot be faulted as "inadequate," but it may be too much. PMs tend to self-correct to some extent. If a belt is not checked or replaced, it will break and the machine will stop—neither a desirable effect nor one indicating control, but obvious. Calibration failures, on the other hand, are not readily visible and are more insidious. Thus, the requirement to take equipment from service if a PM is not completed on time may be mitigated by this consideration. Reported? Absolutely. Remedied? Absolutely. Production stopped? You decide.

Much more can be said about PM, indeed about all the topics in this chapter, but the intent is to provide broad concepts and things to ponder.

Corrective Maintenance: Corrective maintenance differs because it has no preset steps to follow. It depends upon the skill of the technician to troubleshoot a problem and/or replace a part. There is the requirement to have the program defined, but not the specific corrective work. As part of that program definition, there needs to be such items as priority setting, responding to emergencies, prohibiting unauthorized changes to equipment, and recording of the work done.

Setting priorities is straightforward, but has been complicated by the ability to define them to the nth degree in modern CMMS. Basically there are two priorities: Do it now or do it later. Since a modern CMMS will allow 6.02×10^{23} priorities (maybe a slight exaggeration), some feel it necessary to develop that many priorities. Limit the number of priority classifications to those that are really useful.

There should be definition of how to respond to emergencies—not only shift considerations but availability of the CMMS, parts, and so on. There also needs to be regular reinforcement of the prohibition against unauthorized changes to equipment by technicians. Beware of the well-meaning second shift mechanic. Since he/she is well-meaning, he/she wants to get the equipment running. He/she is on the second shift and does not have all the support framework available to the day shift workers. And he/she is a mechanic. He/she has tools. And tools can really mess things up. He/she has to know that no changes are to be made to the equipment, regardless of the yelling of the production folks, regardless of how easy it would be to fix, and regardless of how sure he/she is that the change will not impact anything.

Recording the work done takes on special significance with corrective maintenance. Since it cannot be defined as PM prior to doing the work, its recording is more critical. Technicians need to record four basic things:

- A technical description of the problem (more than the originator's "it does not start")
- A technical description of the solution and testing (if required)
- Who did it and when
- What parts were used

Technicians should be discouraged from incorporating any editorial comments in the closing notes of the WO, such as:
"This equipment never worked and never will!"

"The engineer who installed this got a degree from a cereal box!"

"If the operators knew what they were doing, we would not have such problems!"

Remember, the first person to review the closing notes should be the supervisor. The next person to do so could be a regulatory inspector.

Training: Training for maintenance technicians is peculiar. In the calibration arena, there are usually SOPs for calibrating devices. Although maintenance may have the equivalent in the PM regimens, they are usually much more varied and numerous. As a result, often maintenance technicians are trained on general GMPs, documentation, access, and so on, but not on their specific job functions. Training on 4000 PM regimens would leave no time for actually performing the work. A suggestion is to draw on the commonality of PM activities and train on the general skills required to perform maintenance. For example, regular training sessions could be conducted on such topics as:

- Basics of lubrication
- Shaft and belt alignment
- Electrical contact maintenance

Using this scheme would provide training on the functions required to perform their jobs, but avoid 4000 training sessions. It also has the advantage of being able to use vendors as training resources.

As usual, much more detail could be provided but space limitations allow only some basic considerations for a compliant maintenance program.

Calibration Programs

Calibration is the specialized subset of maintenance. It usually employs the most trained technicians and has more visibility than maintenance. This is because the program usually employs stickers that should be checked regularly by the operators to ensure that the instrument is in calibration, and the devices themselves are often meant to be read and readings recorded. Similar yet different requirements are placed on a calibration program. There still needs to be a methodology to add, modify, and delete instruments. There needs to be a rigorous approval of all such actions. There needs to be documented evidence of all such activities. Different data are required to add (modify or delete) a device to the program in the calibration area than in maintenance.

Classification

Instruments are generally classified according to their potential impact on product. There are varying classification schemes out there. Some with two categories, some with three. Presented here is a logical three-tier scheme. After the initial categorization of an instrument (device or loop) as GMP—used in or to support the manufacture of a drug or related product—the instrument can be placed into one of three the classifications. GMP critical, GMP non-critical, or GMP utility.

A GMP critical instrument is one that monitors or controls a parameter that can have a direct impact on product quality, identity, purity, and so on. The terms "direct impact" and "indirect impact" are currently in vogue in the pharmaceutical and related industry, but still leave room for interpretation. A more complete definition could be:

GMP Critical Instrument: An instrument, which measures, monitors, records, or controls a critical parameter or any parameter with the potential to impact product; any instrument used to test or determine product quality; and/or any instrument used to produce data used in a regulatory document. This includes all calibration standards and all critical instruments referenced by a validation protocol or study.

GMP Non-Critical Instrument: Instruments that may reasonably and frequently be used by maintenance, production, R&D, laboratory, or technical services personnel to troubleshoot or anticipate changes in critical parameters and are not used to determine product quality.

GMP Utility Instrument: An instrument is classified as GMP utility if it monitors or controls a local facility utility supply to a GMP process, an on/off condition, or provides an indication of a non-critical parameter on a piece of GMP equipment.

Other circumstances that may warrant GMP utility classification include:

- Visibility to maintenance personnel only for troubleshooting purposes.
- Very coarse scaling factors relative to the parameter measured (gross readings only).
- Instruments with low accuracy ratings.
- Devices in the output circuit of a control loop.

As an aside, note that the last category is *not* called "for reference only." It is a personal quirk of the author who thinks that a device used for reference should have some accuracy basis and not a category that ignores ongoing calibration. After all, who wants an encyclopedia for reference that is not accurate? It is also worth noting that just because a device is classified as GMP utility (or even non-GMP) it is not exempt from the second law of thermodynamics. It still wears out! There is no use in installing an instrument in a system and then never checking its calibration or operation. If it is never checked, it should never be installed, because the laws of creation tell us that it will wear out. However, the devices in this classification are often very nominal, sometimes a pressure gauge that is color coded red and green with no or very coarse scaling factors. Perhaps it is only used to provide an indication of a gross blockage in a compressed air filter, or the on/off condition or a control solenoid, or the pressure of the water feed, and return of a water-sealed vacuum pump. These devices should not be ignored, but do not require a rigorous calibration. Some may be well served by a simple operation check to make sure that they are not stuck in one position. Stroking a valve to see if it goes through its full range of motion in response to a spanning of the input signal may be the most appropriate check. The bottom line is devices used in the manufacturing, or to support the manufacturing of a drug or related product, are required to be in a maintenance/calibration program. Do not ignore the GMP utility devices because they are "unimportant." If a device is not in the program, perhaps it should not be in the facility!

A statement worth remembering is "Complexity that does not enhance compliance is a compliance risk." Thus, if the effort is expended to classify instruments, there should some rationale for such an activity. Once instruments and devices are classified, that classification can be used as a basis for limiting the number of out-of-tolerance (OOT) events by first applying the OOT requirements to only the GMP critical devices—those that have a direct impact on product—and secondly, by setting limits as triggers for the OOT events that are based on the process—not the instrument—and applying these limits to only the GMP critical devices.

Calibration Limits: There are several different limits to be considered in a calibration program. The most basic limit of accuracy (or uncertainty) is that provided by the manufacturer—the instrument accuracy limits. These are the advertised values that manufacturers are constantly making tighter so that the instrument is seen with a better quality. However, these may not be a good thing for the pharmaceutical industry. Extreme accuracy, like apple pie, is hard to argue against, but the necessity of such accuracy needs to be evaluated. Why would one install a device with an accuracy of $\pm 0.1\%$ on a process that can vary $\pm 5.0\%$ with no impact on output quality? The calibration program should be based on the needs and requirements of your process, and not the best technology of an instrument manufacturer! That leads to another set of limits, the calibration limits.

The calibration limits are those values used to determine if an instrument requires adjustment during a calibration. As such, they are functions of the instrument, though these calibration limits may be set wider than the instrument accuracy limits of the manufacturer. In the hypothetical example given above (the $\pm 0.1\%$ instrument accuracy limits on a process that can vary $\pm 5.0\%$ with no impact on output quality), there is no value to calibrating the device to the manufacturer's limits. The manufacturer probably only makes instruments meeting these rigid specifications, and requests for more reasonable specifications to match the customer's needs will fall on deaf ears. Using the preceding values, calibration limits of $\pm 0.5\%$ or even $\pm 1.0\%$ would make sense and avoid unnecessary adjustment and out-of-calibration events. Thus, the guidance is to set calibration limits as a function of the instrument, but related to the customer's use of the instrument. Note that an inexpensive pressure gauge may have an instrument accuracy limit of 2.0% and enlarging this limit may not be prudent. As a guide, electronic devices often have manufacturer's accuracy tighter than required, but mechanical devices have not seen the same advances in accuracy limits and there may not be an opportunity to increase the instrument accuracy limits when setting calibration limits.

There is a third set of limits. Often these limits are referred to as process calibration tolerances (PCTs). They should not be called process tolerances, since process tolerances are related to the process design and not calibration per se, though there is an indirect relationship. PCTs are those values beyond which the product quality may be compromised. Since only GMP critical instruments can directly impact product quality, only they should have this secondary calibration tolerance applied. PCTs are recognition that the industry uses off-the-shelf instruments in particular applications, and the calibration program should identify limits related to the process, not the instrument manufacturer's technology. PCTs are triggers for OOT events, based on the process needs, and applied only to GMP critical devices. PCTs are significantly less than process tolerances to protect the process, but are yet broader than the calibration limits (or equal to the calibration limits if PCTs have not been established).

How are PCTs determined? In the best of all possible worlds, they would be derived from process development studies and the process tolerances. But often we are not in the best of all possible worlds. We still have hope. The operation folks should know (and are responsible for determining in any case) the limits of the process they operate before the quality of the product suffers. These limits may even be incorporated in an SOP and can be extracted from that SOP. Another source for determining PCTs is past investigations. If an investigation determines that a calibration error of 2% does not have a direct impact on product quality, there is no sense in issuing an OOT the next time when the calibration is out by 1%! Change or apply a PCT of 2%. Do not do the same thing over and expect different results.

Frequency of Calibration: Another calibration parameter to be tied to the classification is the frequency of calibration or the inverse, the calibration period. It stands to reason that if a device is GMP critical and its inaccuracy can impact product quality, it should have its accuracy checked more frequently than devices that cannot directly impact product quality. For example, one scheme would be to set the default calibration frequencies to quarterly for GMP critical, semi-annually for GMP non-critical, and annually for GMP utility device checks. These frequencies can be modified as experience is gained with a particular device, though there should be upper limits, that is, no less frequently than semi-annually for GMP critical devices, no less frequently than annually for GMP non-critical devices, and no less frequently than bi-annually for GMP utility devices.

Out-of-Tolerance Events: In the pharmaceutical and related industry there is a requirement to investigate any impact that a badly out-of-calibration instrument may have had on past production. "Badly out-of-calibration" is often termed as out-of-tolerance (OOT) to distinguish it from the normal adjustments to bring an instrument closer to the "true" value of the standard. Out-of-tolerances are a fact of life. They must be generated by the person finding the OOT immediately, but then they must notify the owner of the process (and quality) of the event. Until there is a formal, documented notification of an instrument found OOT, the burden of this knowledge is with the calibration technician. The monkey is on his/her back. He/she needs to transfer custody of this monkey to the operations and quality folks in an immediate and documented way, and then he/she can back away. The calibration technician is not qualified by education, training, or experience to evaluate the impact of the calibration error for any parameter on product quality. That has to be the purview of the folks who know the process and the chemistry or biology of the product, not the calibration technician. The calibration technician's role is that of the fabled Greek messenger—he or she is the bringer of bad news but he/she is not the one responsible for remediating the bad news.

There are some reactions to an OOT event that are within the purview of the calibration department. First, if an instrument is found out-of-tolerance, it is less than wise to wait a full calibration cycle to check the calibration again. One approach is to schedule a demand calibration halfway through the next normal calibration period. Thus, if a GMP critical device was found OOT during a normal quarterly calibration, an "extra," demand calibration would be scheduled six weeks (one and one-half months) after the OOT event. This OOT remediation calibration would help determine if the OOT event was singular or a sign of the device's continuing failure. If the device was found OOT again at the shortened period calibration, there is less product in jeopardy and the process to replace the device can be initiated (as the device is checked at an ever increasing frequency to ensure production as required).

A second reaction to an OOT event that is within the scope of the calibration department is evaluation of the history of the device's calibration. For example, if a review of the device's history shows it has failed twice in the last four calibrations, the process to replace the device should be initiated. The calibration department should be proactive in identifying bad actors and taking steps to remove them. Few instruments have a monetary value equal to the cost of a product investigation or, worse case, product recall. The calibration department should also conduct a review of the history of all devices that failed calibration—not out-of-tolerances per se, just found out-of-calibration, GMP critical and GMP non-critical devices alike. In these cases, an out-of-calibration can trigger a review of the records and heuristic rules can be established to address the failed calibrations. Perhaps, if a device is found out-of-calibration two of the last four times, the frequency of the calibration should be increased.

There are several more areas of interest in a calibration program. Device calibration SOPs, test points, loop calibrations, test accuracy ratios.

Device Calibration SOPs: Devices that are calibrated need to have a procedure to do so. But a site may have thousands of instruments. Does that require thousands of SOPs? One hopes not. What a training nightmare! Rather, a site can develop core or type SOPs. For example, rather than having one SOP for calibrating a 0 to 100 psig Ajax brand pressure gauge, and another for a 0 to 150 psig Ajax brand pressure gauge, and yet another for a 0 to 100 psig Zenax brand pressure gauge, a single SOP for calibrating pressure gauges can be developed. It would include setting up the calibration apparatus, requirements for getting as-found data. It would also include the checkpoints as percentages (e.g., 10%, 30%, 50%, 70%, 90%) or fixed values of the range. Checks for repeatability would be included, and general adjustment procedures would be specified, such as: adjust the zero, adjust the span, when no more interaction, adjust linearity if adjustment is available. The SOP would continue with the as-left verification and recording of values and any paperwork or OOT requirements. What these core SOPs would not have is specific data such as calibration limits, PCTs, or test points. These data would have to be provided to the technician via an alternative method, apart from the SOP. One method would be to develop and approve the calibration sheet when the device is added to the program and controlled copies would then be provided to the technician when the calibration is performed. Or, as part of the original approval process, the calibration parameters are entered into the CMMS (or the CCMS, Computerized Calibration Management System) if the system is robust enough to accept such data, provided at the time of the calibration, and is properly validated. In any case, developing and maintaining 20 SOPs is easier and less prone to compliance variations than maintaining 4000. Beware that some devices will have calibration procedures too specific to fit into a core SOP, such as those that require pressing certain buttons to enter the programming of a device, but even accounting for these types (e.g., conductivity meters, etc.) the site is well below 4000 SOPs.

Test Points: The points at which a calibration is verified or adjusted, the test points, was alluded to in the preceding paragraph. There are a few things to consider regarding the test points. First, most calibrations are based on a linear model. One cannot verify linearity with a single point. It takes at least three points to determine a straight line. Five points would be a better determination. A single point check is not a calibration and should not be declared as such. Many devices also require a repeatability challenge to ensure that the same input will result in the same output each time. The most popular example cited for the need for a repeatability check is hysteresis, or mechanical binding of moving parts, especially when a reading is approached from a value below the reading and a value above the reading. Hysteresis is not the only repeatability issue, but it is a major consideration and thus mechanical instruments in particular should incorporate a repeatability challenge in their calibration regimen. Electronic devices can also exhibit repeatability errors, though generally to a lesser degree. Depending on the repeatability specification of the device, repeatability may be a negligible consideration for these devices.

The program should also consider how to handle test points at the extreme of the range of the device. Should the zero adjust test point be 0% or 10%? The span adjust test point be 100% or 90%? The answer is based on whether the device has a "live" zero or full-scale reading. If one can input 0% and know that if the reading is less than zero, it will be apparent, then use the 0% value. For example, if one was calibrating an RTD transducer with an input of 0 to 100°C and an output of 4 to 20 ma, it is given that if the device is not calibrated correctly an input of 0°C can generate an output of 3.8 ma, thereby verifying the out-of-calibration condition. Not all devices have a live zero (or full-scale) reading. Pressure gauges may have mechanical stops at the low and/or top ends that inhibit the output from reflecting the input accurately. Programmable logic controllers (PLCs) may have electronic "clamps" at the limits. So, inputting 21 ma would still only result in a 100% reading just as a 20 ma would. In such cases, the best practice would be to use zero and span test points values removed from the extremes, thus using the common 10% and 90% test points.

The preceding paragraph notwithstanding, every effort should be made to avoid limited range calibrations. Yes, the device may not be used over the full range, but the calibration adjustments may be designed to be used as the extremes (within 10%). Someone once challenged this by saying they had a 0 to 120 psig pressure gauge that was used only for reading 30 psig. Why not calibrate it 0 to 60 psig? The response is three-fold. First, the gauge is not correct for the application. In general, instruments should be routinely read around the midpoint of their range, say 35 to 65%. Regularly reading 30 psig on a 120 psig range gauge is not a good practice. Second, if one finds an error at 60 psig, what does one adjust? The zero or the span adjustment? Neither was designed to be used as a midrange adjustment. Third, if one uses an inexpensive pressure gauge with a ±2.0% of full-scale accuracy, the accuracy of the gauge is ±6 psi. Since this accuracy is given as a full-scale value, the ±6 psi is valid throughout the range, thus reading 30 psig provides a potential error of ±6 psi or ±20%—a reading of questionable value to the process with that kind of built in error.

Loop Calibrations: Loop calibrations? Why not? Why not calibrate the instrument system the way it is used? After all, the basic thing the operator wants to know is: "Does the reading on the panel accurately reflect what the parameter value is in the tank?" The best way to verify that is to calibrate the system as a loop. Take the RTD out of the pipe and put it in a bath and see if the temperature of the recorder is correct. The magic in between is of no interest to the operator. If the loop reads correctly at all points and is repeatable, this is a valid calibration check. Many discrete devices are themselves made up of loops, internal to the enclosure of the instrument perhaps, but loops nonetheless. However, a calibration failure of a loop does necessitate disassembling the loop and calibrating each component, re-assembling the loop, and rechecking the loop as a whole. At the end of the day, this methodology provides a calibration that is most related to the use of the system and can save on calibration resources.

Test Accuracy Ratios: The ratio of the accuracy of the unit or system under test (UUT) to the accuracy of the standard used to test is called the test accuracy ratio (TAR) or test uncertainty ratio (TUR). A calibration industry heuristic rule is that a TAR of 4:1 should be maintained. Thus, if one has a standard with an accuracy of ±0.1% and is used to calibrate a device with a calibration limit of ±1.0%, the TAR is 10:1, greater than required but a good ratio. However, if the standard accuracy is only ±0.5%, the TAR is reduced to 2:1, less than generally accepted. If the calibration limit for a device being calibrated with the ±0.5% standard is ±0.4%, the TAR has dropped below 1:1 and the device is being "de-calibrated." An unacceptable situation. The standard used to calibrate a device must be more accurate than the device it is calibrating.

Beware of two situations regarding TARs. First, the TAR of the standard may have to include auxiliary devices used with the standard. For example, if one has a field calibrator with ±0.2% stated accuracy, but an RTD is plugged into the calibrator that has an accuracy of ±1.0%, both accuracies must be considered when establishing the TAR. The most common method for assigning an accuracy to a loop (which includes a standard and auxiliary components) is the root of the sum squared (RSS), but there are other methods. The main point is to be aware of the effect of auxiliary components on the accuracy of the standard train and include it in the TAR calculations.

The second situation occurs when an organization makes the statement in the SOPs that all calibrations with be done with a TAR of 4:1, but fails to verify that this is happening. There is no verification of the TAR on calibration sheets, or no tools or training provided to the technicians to ascertain the TAR. This is certainly a compliance risk. If the program sets a TAR requirement—as it should—but it is never verified, the program has a glaring gap. There must be a way for calibration technicians to verify the TAR they are being required to maintain.

Maintenance and Calibration Metrics

After a program has been established, there should be metrics to monitor the effectiveness of the program. A modern CMMS and CCMS can provide a vast plethora of metrics. Just because they can tell you the color of the eyes of the last technician who worked on the equipment does not mean you should track that data. Just because it can be done does not mean that it should be done! It is useful to divide the data or metrics to be used into two categories: compliance and business. While all compliance-based metrics have a business justification, all business-based metrics do not necessarily have a compliance component. So generate metrics cautiously, lest you get buried in relatively meaningless paper required by wayward statements in SOPS.

The most basic compliance needs are two for each: maintenance and calibration. For maintenance activities, the two basic metrics are overdue PM and open corrective work. For calibration, the two basic metrics are overdue calibrations and out-of-tolerances issued. The first of each of these two sets provide paperwork to tell how well your practices are complying with your procedures. The second item in each set, maintenance and calibration, reports on perturbations in the system and provides insight on how well they are handled. More metrics may be useful to run the business, but it is suggested that they may not be part of the compliance SOPs to avoid compliance constipation.

Summary

All that can be said about maintenance and calibration programs in the pharmaceutical and related industries has not been said in this chapter section. The goal was to provide points to consider when designing a program and a vision of a path forward, not a detailed program for each area. Lots of details can be implemented in various ways. A calibration review team may be established to review calibration parameters, or the parameters can be routed for approvals to different folks without establishing a team. Parts equivalency can be done via a technical review and approval or via major change control. The major considerations are outlined here; the details can be developed as best suits your circumstances.

---

§211.68 AUTOMATIC, MECHANICAL, AND ELECTRONIC EQUIPMENT
Robert Del Ciello and Steven Ostrove

(a) Automatic, mechanical, or electronic equipment or other types of equipment, including computers, or related systems that will perform a function satisfactorily, may be used in the manufacture, processing, packing, and holding of a drug product. If such equipment is so used, it shall be routinely calibrated, inspected, or checked according to a written program designed to assure proper performance. Written records of those calibration checks and inspections shall be maintained.

(b) Appropriate controls shall be exercised over computer or related systems to assure that changes in master production and control records or other records are instituted only by authorized personnel. Input to and output from the computer or related system of formulas or other records or data shall be checked for accuracy. The degree and frequency of input/output verification shall be based on the complexity and reliability of the computer or related system. A backup file of data entered into the computer or related system shall be maintained except where certain data such as calculations performed in connection with laboratory analysis, are eliminated by computerization or other automated processes. In such instances, a written record of the program shall be maintained along with appropriate validation data. Hard copy or alternative systems, such as duplicates, tapes, or microfilm, designed to assure that backup data are exact and complete and that it is secure from alteration, inadvertent erasures, or loss shall be maintained.

As indicated in 211.68(a) above, the use of automated controls is not only allowed, but also expected to be used in the equipment used in the manufacture of pharmaceutical products. While the use of such devices is not mandatory, it is difficult to acquire equipment today that does not have some type of computer/automated control. As stated in section (a), the control unit needs to be calibrated and verified to be working correctly. This last part is the essence of qualification. As with other systems, written records need to be maintained. In paragraph (b) above, the regulation goes on to specify what needs to be verified in the testing, that is, input and output (I/O), backup systems, and calculations.

With a careful reading of the above requirements it can be seen that there are two phases to complete computer system (or control system) validation. These are the equipment qualification and the software validation. The hardware itself needs to be qualified, as does any other process equipment. In addition, Part 11 (electronic signatures and electronic records) and requirements overlap both the hardware and the software aspects of the systems.

Based on the two major aspects of automated controls and computer controlled systems and processes there are again two areas to be considered. These are known as the structural validation and the functional validation stages of qualification/validation.

Structural qualification/validation is primarily set to deal with the development of the software source code, the capabilities and qualifications of the programmer, and the related software functions (e.g., version control). The functional qualification/validation aspect of the program, as the name implies, involves the qualification of the hardware to operate under the conditions it is designated for as well as the actual validation of the entire unit (hardware and software with associated devices/equipment).

Software validation of control systems (hardware and software) has developed to a mature level with the advent of such documents as of GAMP (Good Automated Manufacturing Processes) and the FDAs "Guide for the Validation of Automated Systems." These documents (and others) provide a structured approach to the design, specification acquisition, installation and validation of computer/control systems. The level of qualification required in the GAMP approach is dependent upon the type of control system and the categorization of the software used for the control system as follows:

Category	Example Software/System Type	Basic Level of Validation
1. Operating systems	Windows, DOS, Unix providing operating platform for a system	None
2. Firmware	Micro controllers and discrete devices. Process instruments controlling or recording for example, temperature, flow, pressure, conductivity, level, and pH.	Minimal
3. Commercial off-the-Shelf	Standard application software. Source code not supplied/ established customer base	Basic
4. Configurable software	Customized application software source code not supplied/ established customer base. Generic software configured to client needs for example, DCS	Intermediate
5. Custom (Bespoke) software	Source code developed to client needs for example, PLC	Complete

NEW COMPUTER SYSTEM EXISTING COMPUTER SYSTEM

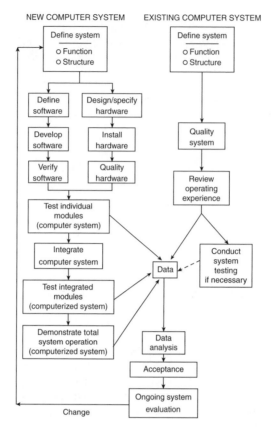

FIGURE 1   The validation life cycle approach.

GAMP approach can be tied directly into the life cycle approach to software development and computer control qualification. It is not meant to be a stand-alone approach, but one that ties all aspects of the qualification process together. Figure 1 shows the life cycle approach to computer system validation. This includes both software and hardware qualification.

The maintenance of the control system is handled within the maintenance/calibration programs as discussed above. Additional skills are required as the complexity of the control system increases, from a simple embedded firmware controller to a distributed control system that operates the entire manufacturing process. Appropriate personnel are needed to support these systems.

§211.72   FILTERS

Filters for liquid nitration used in the manufacture, processing, or packing of injectable drug products intended for human use shall not release fibers into such products. Fiber-releasing filters may not be used in the manufacture, processing, or packing of these injectable drug products unless it is not possible to manufacture such drug products without the use of such filters. If use of a fiber-releasing filter is necessary, an additional non-fiber-releasing filter of 0.22 μm mean porosity (0.45 μm if the manufacturing conditions so dictate) shall subsequently be used to reduce the content of particles in the injectable drug product. Use of an asbestos-containing filter, with or without subsequent use of a specific non-fiber-releasing filter, is permissible only upon submission of proof to the appropriate bureau of the FDA that use of a non-fiber-releasing filter will, or is likely to, compromise the safety or effectiveness of the injectable drug product.

As indicated in the previous version, this subsection has restricted applicability to the manufacture of injectable products for humans. The later introduction of limits on paniculate matter in injectables encouraged the move away from fiber-releasing filters. The ban on use of asbestos filters, without FDA approval, relates not only to reduction in particulate matter in injections, but also to minimization or elimination of worker exposure to airborne asbestos fibers.

More important for sterilizing filters are extractables and validation of filtration effectiveness. Many companies, with the support of the filter manufacturers, have used a matrix approach to address these issues, provided there is a scientific justification.

EXAMPLES OF OBSERVATIONS FROM FDA 483 CITATIONS

1. Worst-case conditions are not undertaken during the validation study.
2. Cleaning failures noted in the ongoing cleaning validation program are not investigated and corrected.
3. The maintenance support group was using an obsolete SOP for maintenance and calibration of equipment.
4. The firm's cleaning validation program has not addressed how long a product can remain in the processing equipment before the equipment must be cleaned.
5. There are no maintenance records for the tableting machines to indicate when routine repair and replacement of parts is performed.
6. Filters used to sterilize bulk drug solutions are not being subjected to a pre-filtration integrity test.
7. There are no written procedures for calibration and PM of laboratory instrumentation.
8. The record generated during the calibration of the fluid bed dryer sensors and chart recorder appear inadequate in which there is no written protocol for this operation; the probes of the original Digistrip recorder used to calibrate the Digistrip thermocouples had not been calibrated since.
9. There are no qualification studies performed on equipment to assure that they perform as intended.

SUGGESTED READINGS

1. Mead WJ. Maintenance: its interrelationship with drug quality. Pharm Eng 1987; 7:29.
2. Branning RC. Computer system validation: how to get started. Pharm Eng 1987; 7:11.
3. Fry EM. FDA regulation of computer systems in drug marketing. Pharm Eng 1988; 8:47.
4. DeRisio R. Equipment design: moist heat sterilizer. Pharm Eng 1987; 7:43.
5. Harris JR et al. Validation concepts for computer systems used in the manufacture of drug products. Pharm Technol 1986; 10:24.
6. Chapman KG, Harris JR, Errico JJ. Source code availability and vendor-user relationship. Pharm Technol 1987; 11:24.
7. FDA. Guide to Inspection of Computerized Systems in Drug Processing. Washington, DC: U.S. Department of Health and Human Services, February 1983.
8. FDA. Guide to Inspections of Validation of Cleaning Processes. Washington, DC: U.S. Department of Health and Human Services, July 1993.
9. Fourman GL, Mullen MV. Determining cleaning validation acceptance limits for pharmaceutical manufacturing operations. Pharm Technol 1993; 17(4):54.
10. Smith JA. A modified swabbing technique for determination of detergent residues in clean-in-place systems. Pharm Technol 1992; 16(1):60.
11. Jenkins KM, Vanderwielen AJ. Cleaning validation: an overall perspective. Pharm Technol 1994; 18(4):60.
12. PhRMA. Computer system validation: auditing computer systems for quality. Pharm Technol 1994; 18(9):48.
13. Agalloco J, Mascherpa V. Validation of a computerized system for autoclave control. Pharm Technol 1995; 19(1):42.
14. Tetzlaff R, Shepherd R, LeBlanc A. The validation story. Pharm Technol 1993; 17(3):100.
15. Rohsner D, Serve W. The composition of cleaning agents for the pharmaceutical industry. Pharm Eng 1995; 15(2):20.
16. Paul BO. Continuous bagging eliminates dust, powder handling. Chem Process 1995; 47.

17. PMA. Validation concepts for computer systems used in manufacturing of drug products. Pharm Technol 1986; 10(5):24.
18. Double ME. The role of quality assurance in the validation process. Pharm Eng 1991; 27(5):50.
19. Chapman KG. A history of validation in the United States: Part 1. Pharm Technol 1991; 15(10):82.
20. Chapman KG. A history of validation in the United States: Part 2. Validation of computer related systems. Pharm Technol 1991; 15(11):54.
21. PMA. Computer system validation—staying current: software development testing strategies. Pharm Technol 1989; 13(9):142.
22. PMA. Computer system validation—staying current: change control. Pharm Technol 1990; 14(1):20.
23. McKinstry PL, Atwong CT, Atwong MK. An application of a life cycle approach to computer system validation. Pharm Eng 1994; 14(3):46.
24. PMA. Computer system validation—staying current: installation qualification. Pharm Technol 1990; 14(9):88.
25. PMA. Computer system validation—staying current: security in computerized systems. Pharm Technol 1993; 17(5):48.
26. Chapman KG, Harris JR, Bluhm AR, Errico JJ. Source code availability and vendor—user relationships. Pharm Technol 1987; 11(12):24.
27. FDA. Computerized drug processing; source code for process control. Compliance Policy Guide (7132a.15), 1987.
28. Levehuk JW. Good validation practices: FDA issues. J Pharm Sci Technol 1994; 48(5):222.
29. PDA. Validation of computer related systems. Technical Report No. 18, 1995.
30. George JM. Lessons from the field, real life experiences in computer system validations. Pharm Technol 1994; 18(11):38.
31. GAMP4. Guide for validation of automated systems. ISPE 2001.
32. Wood JP. Containment in the Pharmaceutical Industry. Marcel Dekker, 2001.
33. Avis KA, Liebermann HA, Lachman L. Pharmaceutical Dosage Forms. Vols 1-3. Marcel Dekker, 1993.
34. deSpautz JF. Automation and Validation of Information in Pharmaceutical Processing. Marcel Dekker, 1998.
35. Signore AA, Jacobs T. Good Design Practices for GMP Pharmaceutical Facilities. Marcel Dekker, 2005.
36. ISPE. ISPE Baseline Guides, Commissioning and Qualification. Vol. 5. ISPE, 2001.
37. GAMP Good Practice Guide, Calibration Management. ISPE, 2001.
38. Nassani M. Cleaning validation in the pharmaceutical industry. J Valid Technol 2005; 11(4).

6 | Control of Components and Drug Product Containers and Closures
Subpart E

Arlyn R. Sibille
Consultant, Harmony, Pennsylvania, U.S.A.

§211.80 GENERAL REQUIREMENTS

(a) There shall be written procedures describing in sufficient detail the receipt, identification, storage, handling, sampling, testing, and approval or rejection of components and drug product containers and closures; such written procedures shall be followed.
(b) Components and drug product containers and closures shall at all times be handled and stored in a manner to prevent contamination.
(c) Bagged or boxed components of drug product containers or closures shall be stored off the floor and suitably spaced to permit cleaning and inspection.
(d) Each container or grouping of containers for components or drug product containers or closures shall be identified with a distinctive code for each lot in each shipment received. This code shall be used in recording the disposition of each lot. Each lot shall be appropriately identified as to its status (i.e., quarantined, approved, or rejected).

The general requirements again emphasize the need for written procedures, which are to be followed. Details on typical procedures and compliance issues are presented later under Section 211.100.

During storage, handling, and sampling, materials are potentially vulnerable to contamination. Warehouses are designed to allow easy access for delivery vehicles; however, this also allows access to rodents, insects, birds, extraneous dust, and vehicle exhaust fumes. This unwanted access can be minimized by separating the actual delivery area from the main storage area by doors or air curtains. Certain raw materials, such as sugar and starch, can attract pests and it is therefore important that any spillages be quickly cleaned up.

Cartons, barrels, and rolls of packaging materials should be stored on pallets or racks to facilitate cleaning, inspections, and pest trap placement. Since these precautions are still unlikely to completely prevent access to insects and other sources of contamination, extermination, or elimination, programs are required to control insects and rodents.

Pest monitoring and maintenance programs typically involve a combination of both chemical and nonchemical prevention and control techniques. Integrated pest management is a more effective program than through the utilization of any one technique. Rodents can usually be controlled by the placement of baitless traps; the use of poison bait is not acceptable in a pharmaceutical establishment. Insects are frequently eliminated by the use of electric exterminators, whereas birds may be trapped and then removed from the premises.

Many companies have implemented procedures that outsource the inspection, monitoring, managed product use, and record keeping requirements for management of insect and rodents. These procedures often require employees to make a notation in a Pest Sighting Log to keep a running record of pest sightings. These procedures compliment the program in order to monitor effectiveness of the program on a continuing basis. This monitoring component of the program ties into the recurring application of pesticides, trappings, or cleaning.

When containers of materials are opened for sampling, the contents have an increased vulnerability for contamination from other materials, microorganisms, or foreign particulate matter (dust). Warehouse conditions may not be suitable for sampling raw materials. Many companies have dedicated sampling areas (see §211.84). Therefore, the sampling techniques used and the sampling environment should be specified and controlled through written procedures. The procedure(s) should specify whether a separate room provided with improved air handling systems, readily cleanable surfaces, and availability of dust extraction hoods (i.e., similar to a dispensing operation) must be utilized during sampling. Sampling techniques to prevent contamination, including acceptable containers and labeling requirements, should also be specified in the sampling procedures.

The requirement for a distinctive code for each lot of components received can be met by retaining the supplier coding where this is adequate. However, most companies establish a material coding system to uniquely identify each material. The coding system allows the manufacturer to trace the material through the receipt, quarantine, sampling, testing, and release/rejection process. This requires that the distinctive number be physically applied to all of the containers within each shipment.

The regulation also requires status identification. This has been interpreted as physical labeling of container during the receipt, quarantine, sampling, testing, and release process. A label with provisions for initialing and dating at each stage from receipt to release to manufacturing can be used. It is now becoming common for manufacturers to apply bar codes to materials, which can be used to effectively monitor the use and movement of the material through its life cycle.

§211.82 RECEIPT AND STORAGE OF UNTESTED COMPONENTS, DRUG PRODUCT CONTAINERS, AND CLOSURES

(a)   Upon receipt and before acceptance, each container or grouping of containers of components, drug product containers, and closures shall be examined visually for appropriate labeling as to contents, container damage or broken seals, and contamination.
(b)   Components, drug product containers, and closures shall be stored under quarantine until they have been tested or examined, as appropriate, and released. Storage within the area shall conform to the requirements of Section 211.80.

Visual examination of materials on receipt is an important quality step. This should confirm that the correct material has been delivered. The delivery information should be checked against the purchase order to ensure that the part number, product code, or descriptions are correct and that the expected quantity for each component or material has been received. The procedure for verifying the receipt of materials should be captured in a written procedure. The procedure should require some notation on the shipping documents to indicate that the inspection has been performed. In some cases, a checklist is used and included with the receiving forms.

If any physical damage has occurred to the container, its potential impact on quality must be considered. Broken seals on containers may indicate that the container has been opened somewhere during transit. An opened container may indicate that the material may have been exposed to unacceptable environmental conditions and may give cause for alert for rejection of the container.

Containers should also be examined for physical deformation and for visible signs of spillage from other materials, as well as for potential rodent attack. Since the possibility of deliberate sabotage exists, the suppliers should be encouraged to use seals with unique designs or logos to minimize the potential for deliberate tampering and to make it easier for the inspection process to document deliberate tampering. Since seals on outer containers are sometimes broken or lost inadvertently during transportation, examination of any inner seals may be required before a final decision can be made. All of these situations will require additional evaluation of the materials and could ultimately result in rejections. A final inspection of material and components should be required prior to use and the inspection documented in the batch record. This additional check is particularly important if each container has not been individually inspected or opened earlier. In fact, low-frequency defects in packaging components are more often detected during the filling/packaging process than by sampling on receipt.

It is also essential to confirm the name of the supplier; this is elaborated upon in §211.84. When materials are purchased through agents, these should be requested to identify the actual producer. The purchase of raw materials and packaging components is usually controlled under a supplier contract. This contract should specify the quality of the material, the manufacturer, manufacturer's part number/material code, and the supplier's code if different. The contract should specify the information to be included in the certificate of analysis (CoA) when one is required to accompany the material shipment. Usually, specific test results with specifications are also specified in the contracts. A supplier contract must also specify prior notification if any material supplier or specification changes, since a change in supplier or specification may have an impact on product quality. A change in product supplier will require additional testing to qualify the supplier. This will eliminate any potential impact on the final product. Accelerated stability studies comparing the current and new materials are required and also, on occasion, accelerated comparative stability on the dosage form itself.

Section (b) refers to storage under quarantine until release. The acceptability of a system as an alternative to physical separation has been indicated previously in §211.42(c), which states that these operations shall be performed within specifically defined areas of adequate size. There shall be separate or defined areas or such other control systems for the firm's operations as are necessary to prevent contamination or mixups during the course of the all procedures from receipt of materials to disposition of the final product. The separation can be accomplished by using fenced areas with controlled access or separate rooms within a warehouse area.

§211.84 TESTING AND APPROVAL OR REJECTION OF COMPONENTS, DRUG PRODUCT CONTAINERS, AND CLOSURES

(a) Each lot of components, drug product containers, and closures shall be withheld from use until the lot has been sampled, tested, or examined, as appropriate, and released for use by the quality control unit.

The need to withhold materials for use until released by quality control is laudable and equates to the normal situation. However, there are occasions when, for a variety of reasons, materials arrive at a plant site and are required for immediate use. These situations could be exacerbated by lengthy component testing such as microbial evaluation. If a situation such as this occurs, the drug product manufacturer must make the decision to "manufacture at risk" while a material is undergoing testing. The product must be placed in quarantine immediately upon completion of manufacturing until the final release of the material in question. Manufacturing at risk runs the possible risk of a product rejection. Obviously, the manufacturer would be cognizant of these potential financial risks and before initiating such an action would evaluate material and product history and the magnitude of the added value. However, the Food and Drug Administration (FDA) has specifically stated that such an approach is not acceptable since this "increases the risk to the consumer that an unsatisfactory lot might erroneously be released." If a manufacturer were to choose this route, the procedures to ensure that the lot is not released until all testing is complete must be written and followed closely.

Deliveries of material to bulk storage require special mention. It is frequently impractical to hold a delivery vehicle until material can be fully evaluated. In such cases, it is usual to ensure that the CoA accompanies the delivery and that the more sensitive tests are performed before the material is discharged into the bulk storage system. In the event that the full analysis identifies a problem, it may be necessary to quarantine the contents of the storage tank until a comprehensive evaluation has been performed.

When a second delivery of a previously released material is received, it is still necessary to sample and evaluate. The conditions to which the later delivery may have been exposed could have differed from the original delivery. The same level of evaluation may not be necessary, but any parameter that might be affected by shipping and storage conditions should be examined.

(b) Representative samples of each shipment of each lot shall be collected for testing or examination. The number of containers to be sampled, and the amount of material to be taken from each container, shall be based upon appropriate criteria such as statistical criteria for component variability, confidence levels and degree of precision desired, the past quality history of the supplier, and the quantity needed for analysis and reserve where required by §211.70.

The quantity of material that should be sampled will depend upon the testing required. Manufacturers usually sample a quantity of material at least twice the size needed for testing when the cost of the material is not prohibitive. This allows for retesting if out of specification results are obtained or material is spilled or contaminated during testing. During the qualification process for new suppliers, the qualification procedure may require more extensive sampling and evaluation until consistency in the product is demonstrated. This could also be the case for a new dosage form, which, although validated using a minimum of three batches, could still undergo process improvement/optimization. Additional amounts of sample could allow more extensive evaluation of the material in relation to these optimization studies.

(c) Samples shall be collected in accordance with the following procedures:
   (1) The containers of components selected shall be cleaned, where necessary, by appropriate means.
   (2) The containers shall be opened, sampled, and resealed in a manner designed to prevent contamination of their contents and contamination of other components, drug product containers, or closures.
   (3) Sterile equipment and aseptic sampling techniques shall be used when necessary.
   (4) If it is necessary to sample a component from the top, middle, and bottom of its container, such sample subdivisions shall not be made into composites for testing.
   (5) Sample containers shall be identified so that the following information can be determined: name of the material sampled, the lot number, the container from which the sample was taken, the date on which the sample was taken, and the name of the person who collected the sample.
   (6) Containers from which samples have been taken shall be marked to show that samples have been removed from them.

The process of sampling can itself pose risks of contamination. For this reason, containers may need to be cleaned prior to sampling—a vacuum system is very effective for a large container. Generally, an air-blowing system should be avoided because this is more likely to spread a potential problem. Small containers may be wiped down with an appropriate solvent or distilled water. Containers should be opened for sampling in an acceptable environment that will not expose the material to further risk of contamination. For drug substances and excipients, it is preferable to provide a sampling area with environmental conditions similar to those in manufacturing. This sampling area may be a designated room near or adjacent to the warehouse. Containers and closures can usually be sampled in the warehouse, but any outer protective coverings should be securely replaced.

Materials requiring microbiological evaluation need to be sampled under more rigorous conditions involving the use of sterile equipment and a sterile environment for such sampling. The procedures used for sampling should be specified in written procedures. Employees must be properly trained in such sampling techniques and the training must be documented prior to sampling.

The regulations do not preclude the composing of samples for testing, except as indicated in (c)(4). If there is some doubt about the homogeneity of a component, it may be advisable to evaluate this by taking samples from various positions in the container. Obviously, the compositing of these samples would be scientifically invalid.

Upon completion of sampling of a container, the container must be labeled to indicate that material was removed from the container. The quantity of material removed should be included on the label as well as the date and the person sampling the material. It is very important to ensure that the container from which the material is removed is resealed in a manner to prevent contamination when returned to the quarantine area. Barrels may be taped, bottles may have a stretch plastic film placed securely around the top, boxes should be taped, or other appropriate measures should be undertaken to ensure that the container is resealed. Some containers may include closures that prevent contamination and therefore no further sealing is required.

(d) Samples shall be examined and tested as follows:
   (1) At least one test shall be conducted to verify the identity of each component of a drug product. Specific identity tests, if they exist, shall be used.
   (2) Each component shall be tested for conformity with all appropriate written specifications for purity, strength, and quality. In lieu of such testing by the manufacturer, a report of analysis may be accepted from the supplier of a component, provided that at least one specific identity test is conducted on such component by the manufacturer and provided that the manufacturer establishes the reliability of the supplier's analyses through appropriate validation of the supplier's test results at appropriate intervals.
   (3) Containers and closures shall be tested for conformance with all appropriate written procedures. In lieu of such testing by the manufacturer, a certificate of testing may be accepted from the supplier, provided that, at least, a visual identification is conducted on such containers/closures by the manufacturer and provided that the manufacturer establishes the reliability of the supplier's test results through appropriate validation of the supplier's test results at appropriate intervals.
   (4) When appropriate, components shall be microscopically examined.
   (5) Each lot of a component, drug product container, or closure that is liable to contamination with filth, insect infestation, or other extraneous adulterant shall be examined against established specifications for such contamination.
   (6) Each lot of a component, drug product container, or closure that is liable to microbiological contamination that is objectionable in view of its intended use shall be subjected to microbiological tests before use.

Components, containers, and closures used for production of pharmaceuticals must obviously comply with their quality specifications. As suppliers introduce effective procedures and embrace the principles of vendor certification, the need for customer testing is reduced. However, basic caution and the Current Good Manufacturing Practice (CGMP) regulations do require that some testing be performed. For a new supplier, it will be necessary for the customer to perform full testing for a minimum of the first three lots of materials received from the supplier or the number of lots designated in the manufacturer's supplier/vendor qualification procedure. Manufacturers should require that suppliers provide certificates of analysis, which can be used to compare supplier and customer results as a requirement of the ongoing maintenance of a qualification of a supplier process. The vendor approval status indicates that the supplier has submitted a predetermined number of lots of materials, which has been tested by the manufacturer with all tests meeting predetermined specifications.

After a supplier has been qualified, the certificates of analysis provided with each lot should refer to all of the agreed tests in the specification. Actual numerical values should be given for quantitative tests and limit tests should include the specification limit. Upon receipt of the material CoA, the manufacturer must compare customer and supplier results and ensure that all of the specifications are within the contracted limits. The CoA must be signed and dated by a competent person from the supplier. Any significant differences in results between supplier and customer must be investigated and the cause identified and corrected.

Once a supplier confidence has been established (approval or certification) certified or qualified, it may be possible for the customer to perform minimal testing on the material or container/closure. The regulations still insist on a minimum of one identity test, which for containers and closures may be visual. For chemical components, the identity test should be specific, such as an infrared spectrum or chromatographic procedures. For compendial materials, the identity test should include the identity test specified in the Compendia. Any parameter that could be subject to change during shipping and that could have a significant impact on quality needs to be included in a reduced testing protocol. These protocols must clearly define the conditions to be met before reduced testing can be introduced. The protocols should indicate that the procedure is material *and* supplier related, the frequency of full testing, and the circumstances that require a return to full testing. It must be emphasized that the product manufacturing Quality Control Department is responsible for the release of purchased materials into production, and consequently, reduced testing should only be introduced when there is proven confidence in the supplier.

VENDOR QUALIFICATION

In recent years, emphasis has been on the application of validation techniques to increase the level of quality assurance. This has extended to suppliers and is most usually referred to as vendor qualification or certification. Vendor qualification is a system that assures that a supplier's product is produced under controlled conditions, resulting in consistent quality conformance. Being based on the principle of defect prevention, rather than defect detection and inspection, it significantly reduces the need for customer inspection. Vendors are qualified when there is evidence to support their ability to supply a material that consistently meets all quality specifications.

Vendor qualification is a supplier–customer partnership and can only be successful with the full involvement and agreement of both partners. Several key steps are involved in the qualification process.

1. The manufacturer must have a Vendor Qualification Procedure that outlines the steps required to qualify or certify a vendor.
2. The vendors must be alerted to the procedure and the steps for qualification and there must be a mutual agreement to the process and the benefits for both. The initial selection of potential partners should take into account the supplier's history in terms of quality, delivery, and support service as well as the importance of the specific material to the business. Vendor qualification has a higher chance of success with a supplier who already has a high commitment to quality and customer service.
3. The following elements should be included in the qualification process. Audit of the supplier's manufacturing facility, where practical or possible. The drug manufacturer should determine whether a) quality systems are in place at the supplier end that meet CGMPs, b) the product quality manufactured will meet your quality specifications, and c) the plant has the capacity to meet your production needs. Where a material may be supplied from more than one plant of the supplier, each plant must be treated as a separate entity for certification purposes.
4. *Specifications*—A detailed review should be made of product specifications with particular reference to legal requirements (Compendia, FDA, etc.) and fitness for use. This latter point is likely to require a supplier understanding of the customer's process; in this way, it may be possible to relax certain less critical specification parameters while tightening or increasing the level of assurance on more critical parameters. This can be particularly important with packaging components where improvements in some areas can dramatically impact on line speeds and efficiency and with particle size of powder ingredients. Obviously, test methods should be identical and agreed upon. Where this is not possible, equivalence must be demonstrated.

5. *Process and Specification—Changes.* Another important element in the vendor certification process is the procedure for handling any changes to the vendor's or drug manufacturer's process or product specification. Any proposed changes must be clearly documented, with reasons and supporting data, and be reviewed and accepted by the drug manufacturer prior to introduction. Most changes will require customer evaluation and/or FDA approval before acceptance. If the drug manufacturer is changing a process parameter that may impact the use of an approved material, this change should be discussed with the material supplier to ensure that the material is best suited for the process change. Changes in the process may also require a change in the raw material or a packaging component.

6. *Process Evaluation*—Vendors are being requested to demonstrate that their production processes are validated, especially for the manufacture of bulk pharmaceutical chemicals. If the supplier's process is under control, any evaluation by the customer should only have value with respect to any changes during shipment. Sections 211.84(a) and (d) do allow for reduced testing, but the elimination of incoming material testing by the customer is precluded by 211.84(d)(2) and (3).

   The customer should perform audits of the supplier's process at appropriate intervals. There are several factors to consider in determining the frequency of auditing. These include a) the capability of location management, b) degree of change that has documented at manufacturing location, c) management turnover, d) history of problems and issues, e) indication of performance measures that suggest problems, and, more importantly, f) the inherent risk in an operation or location to your production. If the vendor is purchased by another company, an out-of-schedule audit may be prompted.

7. A similar procedure should be in place in the event the customer intends to change the specification. Any proposed changes to the customer's process that could impact on the usability or performance of the supplier's material also require prior review and agreement with the supplier. For example, if the customer were contemplating replacement of a packaging line, there would need to be discussions with the supplier of the packaging components. Having established a working partnership that can manage change, it should be possible to work together to identify areas of improvement.

8. *Supplier Reporting*—It is important that both supplier and customer are kept informed of each other's difficulties. The supplier must notify the customer of any atypical situations or process deviations prior to shipping material so that any additional testing or evaluations may be performed. The supplier should also provide certificates of compliance or certificates of analysis for every batch—formatted in a manner that is acceptable to the customer. The customer should also provide feedback to the supplier with respect to compliance with specification, performance in use, and delivery service.

Vendor qualification results in a high level of reliance on the supplier: reduced incoming inspection, reduced inventories, and higher output. The main result of vendor qualification is an assured reduction in quality variability, which provides several benefits. Reduced testing eliminates some testing costs, but more importantly can make materials available to production more quickly. This allows further inventory reductions and is also of benefit when materials are urgently required for unexpected production.

As long as a vendor's product performs to predetermined specifications and delivery schedules, the vendor usually remains a qualified vendor. Any failure by the supplier can have serious consequences. These failures may result in disqualification of the vendor. Depending on the nature of the problem, it may be possible to work with the supplier to re-establish a qualification status. It is a serious action for a manufacturer to disqualify a vendor. The manufacturer must have an alternate supplier already qualified or be able to quickly qualify a second vendor to ensure that there is no disruption in the manufacturing process due to the disqualification. This emphasizes the serious nature of the relationship between the manufacturer and the supplier. Open communication is very important, as is the need for identified contacts in each facility.

> (e) Any lot of components, drug product containers, or closures that meet the appropriate written specifications of identity, strength, quality, and purity and related tests under paragraph (d) of this section may be approved and released for use. Any lot of such material that does not meet such specifications shall be rejected.

Any lot of components, containers, or closures not meeting the specification is to be rejected. This does not preclude recovery by an appropriate rework or inspection procedure, provided the material after this rework meets the specification. In such instances, it may be advisable to include certain noncritical parameters as action levels, provided the procedures clearly define who makes the decision. There is an obvious need for the supplier and customer to agree on specifications. The manufacturer must ensure that incoming materials meet the predetermined, agreed-upon specifications, that they are quickly released to manufacturing, and that rejected material is reduced to a minimum. For information on Vendor Qualification.

Refer to Chapter 16 on Contracting and Outsourcing for related information.

§211.86 USE OF APPROVED COMPONENTS, DRUG PRODUCT CONTAINERS, AND CLOSURES

> Components, drug product containers, and closures approved for use shall be rotated so that the oldest approved stock is used first. Deviation from this requirement is permitted if such deviation is temporary and appropriate.

Using oldest stock first helps to reduce the possibility of contamination and to assure that material conforms to appropriate requirements. Because it may be desirable to package using a single lot of components, containers, or closures, an exemption from strict use of oldest stock first is provided. Other legitimate uses of the exemption are for evaluation of a new supplier, or new equipment or processes with respect to a preferred lot of materials, or the temporary physical inaccessibility of the oldest stock.

Materials management systems now include a need to re-evaluate material after a predetermined time and prior to use. This will further minimize the possibility of materials in an unsuitable condition being used (see also §211.87).

§211.87 RETESTING OF APPROVED COMPONENTS, DRUG PRODUCT CONTAINERS, AND CLOSURES

> Components, drug product containers, and closures shall be retested or re-examined, as appropriate, for identity, strength, quality, and purity and approved or rejected by the quality control unit in accordance with §211.84 as necessary, e.g., after storage for long periods or after exposure to air, heat, or other conditions that might adversely affect the component, drug product container, or closure.

The release of components, containers, and closures for use cannot be for an indefinite time. During storage, degradation may occur, moisture may be absorbed, or materials may simply become contaminated during the storage process. Re-evaluation time scales should be developed from historical data, where possible. Except for particularly sensitive materials, a one-time period, often one year, has been established by many manufacturers. Either the product release label or the system should clearly indicate when materials are to be re-evaluated. This re-evaluation will not usually require full testing, but only examination of those parameters known to be subject to change. For infrequently used materials, re-evaluation coincides with just prior to the use of the material.

Under normal circumstances, materials should be used before they become eligible for re-evaluation. Consequently, when re-evaluation is necessary, the reason for the material still being around should be investigated. The usual reasons include minimum purchase quantities and changes in forecast, but occasionally, this can identify a flaw in the purchasing or planning processes. A further point to be considered when using older materials is the impact on the stability of the dosage form. Degradation is not always linear, and, in some instances, a limited accelerated stability study may prove advisable.

An important factor to consider in the re-evaluation process is the comprehensiveness of the material specification. For some materials, the specifications may have been derived many years ago and the evaluation of ingredients may not meet current expectations. This applies to some of the older bulk drugs in the USP. In these cases, it may be necessary to supplement the existing specification with additional degradation and impurity evaluations. There is a shortcoming in requiring additional testing on older material, i.e., there is no zero point data for comparison. Also, when re-evaluating materials, the extent of changes should be considered even for parameters still within specifications.

For sensitive materials, care should be taken to store them under the appropriate conditions where these are specified. Where not specified, it may still be advisable to identify and use positions in the warehouse that are least susceptible to adverse climatic changes.

§211.89 REJECTED COMPONENTS, DRUG PRODUCT CONTAINERS, AND CLOSURES

Rejected components, drug product containers, and closures shall be identified and controlled under a quarantine system designed to prevent their use in manufacturing or processing operations for which they are unsuitable.

This section reaffirms what was reviewed earlier in Section 211.42(c)(2). Although a segregated reject area is not required, if adequate control systems exist, many companies do segregate reject materials. This is an added precaution against inadvertent use.

FDA investigators frequently use a visit to the reject area as a potential source of identifying deficiencies. If rejections occur, it is possible to assume that the vendor process is not adequately under control, and an evaluation of the cause should have been performed and documented. If the rejection is based on an out-of-specification test result generated during quality control testing and/or inspection, the ensuing investigation should be conducted as described in a written procedure based on FDA *Guidance for Industry: Investigating Out of Specification (OOS) Test Results for Pharmaceutical Production*, published October, 2006.

§211.94 DRUG PRODUCT CONTAINERS AND CLOSURES

(a) Drug product containers and closures shall not be reactive, additive, or absorptive so as to alter the safety, identity, strength, quality, or purity of the drug beyond the official or established requirements.
(b) Container closure systems shall provide adequate protection against foreseeable external factors in storage and use that can cause deterioration or contamination of the drug product.
(c) Drug product containers and closures shall be clean and, where indicated by the nature of the drug, sterilized, and processed to remove pyrogenic properties to assure that they are suitable for their intended use.
(d) Standards or specifications, methods of testing, and, where indicated, methods of cleaning, sterilizing, and processing to remove pyrogenic properties shall be written and followed for drug product containers and closures.

The assessment of suitability of containers and closures is described in Section 211.166 (stability testing). The U.S. Pharmacopeia, ⟨661⟩, provides information on specifications and test methodology for a range of container materials. These include:

1. Light transmission for glass and plastics;
2. Chemical resistance for glass;
3. Physicochemical tests on plastics;
4. Biological tests on plastics and other polymers;

5. Chemical tests on polyethylene containers for dry oral dosage forms;
6. Polyethylene terephthalate (PET) and PET G containers;
7. Polypropylene containers;
8. Repackaging into single unit containers.

These tests may also be modified to apply to the use of plastics other than polyethylene, polypropylene, and PET and to the use of plastics with liquid dosage forms. Where other plastics are involved, any specific signal compounds may need evaluation, e.g., vinyl chloride monomer levels from polyvinylchloride (PVC) containers as well as antioxidants and antiozonants and other additives that are susceptible to leach into liquid drug formulations.

U.S. Pharmacopeia, ⟨381⟩, provides information on specifications and test methodology for testing, applicable to container permeation. Section ⟨381⟩, provides information on the test methodology and specifications required to qualify an elastomeric closure (a packaging component that is or may be in direct contact with the drug). The tests are broken down into biological and physiochemical test procedures. The monograph also requires the customer to review all the ingredients of the closure formulation to assure that no known or suspected carcinogens or other toxic substances are included. Because many elastomeric formulations are proprietary, the manufacturer (user) should request the supplier to confirm the absence of carcinogenic and toxic additives. This confirmation should be in writing and a copy should be maintained in the product files. The supplier must be required to notify the drug manufacturer of any changes in formulation whether they impact the suspected carcinogenic or toxic ingredient information.

There is an expectation of the regulator agencies that the manufacturer will evaluate the potential impact of storage, cleaning procedures, and product on the suitability for use. Section ⟨1151⟩ Pharmaceutical Dosage Forms—Aerosols states "Since pressurized inhalers and aerosols are normally formulated with organic solvents as the propellant or the vehicle, leaching of extractable from the elastomeric and plastic components into the formulation is a potentially serious problem." The USP further states "The extractable profiles of a representative sample of each of the elastomeric and plastic components of the valve should be established under specified conditions and should be correlated to the extractable profile of the aged drug product or placebo, to ensure reproducible quality and purity of the drug product. Extractables, which may include polynuclear aromatics, nitrosamines, vulcanization accelerators, antioxidants, plasticizers, monomers, etc., should be identified and minimized wherever possible."

A container closure system usually serves several roles: to prevent egress (leakage) of the contents, especially for liquids; to prevent ingress of microorganisms, especially for sterile products; and to provide access by the consumer to the contents. The evaluation of container closure performance during storage and transportation will be addressed in the section on stability (§211.166). Consumer acceptability, with the exception of child-resistant closures for which there are defined test methods and acceptance criteria, is often given minimal attention. The United States requires that over-the-counter (OTC) medicines be packaged with at least one tamper-resistant (tamper-evident) feature. Such features may be part of a container closure system, as in the case of seals over the mouth of a bottle, and require stability evaluation. Others, such as neck seals and carton overwraps, do not impact on stability. The initial CGMP regulations did not address product tampering specifically, but the wording in (b), "protection against foreseeable external factors in storage and use," could be considered to apply. Tampering, although rare, has become a "foreseeable" possibility resulting in the addition of §211.132—tamper-resistant packaging requirements for OTC human drug products. Five incidents of drug product package tampering have resulted in death. These five involved cyanide-laced capsules placed in packages of the following OTC drugs: Tylenol in 1982 and 1986, Excedrin in 1986, Sudafed in 1991, and Goody's Headache Powder in 1992. Incident reports have also emerged for nondrug products.

In addition to confirming the suitability of containers and closures, their container/closure specifications defining composition and dimensions must also be established and monitored. The compositions or formulation of many plastic and elastomeric materials is considered proprietary by suppliers. Consequently, close working relations should be established with these suppliers to ensure that they use only acceptable additives and that no changes are made without prior notification with adequate time for evaluation and FDA approval where required.

The cleanliness of container closures is assessed through visual inspection, microbial testing, and validated cleaning procedures. All these assessments of cleanliness are performed under written procedures and are documented. Sterilization process must be validated and re-evaluated on a regular basis.

EXAMPLES OF RECENT OBSERVATIONS FROM FOOD AND DRUG ADMINISTRATION 483 CITATIONS

1. Failure to withhold each lot of component, drug product containers, and closures from use in the manufacture of ophthalmic herbal tinctures until the lot has been sampled, tested, or examined as appropriate [21 CFR 211.84(a)].
2. Failure to establish written procedures that describe in sufficient detail the receipt, identification, storage, handling, sampling, testing, and approval or rejection of components and drug product containers and closures [21 CFR 211.80(a)]. Specifically, your firm does not have any written procedures describing the receipt, identification, storage, handling, sampling, testing, and approval or rejection of components and drug product containers and closures.
3. You have not conducted a visual identification on each lot of container and closure or established the reliability of the supplier's test results, as required under 21 CPR 211.84(d)(3). For example, although you receive a CoA, you do not perform a visual examination on containers and closures and you have not established the reliability of the supplier's test results.
4. Failure to properly identify each lot of drug components of its status in terms of being quarantined, approved, or rejected. Specifically, several containers of raw materials were observed in use that bore only the "Quarantine" sticker [21 CPR 211.80(d)].
5. You have not performed at least one specific identity test on each component or established the reliability of the supplier's test results, as required under 21 CFR 211.84(d)(2). For example, although you receive a certificate of analysis, you do not perform identity testing on incoming raw materials and you have not established the reliability of the supplier's test results.
6. Failure to establish and follow written procedures describing in sufficient detail the receipt, identification, storage, handling, sampling, examination, and/or testing of labeling and packaging materials [21 CFR 211.122(a)]. Specifically, your SOP on creating labels does not address the processes of controlling label receipt, storage, handling, sampling, issuance, and reconciliation.
7. The reliability of the suppliers' CoA was not established in that a complete analysis was not performed and compared with the CoA at the appropriate intervals.
8. Failure to have written procedures describing the receipt, identification, quarantine, sampling, release, and handling of labeling material. Furthermore, incoming labels received from the vendor are not proofed against the master label.
9. Failure to withhold from use each lot of components, drug product containers, and closures until the lot has been sampled, tested, examined, and released by the quality control unit. Failure to conduct at least one specific identity test and establish the reliability of the supplier's analyses. Specifically, your firm does not complete any raw material testing for any raw material received, including both active and inactive ingredients. In addition, although your firm receives a CoA from the component supplier, you do not conduct a specific identity test on components and have not verified the supplier's CoA [2I CFR 211.84(a) and (d)].
10. Failure to properly identify each lot of drug components of its status in terms of being quarantined, approved, or rejected. Specifically, several containers of raw materials were observed in use that bore only the "Quarantine" sticker [21 CPR 211.80(d)].
11. Failure to have operations performed within specifically defined areas of adequate size as necessary to prevent contamination and to have operations relating to the manufacturing,

processing, and packing of penicillin and cephalosporin performed in facilities separate from those used for other drug products for human use [21 CFR 211.42(c) and (d)].

12. Sampling of containers/closures is not based on appropriate statistical criteria.
13. The firm has not included a pyrogen and/or bacterial endotoxin specification for active drug substance raw material.
14. Several batches of tablets were rejected because the active raw material did not meet the firm's established bulk density specifications. No explanation was given in the process validation report as to how bulk density affects the finished product.
15. The firm is aware that . . . has shown marked degradation over time, but no testing was performed on current lots in order to justify the one-year material storage time limitation.

SUGGESTED READINGS

1. Parenteral Drug Association. Technical Report 27, Pharmaceutical Package Integrity, July, 1998.
2. Bossert JL. The Supplier Management Handbook. ASQ, 2004.
3. Mass RA, Brown JO, Bossert JL. Supplier Certification: A Continuous Improvement Strategy. ASQ, 1990.
4. Pannella CR. Managing Contract Quality Requirements. ASQ, 2006.
5. Beagley KG. Should you certify your vendors? Pharm Med Pack News 1994; 34.

7 | Production and Process Controls
Subpart F

Joseph D. Nally
Nallianco LLC, New Vernon, New Jersey, U.S.A.

Michael D. Karaim
Brewster, Massachusetts, U.S.A.

§211.100   WRITTEN PROCEDURES; DEVIATIONS

(a)   There shall be written procedures for production and process control designed to assure that the drug products have the identity, strength, quality, and purity they purport or are represented to possess. Such procedures shall include all requirements in this subpart. These written procedures, including any changes, shall be drafted, reviewed, and approved by the appropriate organizational units and reviewed and approved by the quality control unit.

(b)   Written production and process control procedures shall be followed in the execution of the various production and process control functions and shall be documented at the time of performance. Any deviation from the written procedures shall be recorded and justified.

These two subsections embody the basic underlying concept of current good manufacturing practice (CGMP): there shall be adequate written procedures that have been approved by responsible persons, and for which there is documentation that the procedures have been followed.

The drafting and approval of important procedures such as these cannot be left to chance. Written standard operating procedures (SOPs) define how things are to be done and provide a basis for the training of new or relocated personnel. SOPs are a fundamental extension of the CGMPs—the latter ideally should define what is to be achieved, whereas the SOPs provide company-specific approaches on how to meet these requirements. There should be a master SOP, which describes the overall procedure (SOP on SOPs). This procedure defines how to initiate/revise an SOP, format, who should review and approve (with defined areas of responsibility), frequency of routine review (often every two years), mechanism of issue and replacement of outdated versions, training, archiving, and destruction. Document control is essential for the SOP system (and all manufacturing and control documents). The life cycle of a document (creation, distribution, use, archiving, and destruction) must be considered in establishing the control system. Many firms now use automated document systems that provide a higher level of document control in the review, approval, and distribution processes. For example, current versions of SOPs are available electronically for review at designated terminals and only to trained and qualified personnel. Hard copy printouts are date stamped and accompanied by a statement such as copy only valid for XX/XX/XX (that day only). Manual or paper systems are also common and can be very effective. However, version and distribution control (especially forms and attachments) can provide additional challenges. The SOP supplier [Usually Quality Assurance (QA)] should ideally issue numbered copies to identified recipients and copying of distributed SOPs should not be permitted; this can be monitored by printing the number (or other confirmation) in a different color so that photocopying would be obvious. The use of paper stock with different colors, borders, or unique markings is another way of assuring original distribution and avoiding/prohibiting photocopying.

Many SOP systems are a combination of electronic and paper system and require both electronic and paper controls (numbering system, version control and incrementing, referencing related documents, distribution control of draft and official document copies, and change control).

It is important to note that *any* process or system that does not work in a manual (paper) system will probably also fail in an electronic system.

A typical SOP format includes sections for the following:

- Purpose and Scope
- Definitions
- Equipment/Materials/Supplies and Precautions
- Responsibilities
- Procedure (work process)
- References and Linkages (other related procedures/systems)
- Attachments (forms)
- Change History

SOPs are usually written by department and function (e.g., Manufacturing, Packaging, QA) and describe work activities, parts of systems, or in some cases a complete departmental system. This approach is somewhat limited when considering quality systems that cross multiple departments [e.g., investigations, corrective/preventive action (CAPA), Change Control, Annual Product Review, Validation, Technology Transfer, etc.]. Multi- or cross-functional SOPs are often needed to define the system or work process. Refer to chapter 14 on quality systems approach.

Considerations for designing and implementing effective SOPs and sustainable systems include the following.

Design and Definition Elements

- Logical/understandable flow of inputs, activities, and outputs.
- Language and/or reading level appropriate to the user.
- Clear scope (individual activity/process, departmental process, multi departmental process).
- Responsibilities for tasks, decisions, and results are clearly defined.
- All necessary routine tasks/activities are clearly defined in enough detail to be reproduced consistently.
- How to handle and report exceptions (what-ifs) are defined/covered.
- Decision points, processes, and criteria are identified and defined.
- The process for determining, measuring, and reporting system/procedure performance or metrics is defined (where needed).
- All related SOPs and systems are listed and integrated where needed.
- Interactions, hand-off documents and/or criteria are defined.
- Interaction responsibilities or decisions are defined.
- Templates, forms, or records for complete documentation are provided or linked.
- Change control system linkage is included where needed.

Implementation and Maintenance Elements

- Enough personnel to execute the activities, decisions, and documentation.
- The appropriate facilities and/or equipment are available.
- Equipment is the right design for the intended use.
- The right tools (hardware, problem solving, etc.) are available.
- Personnel training is complete and effective.
- Periodic retraining or reinforcement is conducted.
- The learning capacity or skill level of personnel meets the procedural needs/requirements.

- Procedural communication exists between operators, decision makers, supervisers, and the customer of the output.
- Management/supervision:
  - Provides the right environment (physical conditions and working atmosphere).
  - Provides enough time or information (planning/scheduling) to complete the procedure.
  - Accepts responsibility and accountability for the results.

The master batch record, which provides full details of how a product is to be manufactured, could be considered a very important SOP. As with all SOPs, the amount of detail provided should be adequate to assure that different individuals will be consistent in following the process. This not only enhances the potential for consistent product quality, but it also allows more effective evaluation of the causes of any quality deviations and provides a firm basis for process optimization.

The procedures for production and process control are to be reviewed and approved by the quality control unit or more typically QA. This does not mean that QA is to be considered expert in each area of the operation. For example, the production document would normally be reviewed initially by production and/or technical services; the role of QA would be to confirm this review by a responsible person and to further review the document for possible adverse impacts on quality and safety. When reviewing the production and process control documentation, it will be essential to check that:

a. The various requirements referenced in the CGMP regulations (especially the other subsections of Subpart F) have been adequately addressed.
b. The documents are in compliance with the relevant sections of any new drug application (NDA) and approved NDA (ANDA).
c. When applicable, there are appropriate supporting data such as process validation, analytical method validation, and product stability.
d. The reasons for any proposed changes from previous procedures are clearly defined and supported.
e. The appropriate functions, such as production and technical services, have been reviewed and signed off.
f. The procedures are compatible with any compendial requirements [such as United States Pharmacopeia (USP)].
g. The procedure is of the appropriate design and level of definition (refer to SOP considerations above).

Having provided written and approved procedures, the next (and more difficult) stage is to ensure that they are followed. This involves training and verification steps. Employees must be given training in all relevant procedures. This should include an understanding and awareness of the purpose of the procedures and why they need to be followed. As with all training, it should be confirmed that the employee has actually learned the relevant information and there should be a record of the successful completion of the training (Chap. 3). Next comes the verification step. A combination of some, or preferably all, of the following approaches provides data on compliance.

1. Regular monitoring of compliance by supervisors and managers as they do their daily work. This can be informal but it allows immediate identification and correction of potential compliance problems. It further demonstrates to employees that management does consider compliance to be important.
2. A more systematic review of compliance can also be performed by supervisors and managers on a less frequent basis—perhaps monthly. This again would be done by comparing actual activities with written procedures. This more systematic approach ensures that no department, process, or shift is ignored.

3. Quality assurance, along with departmental management, performs an audit of each function. A written report should be issued and if possible any deficiencies should be quantified thereby allowing trends to be monitored. Quantification can be relatively simple, such as classification of deficiencies into critical, major, and minor and recording the number and percentage of each. Alternatively, a numerical weighting system can be used. Management would be expected to evaluate the audit report, identify the root causes of any noted deficiencies, and to specify appropriate corrective action.
4. Independent audit from outside of the plant adds another level of review. This is similar to three above but may involve personnel from another facility or corporate staff. Also included in this category are regulatory audits such as those by the Food and Drug Administration (FDA). One effective approach to independent audit is to adopt the process assessment approach as used by the Malcolm Baldrige National Quality Award. This has the advantages that it can be focused toward specific processes, allows clear identification of causes of deficiencies, gives credit for positive achievement, identifies centers of excellence, and can provide numerical trend data that can stimulate top management to action.
5. The routine quality assurance check of batch records also provides basic information on compliance. This should also be used to review deviation frequency, evaluation, and corrective action and to confirm compliance with FDA registration data.

It should be emphasized that if the data generated in 1, 2, and 5 above are used to identify and correct basic problems, then the audits described in 3 and 4 should simply provide confirmation of compliance. Traditionally, independent audits (4 above) were used to identify areas of noncompliance. Since they are performed relatively infrequently and can only examine small parts of a production operation, they are ineffective as a basis for identifying *all* noncomplying activities. The emphasis should be on self-evaluation (1 and 2), which is more likely to be successful than the utilization of a police-type activity. The persistent finding of noncompliance issues by QA should signal that management is not giving enough attention to the subject, that training is not adequate, or that procedures are too complex. QA should then work with the appropriate managers to identify the causes and to initiate corrective action.

In order to encourage self-audit, the FDA had agreed not to ask for copies of internal audit reports (Compliance Policy Guides 7151.02). They may, however, ask for evidence that audits are performed. Also, in the event of litigation requests may be made to see such records.

The computerization of process documentation can improve the effectiveness of compliance. The production system can be designed so that one stage has to be completed and any relevant data entered into the computer before the next stage can be initiated. Process control limits can also be included and any atypical results can be made to automatically initiate managerial review. The subject of validation of computer systems is included in Chapter 5.

It would be difficult, if not impossible, to draft an operating procedure that will meet all circumstances. On occasion, deviations from the defined procedure will occur or will be necessary (for example, to evaluate a change). There are usually two types of deviations: planned or unplanned. Planned deviations may be permitted provided there is a planned deviation system that allows for the appropriate documentation and approvals (including QA) prior to execution. Unplanned deviations can be accidental or deliberate. In either case, an investigation is typically performed, potential impact on involved batches is determined, and CAPA generated. CAPA could include procedural modification or personnel discipline. Obviously, in an effective compliant operation, deviations should not be a common occurrence.

Section 211.100(a) requires the review and appraisal of changes to production and process control procedures. However, this review and approval should be considered in the broader concept of change control.

CHANGE CONTROL PROCESS

Inputs	Activities	Output
Changes to:	Change request	Request approval
Bill of materials	initiation (forms and	↓
Manufacturing process	supporting docs)	
Packaging process	↓	Change impact
Shipping process		evaluation report
Product labeling	Request review and	↓
Test methodology	approval	
Standard operating	↓	Final close out report
procedures		
Registration documents	Change	
Computer systems	implementation	
Facilities/utilities	↓	
Equipment		
New product intro	Change results and	
Product discontinuation	evaluation	
	↓	
	Final close out	

A consistent achievement of product quality is dependent on the availability of defined/approved/validated procedures and the application and adherence to these procedures by trained personnel. In the event that any change is to be introduced into the production operation, it is important to evaluate its potential impact and where necessary provide appropriate evaluation and/or actions. The procedure that controls change is, not unexpectedly, called "change control." This should be a defined, proactive management system that facilitates a review of any proposed change and monitors the impact of the change. The system should be fail-safe by preventing changes that could adversely affect product quality or conflict with registration or regulatory requirements. The procedure should have identified ownership with responsibility for maintenance, monitoring, and improvement of the procedure along with training.

The procedure will involve multiple disciplines including sales and marketing, medical, legal, manufacturing, regulatory affairs, R&D, technical services, and maintenance, as well as QC/QA. Not all functions will need to be involved with all changes. The evaluation of the change, which must be documented, should include the following.

- Clear definition of the proposed change with the reason for the change.
- Identification of potential impact and the evaluations to be performed, such as accelerated stability, revalidation, and retraining.
- Regulatory impact (all countries involved) and approvals required.
- Schedule for implementation.
- Definition of who needs to approve the change and a record of their concurrence.
- Post introduction review to confirm that the change did not have any adverse impact.

The change control procedure is possibly the most important SOP in a plant operation. It is also one of the broadest ranging and most complex. Consequently, the management of the process must be delegated to someone with the necessary knowledge and skills to understand and manage this complexity. In larger facilities, there may be separate change control procedures for different types of change. For example, Labeling, Equipment, Computer Systems, and Procedure/Documents can have individual change control systems but QA must be involved in oversight and approval and the outcome must be the same-control and management of change.

Evaluation of change control should be part of the routine QA plant audit.

§211.101 CHARGE-IN OF COMPONENTS

Written production and control procedures shall include the following, which are designed to assure that the drug products produced have the identity, strength, quality, and purity they purport or are represented to possess.

(a) The batch shall be formulated with the intent to provide not less than 100% of the labeled or established amount of active ingredient.
(b) Components for drug product manufacturing shall be weighed, measured, or subdivided as appropriate. If a component is removed from the original container to another, the new container shall be identified with the following information:
    (1) component name or item code,
    (2) receiving or control number,
    (3) weight or measure in new container,
    (4) batch for which component was dispensed, including its product name, strength, and lot number.
(c) Weighing, measuring, or subdividing operations for components shall be adequately supervised. Each container of component dispensed to manufacturing shall be examined by a second person to assure that:
    (1) The component was released by the quality control unit.
    (2) The weight or measure is correct as stated in the batch production records.
    (3) The containers are properly identified.
(d) Each component shall be added to the batch by one person and verified by a second person.

The requirement to "formulate with the intent to provide not less than 100% ... of the active ingredient" requires some explanation. This certainly makes it unacceptable to add only sufficient material to meet the lower end of the specification, although no reputable manufacturer would ever use this approach. It is not the intention of the regulations to require calculation of an exact amount of active ingredient based on the assay value of the material for each batch of product. Most active ingredients show assay results that are not exactly 100%. With inherent errors in analytical methodology in the order of 1% to 2%, it is not possible to precisely determine the "true" assay value. Consequently, it is acceptable to use material that is within the acceptable specification without specific adjustment to accommodate batch analytical variations. This may not be adequate for materials with a significant, and variable, loss on drying. When it is necessary to calculate a specific quantity, this requirement should be specified in writing by QC or QA and not be the subject of telephone or other verbal communication. For a product that is known to show some inherent loss of potency during the production process, it may be advisable to take the assay value into account for each batch. It may also be necessary to add an overage to allow for this potency loss.

The dispensing step is a critical stage of the manufacturing operation. It ensures that the right amounts of the correct material, released by QC/QA, are allocated to the specified batch of product. The labeling of the component containers [(b)(1)—(4)] makes the later checking at production usage more effective. As written, §211.101(b) could be interpreted that it is only necessary to include the labeling requirements if material is transferred from its original container. However, such a literal interpretation would be illogical and would weaken the system since the original container will not reference the drug product name, strength, or lot number.

The dispensing operation also provides an ideal opportunity to visually examine containers for damage and contents for atypical appearance or foreign matter. Dispensing operators are critical (see Critical Process Parameters, p. 94), and all employees should be made aware of the importance of this step and role.

The requirement that "each container of component dispensed to manufacturing shall be examined by a second person" [§211.101 (c)] is usually interpreted to mean that a second person should be available in the dispensary to perform this duplicate check. Several alternatives would also appear to achieve the same result. A single check could be performed in the dispensary with the second independent check being done on receipt by production. With some manual systems, the dispensary label can be removed at the production stage and become part of the batch record. Either routinely, or in the event of a problem, the individual labels can be examined. A second effective alternative is the replacement of the second check by the availability of a suitable computer system. Computer systems that will prevent the weighing of an incorrect or unreleased component are available; they will also disallow completion of the dispensing step if the amount of material being weighed is outside of the defined operational tolerances. Such systems can also be designed to allow only designated individuals to weigh out specified materials, as in the case of controlled substances or materials to which an individual employee may be allergic. Another alternative is the use of bar codes, applied to the incoming materials, and the monitoring of their use and disposition throughout the production operation by scanning equipment. These systems provide much more effective control over the dispensing function than does a second human check and should be introduced whenever possible. An added advantage of such systems is that they allow immediate reconciliation; in the event that a raw material being weighed did not correlate with the records, the dispensing could be automatically put on hold until an investigation had been initiated.

The requirement for a second person, in production, to verify the addition of components to a process is subject to the same argument as that used above for dispensing. For example, if component containers are provided with a bar code, the scanning of this bar code on addition to the batch would provide the assurance required.

However, the original intent of the (manual or witnessed) double check is valid and well proven over time. For example, it is routinely used by the Pilot and Co-Pilot in airline flight operations (flight checklist). An independent second check of an operation can increase the assurance level from ~80% to 99% based on studies of human errors and interactions. Even a second check by the same individual can increase the level of assurance. Experienced carpenters will always preach, "measure twice, cut once" (as opposed to measuring once and cutting twice).

§211.103   CALCULATION OF YIELD

Actual yields and percentages of theoretical yield shall be determined at the conclusion of each appropriate phase of manufacturing, processing, packaging, or holding of the drug product. Such calculations shall be performed by one person and independently verified by a second person.

Theoretical yield is defined in Section 210.3(b)(17) as the maximum quantity that could be produced, based on the quantities of components used, in the absence of any loss or error in production. Theoretical yield consists of the summation of the weights of all raw materials entering the production cycle. For granulations, powders, and tablet coatings, an amount equal to evaporated solvent should be subtracted.

In practice, appropriate phases would include, for example, granulation/mixing, compression, coating, and packaging for a solid dosage form manufacturing operation.

On the basis of historical data, an acceptable range for actual yield at each appropriate stage can be calculated. This range is sometimes set so that 95% of batches produced will fall within the range when the process is operating correctly. The purpose of this is to alert management to atypical situations that may require investigation. Low yields may not only signal potential problems, but also indicate opportunities for process improvement with subsequent cost benefits. Process losses can occur for a variety of reasons including dust extraction, spillage of components or product, machine losses such as in compression, machine adjustments, samples, or residue in equipment. The regulations again require that a second person verify independently the yield calculations. The availability of an automated and validated calculation procedure would seem to be a viable and preferable alternative.

In practice, yield calculations involve reporting the following data/calculations at each appropriate phase:

- actual yield (kg),
- % of theoretical yield = (actual yield/theoretical starting quantity) × 100,
- % accountable yield = [(actual yield + accountable waste)/ theoretical starting quantity] × 100.

The latter provides additional information as to the source/reason for losses that can be useful in investigations of percentage of theoretical yield failures.

§211.105   EQUIPMENT IDENTIFICATION

(a)   All compounding and storage containers, processing lines, and major equipment used during the production of a batch of a drug product shall be properly identified at all times to indicate their contents and, when necessary, the phase of processing of the batch.

This regulation requires that all equipment and lines always bear a label showing their status: clean, to be cleaned, or with the product name and lot number and, if necessary, the phase of processing. If equipment is permanently installed and used for only one batch of product at a time, it may be acceptable to status-label the complete suite. This approach is economical with respect to the application of status labels but individual pieces of equipment tend to be cleaned separately. It may still be advisable to status-label individual items after cleaning to ensure that no uncleaned equipment is allowed to be used.

Some recording system should be introduced to allow back reference to the status data in the event of a problem. Alternative approaches include the retention and filing of status labels or the use of logbooks. The former may be preferred since it is a record of the actual documentation rather than a transcription into a logbook.

The labeling of containers of material in process should clearly define the product, batch number, and state of processing (e.g., granule, bulk tablet, etc.). Where several containers are involved, they should be numbered sequentially. This is of particular value in the event that a problem is later identified and needs to be investigated.

Where materials are to be transported to other sites, it may also be appropriate to place a label inside the container as an extra precaution in case the outer label gets lost or defaced.

(b)   Major equipment shall be identified by a distinctive identification number or code that shall be recorded in the batch production record to show the specific equipment used in the manufacture of each batch of a drug product. In cases where only one of a particular type of equipment exists in a manufacturing facility, the name of the equipment may be used in lieu of a distinctive identification number or code.

The intent of this subsection is to allow identification, at some future date, of the specific piece of equipment involved. This is particularly appropriate where a manufacturer may have several different pieces of the same equipment, which may not behave identically. If the manufacturer has only one piece, then reference by name alone will suffice.

§211.110   SAMPLING AND TESTING OF IN-PROCESS MATERIALS AND DRUG PRODUCTS

(a)   To assure batch uniformity and integrity of drug products, written procedures shall be established and followed, which describe the in-process controls, and tests or examinations to be conducted on appropriate samples of in-process materials of each batch. Such control procedures shall be established to monitor the output and to validate the performance of those manufacturing processes that may be responsible for causing variability

in the characteristics of in-process material and the drug product. Such control procedures shall include, but are not limited to, the following, where appropriate:
(1) tablet or capsule weight variation,
(2) disintegration time,
(3) adequacy of mixing to assure uniformity and homogeneity,
(4) dissolution time and rate,
(5) clarity, completeness, or pH of solutions.

(b) Valid in-process specifications for such characteristics shall be consistent with drug product final specifications and shall be derived from previous acceptable process average and process variability estimates where possible and determined by the application of suitable statistical procedures where appropriate. Examination and testing samples shall assure that the drug product and in-process material conform to specifications.

(c) In-process materials shall be tested for identity, strength, quality, and purity as appropriate, and approved or rejected by the quality control unit, during the production process, e.g., at commencement or completion of significant phases or after storage for long periods.

Valid statistical sampling plans are the topic of much debate and there is a plethora of plans, calculations, probabilities, etc., that can be used. However, what is important is the documented rationale for using a particular sampling plan and that plan should be based on sound scientific rationale and consider consumer or user risk (as opposed to manufacturer's risk). Small sample sizes such as 10 or 20 units provide little assurance of batch quality (by test alone).

The first stage in establishing appropriate process control criteria is the identification of the key factors that impact on quality and the evaluation of acceptable operational ranges for these. This is referred to as process validation.

PROCESS VALIDATION

The FDA in "Guidelines on General Principles of Process Validation" defines process validation as "establishing documented evidence which provides a high degree of assurance that a specific process will consistently produce a product meeting its predetermined specifications and quality characteristics." The designing of quality into a product and its production processes, coupled with supporting validation data, increases the potential for consistently achieving quality standards and reduces dependence on both in-process and end-product testing.

When the concept of validation was introduced, the FDA recognized two approaches: retrospective and prospective. Retrospective validation involved an in-depth evaluation of a large number of consecutive batches of product to correlate processing conditions and analytical results. Provided materials quality and processing conditions are adequately controlled and reported, this approach can clearly demonstrate whether a process is under control. The range of processing conditions resulting in satisfactory product quality can then be used to define acceptable ranges in master production documentation. The advantage of this approach was that it allowed identification of processes that were in control without the need for new additional testing. For those processes that yielded variable product quality, further work was required to make the processes reproducible. For new products, this approach is not acceptable, since many batches would need to be produced before sufficient data were generated. This would mean that in the interim period product quality had to be assured by a heavy reliance on test results. A further disadvantage of the retrospective approach was that in many instances the process variables were inadequately controlled or reported, making the evaluation suspect. The FDA now expects validation to be prospective—completed before commercialization. However, as validation requirements began to be expected for bulk pharmaceutical chemicals, many companies reactivated the retrospective approach—with the same benefits and constraints noted for dosage forms.

The subject of process validation gained new attention by the introduction of the FDA Compliance Program 7346.832 in 1990. The subject was Pre-Approval Inspections/

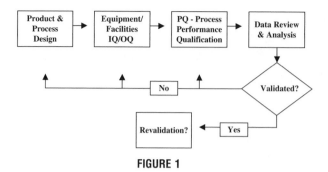

Investigations (PAI). The impact of PAI on the entire compliance program has been significant (Chap. 22).

Prospective validation considers equipment and process and involves several stages: product and process design, equipment installation and operational qualification (IQ and OQ), services qualification, process performance, performance evaluation, and a system to assure timely revalidation (Fig. 1).

For each phase of validation, protocols should be compiled and approved by the relevant functions (e.g., manufacturing, engineering, and QA). These protocols should clearly define the work to be performed and the acceptance criteria. On completion, the data should be evaluated against the acceptance criteria. One very important prerequisite (found by experience) is that the personnel involved directly with the validation exercises must be trained/competent in validation theory, practice, and operational SOPs.

During the product development phase, there should be interactive involvement of all appropriate technical functions, usually R&D, engineering, manufacturing, technical services, and QA. This should ensure that the product as designed by R&D has a high probability of manufacturing success/consistency when transferred into the production operation. This technology transfer process needs to be clearly elaborated with responsibilities and interactive collaboration defined. Since the required involvement by plant operations usually occurs many months (sometimes years) before the product is approved for commercial launch, there can be reluctance to this involvement. However, the potential benefits should override any reluctance. An example that illustrates this point was a situation where R&D developed a tablet product that could not be produced without significant capital expenditure in the plant to provide special environmental conditions—the equipment was on a long lead time, which delayed introduction of the product. Earlier involvement of production personnel could have encouraged modification of the product or process or, alternatively, the required equipment could have been ordered earlier. Early involvement of QC can also be used to evaluate the robustness of analytical methods.

In May 2006, FDA published the "Guidance for Industry Q8 Pharmaceutical Development." "The Pharmaceutical Development section is intended [to] provide a comprehensive understanding of the product and manufacturing process for reviewers and inspectors." Refer to Appendix B for the list of ICH documents. This document is must reading for personnel involved with product transfer and validation as well as development personnel. Of special interest in this Guidance is the recognition of some new concepts that can be applied to pharmaceutical product development, validation, manufacturing and control.

- *Continuous process verification*: An alternative approach to process validation in which manufacturing process performance is continuously monitored and evaluated.
- *Design space*: The multidimensional combination and interaction of input variables (e.g., material attributes) and process parameters that have been demonstrated to provide assurance of quality. Working within the design space is not considered as a change. Movement out of the design space is considered to be a change and would normally initiate a regulatory postapproval change process. Design space is proposed by the applicant and is subject to regulatory assessment and approval.

- *Formal experimental design*: A structured, organized method for determining the relationship between factors affecting a process and the output of that process, also known as "Design of Experiments."
- *Life cycle*: All phases in the life of a product from the initial development through marketing until the product's discontinuation.
- *Process analytical technology (PAT)*: A system for designing, analyzing, and controlling manufacturing through timely measurements (i.e., during processing) of critical quality and performance attributes of raw and in-process materials and processes with the goal of ensuring final product quality.
- *Process robustness*: Ability of a process to tolerate variability of materials and changes in the process and equipment without negative impact on quality.

These concepts are compatible with FDA's recent scientific and quality risk management approaches (Guidance for Industry, ICH 9 Quality Risk Management) that are designed to provide a basis for more flexible regulatory methods. Time will tell how these approaches and concepts roll out to acceptable industry practices but at the very least it should result in a more comprehensive understanding of pharmaceutical products and manufacturing processes. This understanding is essential to achieve a compliant process validation.

Equipment. When new equipment is purchased, the first step after receiving is to ensure that what is delivered is what was ordered and then to confirm its proper assembly and installation. This is followed by operation of the equipment to confirm that it does function in accordance with the design and purchase specifications. This includes parameters such as speeds of mixers, heat distribution of ovens, and calibration of monitoring instruments.

Utilities. The specific unit operations are supported by services such as water, electricity, environmental air, and compressed gases. These too must be qualified. This will involve installation/operational qualification, which includes as-built drawings; weld certification for pipework; airflow volumes, pressure differentials, and paniculate levels (viable and nonviable); microbial evaluation of water quality; effectiveness of filters (air, water, gases); validation of computerized systems for environmental control; and temperature and humidity controls for Heating Ventilation Air Conditioning (HVAC) systems.

When the facilities, the supporting systems, and the equipment are qualified, it is time to qualify (validate) the specific production processes.

The process validation protocol should include:

a. The facilities, services, and equipment to be used.
b. The key variables likely to impact on quality. These are usually identified during the product development process or from experience with similar products.
c. The range of conditions to be evaluated for each variable. The range of conditions to be evaluated should extend beyond the anticipated operational ranges of the process. It has been suggested that the validation ranges should extend to the point where the process fails, "worst case." There would seem to be no need for this, provided operational ranges are maintained inside the values evaluated during validation.
d. All test equipment, gauges, and recorders used in the validation process should be calibrated immediately prior to and after each validation experiment.
e. The samples to be taken; location, size, number, and frequency.
f. The tests to be performed and the methodology to be used. Analytical methods must be validated, otherwise any data generated will be of doubtful value.
g. The number of replicate process runs to be performed.
h. Analytical methods validation data for active and detergent cleaning levels
i. The acceptance criteria.
j. Details of who must review the data and where it is to be retained for permanent reference.

The possible interrelating effects of different process variables could require an extensive number of evaluations. The number of experiments can be reduced by the application of suitable statistical methods such as design of experiments. Information from experimental designs can often identify critical and interacting variables.

A critical process step or process control is a parameter that must be controlled within predetermined criteria to ensure that the product meets its specification. Potential critical process steps and parameters for a solid dosage product could include:

- Drug substance properties,
- Excipient properties,
- Weighing/dispensing operations,
- Order of material addition,
- Dry mixer speed and time,
- Quantity, temperature of water, spray rate (wet granulation),
- Granulator impeller and chopper speed,
- Granulation endpoint (time or amperage),
- Milling screen size and mill speed,
- Drying inlet air flow temperature, dew point,
- Drying outlet temperature,
- Drying product temperature,
- Tablet press or encapsulation machine setup,
  - Pre and/or compression forces
  - Machine speed
  - Rejection system air speed
- Coating suspension mixing time,
- Coating pan spray nozzle size, spray gun distance to bed, spray rate,
- Coating pan inlet and exhaust temperature, pan speed,
- Cure temperature and time.

Typical in-process tests or controls (Quality Attributes) for a solid dosage product could include:

- Particle size and distribution,
- Moisture content,
- Blend uniformity,
- Content uniformity,
- Tablet weight, hardness, thickness, friability, disintegration,
- Capsule weight, disintegration,
- Coating weight gain.

Each product and process can have different parameters that are critical. For example, the order and rate of addition of materials during mixing and granulation may be essential for reproducibility on certain formulations and have no significant impact on other formulations. In either case, comprehensive product development data is the key to determining what is critical and must be controlled within a tight range or set point, and what is not critical.

In November 2003, FDA issued the draft guidance on "Powder Blends and Finished Dosage Units—Stratified In-Process Dosage Unit Sampling and Assessments": "This guidance is intended to assist manufacturers of human drug products in meeting the requirements of 21 CFR 211.110 for demonstrating the adequacy of mixing to ensure uniformity of in-process powder blends and finished dosage units. This guidance describes the procedures for assessing powder mix adequacy, correlating in-process dosage unit test results with powder mix test results, and establishing the initial criteria for control procedures used in routine manufacturing."

This guidance emphasizes the importance of a good understanding of mixing and blending dynamics and is in agreement with the FDA general expectations for validation.

- For solid dosage validation, the granule or powder mix must be evaluated for uniformity in addition to the tablets/capsules themselves.

- Sample size from the blend should be small and should resemble the dosage size, preferably not more than three times the active ingredient dosage size.
- Compositing of samples should not occur.
- Samples should be representative of all parts of the batch. This means that samples should be taken from places that might be problem areas.
- Sampling from drums of blended powder or granule should only occur if it could be demonstrated that this was representative.
- Concurrent and prospective validation requires at least three consecutive batches.
- Particle size distribution specifications should be defined and evaluated.

Revalidation. A system must be established that initiates a review of the need for revalidation whenever there is a change in the equipment, facilities, process, services, formulation, or source of components. In some cases, atypical product behavior could generate revalidation activities. Additionally, because some changes may be made without notification and time itself can result in change, it may be advisable to consider revalidation at predetermined frequencies.

On completion of process validation, it is then possible to define the operational parameters for the process which if followed should assure compliant product. The batch record defines these operational parameters and the in-process controls provide confirmation that the process has remained under control. The use of statistical control charts will identify trends and the need for any process adjustment; they will also make it easier to pinpoint and contain substandard product if a process goes temporarily out of control. An important element in the validation/revalidation process is the evaluation of process data from production batches. If process modifications are required to keep the quality consistent or if reworking/rejections occur, then it is possible that the process has not been adequately validated. Some process variables have not been fully identified and revalidation may be required (Fig. 1).

It has been common practice for production personnel to perform many of the in-process tests, usually with QC repeating some of these at less frequent intervals. The current trend in managing quality is to transfer accountability for quality to the individual performing the job. This is much more likely to be successful than by the operation of a police-style QC/QA department trying to play "catch-it." However, QC/QA still do retain the overall responsibility of ensuring that released product meets the required standards.

(a) Rejected in-process materials shall be identified and controlled under a quarantine system designed to prevent their use in manufacturing or processing operations for which they are unsuitable.

Since production processes are designed and operated to produce complying material, a rejection should be a cause for concern. An investigation of the problem must be initiated and must be of sufficient depth to identify the root cause and not just the symptom. Unless the real basic cause is identified, the problem is likely to re-occur.

Identification of root cause should be followed by appropriate corrective action to prevent a future recurrence. During the investigation, it should be established whether there has been a previous similar occurrence; this could indicate that the root cause probably had not been identified on the earlier occasion. Refer to Chapter 11 for more details on investigations.

Having identified the cause of the problem, the disposition of the affected material must be addressed. Rejected materials need not necessarily be destroyed. The material may be suitable for reprocessing (see §211.115). While retained in a reject status, materials should be controlled, either physically or by way of a system, to preclude their inadvertent further processing. Rejected materials should be disposed of as quickly as possible to further minimize the risk. FDA investigators also tend to visit the reject area and use the information obtained to identify potential GMP deviations.

PROCESS ANALYTICAL TECHNOLOGY

What has just been presented is the traditional approach to validation that should still be around for a number of years. PAT application has the potential to replace process validation as it is known today but this will be an evolutionary process.

PAT The concept is not new. It is practiced in other industries and the device GMPs allow for full end testing for devices in lieu of process validation if the diagnostic testing assures 100% functionality of each device. An example of modern process measurement control is in the commercial flooring industry where the dimensions of the Snap&Fit composite flooring systems are continuously monitored, to thousands of an inch, as it is produced. For anyone who has installed one of these floors, the manufacturing process precision is much appreciated.

PAT has been applied slowly by industry and is now an initiative from FDA. The FDA Guidance for Industry document is: PAT—A Framework for Innovative Pharmaceutical Development, Manufacturing and Quality Assurance, September 2004. This is available at: http://www.fda.gov/cder/guidance/6419fnl.htm. Their goal is to promote greater product and process understanding that results in high levels of product assurance while providing effective mechanisms to demonstrate validation. A manufacturing process that utilized PAT could be described as being continuously monitored, evaluated, and adjusted using validated in-process measurements, tests, controls, and process endpoints.

The highlights of this guidance are presented:

Introduction

"This guidance is intended to describe a regulatory framework (Process Analytical Technology, PAT) that will encourage the voluntary development and implementation of innovative pharmaceutical development, manufacturing, and quality assurance. This guidance is written for a broad industry audience in different organizational units and scientific disciplines. To a large extent, the guidance discusses principles with the goal of highlighting opportunities and developing regulatory processes that encourage innovation. In this regard, it is not a typical Agency guidance."

The scope of the guidance includes application drug products (NDA, NADA, ANDA) and non-application drug products. It also includes specified biologics regulated by Center for Drug Evaluation and Research (CDER) and CVM. There is an important statement in the scope section that points out that the development and implementation of PAT is a *voluntary* one and that one can also choose to apply PAT to selected products.

Background

"Unfortunately, the pharmaceutical industry generally has been hesitant to introduce innovative systems into the manufacturing sector for a number of reasons. One reason often cited is regulatory uncertainty, which may result from the perception that our existing regulatory system is rigid and unfavorable to the introduction of innovative systems."

"In August 2002, recognizing the need to eliminate the hesitancy to innovate, the Food and Drug Administration (FDA) launched a new initiative entitled "Pharmaceutical CGMPs for the 21st Century: A Risk-Based Approach." The goals are intended to ensure that:

- The most up-to-date concepts of risk management and quality systems approaches are incorporated into the manufacture of pharmaceuticals while maintaining product quality.
- Manufacturers are encouraged to use the latest scientific advances in pharmaceutical manufacturing and technology.
- The Agency's submission review and inspection programs operate in a coordinated and synergistic manner.
- Regulations and manufacturing standards are applied consistently by the Agency and the manufacturer.
- Management of the Agency's Risk-Based Approach encourages innovation in the pharmaceutical manufacturing sector.
- Agency resources are used effectively and efficiently to address the most significant health risks.

"The approach is based on science and engineering principles for assessing and mitigating risks related to poor product and process quality. In this regard, the desired state of pharmaceutical manufacturing and regulation may be characterized as follows:

- Product quality and performance are ensured through the design of effective and efficient manufacturing processes
- Product and process specifications are based on a mechanistic understanding of how formulation and process factors affect product performance
- Continuous *real-time* quality assurance
- Relevant regulatory policies and procedures are tailored to accommodate the most current level of scientific knowledge
- Risk-based regulatory approaches recognize
  - the level of scientific understanding of how formulation and manufacturing process factors affect product quality and performance
  - the capability of process control strategies to prevent or mitigate the risk of producing a poor quality product"

PAT Framework

"The Agency considers PAT to be a system for designing, analyzing, and controlling manufacturing through timely measurements (i.e., during processing) of critical quality and performance attributes of raw and in-process materials and processes, with the goal of ensuring final product quality. It is important to note that the term *analytical* in PAT is viewed broadly to include chemical, physical, microbiological, mathematical, and risk analysis conducted in an integrated manner. The goal of PAT is to enhance understanding and control the manufacturing process, which is consistent with our current drug quality system: *quality cannot be tested into products; it should be built-in or should be by design*. Consequently, the tools and principles described in this guidance should be used for gaining process understanding and can also be used to meet the regulatory requirements for validating and controlling the manufacturing process.

Quality is built into pharmaceutical products through a comprehensive understanding of:

- The intended therapeutic objectives; patient population; route of administration; and pharmacological, toxicological, and pharmacokinetic characteristics of a drug
- The chemical, physical, and biopharmaceutic characteristics of a drug
- Design of a product and selection of product components and packaging based on drug attributes listed above
- The design of manufacturing processes using principles of engineering, material science, and quality assurance to ensure acceptable and reproducible product quality and performance throughout a product's shelf life"

"A desired goal of the PAT framework is to design and develop well understood processes that will consistently ensure a predefined quality at the end of the manufacturing process. Such procedures would be consistent with the basic tenet of quality by design and could reduce risks to quality and regulatory concerns while improving efficiency. Gains in quality, safety and/or efficiency will vary depending on the process and the product, and are likely to come from:

- Reducing production cycle times by using on-, in-, and/or at-line measurements and controls
- Preventing rejects, scrap, and re-processing
- Real-time release
- Increasing automation to improve operator safety and reduce human errors
- Improving energy and material use and increasing capacity
- Facilitating continuous processing to improve efficiency and manage variability"

Process Understanding

"A process is generally considered well understood when (1) all critical sources of variability are identified and explained; (2) variability is managed by the process; and, (3) product quality attributes can be accurately and reliably predicted over the design space established for materials used, process parameters, manufacturing, environmental, and other conditions." "Although retrospective process capability data are indicative of a state of control, these alone may be insufficient to gauge or communicate process understanding."

"A focus on process understanding can reduce the burden for validating systems by providing more options for justifying and qualifying systems intended to monitor and control biological, physical, and/or chemical attributes of materials and processes."

"Transfer of laboratory methods to on-, in-, or at-line methods may not necessarily be PAT. Existing regulatory guidance documents and compendial approaches on analytical method validation should be considered."

Principles and Tools

"Currently, most pharmaceutical processes are based on time-defined end points (e.g., blend for 10 minutes). However, in some cases, these time-defined end points do not consider the effects of physical differences in raw materials. Processing difficulties can arise that result in the failure of a product to meet specifications, even if certain raw materials conform to established pharmacopeial specifications, which generally address only chemical identity and purity."

There are a number of tools available to enable better process understanding. The PAT Framework defines four categories:

- *Multivariate tools for design, data acquisition and analysis.* Pharmaceutical products and processes are complex multifactorial systems. The use of multivariate mathematical approaches such as design of experiments, response surface methodologies, process simulation, and pattern recognition tools can greatly enhance process understanding and control.
- *Process analyzers.* It is now more common to see a variety of process analyzers being used to monitor pressures, temperatures, pH, etc. The initial needs came from safety concerns and yield optimization but the use for process control and validation is well accepted. Advances in process analyzers make real-time control feasible. However, they typically generate a lot of data. Sorting that data for batch record review can be a challenge. There are three types of measurements:
  - At-line: sample is removed from process stream and analyzed close by.
  - On-line: sample is diverted from process stream and analyzed but could return to the process stream.
  - In-line: process stream is analyzed.
- *Process control tools.* The design and application of process controls should be done at the process development stage. Process controls should monitor the state of the process and actively manipulate it to maintain a state of control. Controls should accommodate raw material attributes, monitoring of critical process parameters, and achievement of process endpoints.
- *Continuous improvement and knowledge management tools.* Continuous learning from data collection and analysis over the life cycle of the product can greatly benefit the manufacturer and regulators. Sound scientific approaches and data can make postapproval product changes a more pleasurable experience.

Risk-based approaches (see Chap. 14) can and should be used in a PAT framework. For well-understood processes, the opportunity exists to demonstrate that less restrictive regulatory approaches can be applied.

Another benefit of PAT application is the possibility of *real-time release* as long as the process data convincingly represent the quality of the in-process and finished product. Real-time release is similar to the *parametric release* used in some terminally sterilized product manufacturing. However, real-time release measures material attributes as well as process parameters.

§211.111 TIME LIMITATIONS ON PRODUCTION

When appropriate, time limits for the completion of each phase of production shall be established to assure the quality of the drug product. Deviation from established time limits may be acceptable if such deviation does not compromise the quality of the drug product. Such deviation shall be justified and documented.

The main purpose of this regulation is to indicate that certain processes are sufficiently sensitive that time limits need to be established for their completion. This could be especially important for:

a. Material vulnerable to microbial attack. Bulk injections are usually required to be filled into the final container within 48 hours; otherwise, any microbial contamination could result in high levels of pyrogenic material.
b. Materials subject to oxidation may be protected with nitrogen. Effective nitrogen protection may be difficult at the bulk stage; also, failure of the nitrogen system could result in batch rejection.
c. Tablet granulations or other bulk solids may absorb or release moisture on storage, making them more difficult to process or even accelerating decomposition. Batch records should clearly indicate any time-scale restriction and dates and times should be recorded. In the event that a defined time scale is exceeded, this may not necessarily result in batch rejection. But an investigation must be initiated to identify the cause and the possible implications of the changed time scales. Only if adverse effects are unlikely can the material be used. For example, if pyrogenicity is the concern, this could be measured and if results are not atypical it may be possible to use the material.

Extension of established and validated time scales may be used as a basis for extending these times. For example, if an accepted holding time of 48 hours has been extended to 60 hours with no adverse impact, it may be possible to change the accepted holding time to 60 hours. In most instances, data from more than one such extension will be required before a permanent change can be implemented. Many firms include holding times in their validation protocols so that varying or maximum conditions can be evaluated that mimic production conditions.

Another important timing parameter relates to the actual time scales for unit processing such as mixing and drying. This is normally addressed at the process validation and process control stages.

---

§211.113 CONTROL OF MICROBIOLOGICAL CONTAMINATION

Michael D. Karaim

(a) Appropriate written procedures, designed to prevent objectionable microorganisms in drug products not required to be sterile, shall be established and followed.

Every day, the human body is invaded by countless numbers of microorganisms, which are found in the food we eat, the air we breathe, and the water we drink. Consequently, for most products other than injections, there is no need for sterility. For products that are not required to be sterile, the presence of microorganisms could still constitute a problem. Certain microorganisms are associated with human illness and should be absent. For example, oral suspensions and solutions should be tested for freedom from *Escherichia coli*; products for topical application should be tested for freedom from *Pseudomonas aeruginosa* and *Staphylococcus aureus.*

Some products may also be prone to microbial degradation resulting in loss of active ingredient or breakdown in physical characteristics, such as emulsions. In such cases, it may be necessary to have a specification for total viable microorganisms. The end use of the product may also make it appropriate to have such limits: for example, for product used around the eyes or on mucous membranes.

The USP 29 General Information chapter ⟨1111⟩ *Microbiological Examination of Nonsterile Products: Acceptance Criteria for Pharmaceutical Preparations and Substances for Pharmaceutical*

Use addresses total plate count criteria for bacteria and yeasts/molds. In addition, it discusses evaluation of other microorganisms recovered in terms of the following:

- The use of the product (e.g., route of administration),
- The nature of the product as it relates to supporting growth,
- The method of application,
- The intended recipient (e.g., neonates, infants, the debilitated),
- The use of immunosuppressive agents,
- The presence of disease, wounds, organ damage.

Chapter ⟨1111⟩ gives examples of total plate count limits as well as the absence of certain microorganisms based on route of administration. For example, as taken from Table 1. Acceptance Criteria for Microbiological Quality of Nonsterile Dosage Forms, the total plate count criteria for nonaqueous preparations for oral use are less stringent than those for oromucosal use products.

Route of administration	Total aerobic microbial count (cfu/g or cfu/mL)	Total combined yeasts/molds count (cfu/g or cfu/mL)	Specified microorganism(s) (1 g or 1 mL)
Nonaqueous preparations for oral use	10^3	10^2	Absence of *E. coli*
Oromuosal use	10^2	10^1	Absence of *S. aureus* Absence of *P. aeruginosa*

NOTE: *These USP-NF General Chapters were originally scheduled to be effective in August, 2007. While this reflects current thinking, worldwide harmonization efforts may delay implementation until 2009.*
     The procedures and conditions required to assure adequate microbial quality will vary according to the specific products but are likely to include some, or all, of the following:

1. Microbial monitoring of potentially susceptible raw materials. This may require special negotiation with the supplier if a microbiological specification is not a normal requirement for his other customers. Current practices involve the setting of microbial specifications for materials of natural (animal, vegetable, or mineral) origin, those likely to support microbial growth and materials to be used in product formulations with rigorous microbial specifications, such as injections.
2. Equipment sanitation procedures that have been proven effective, especially for any specific known deleterious or objectionable microorganisms.
3. Processing conditions that minimize the potential for microbial growth.
4. Environmental control including covers over equipment; laminar flow at susceptible points, wearing of protective clothing such as gloves and masks, clearing filling lines at breaks.
5. Formulations to include preservatives.

     Although there may be a need for limits for liquid products, especially aqueous products, there would seem to be less value for solid oral dosage products. As far as we are aware, microbial contamination has not been a problem except for products involving materials of natural origin. In fact, USP General Information Chapter ⟨1112⟩ *Application of Water Activity Determination to Nonsterile Pharmaceutical Products* addresses water activity in relationship to Microbial Limit Testing strategies. This chapter notes that determination of water activity of nonsterile pharmaceutical dosage forms aids the decision-making process regarding optimized product formulations, susceptibility to microbial contamination, frequency of microbial limit testing, etc. It points out that nonaqueous liquids and dry solid dosage forms will not support microbial growth due to low water activity. Pharmaceutical drug products with water activities well below 0.75 are noted as being excellent candidates for reduced microbial limit testing for product release and stability evaluation.

> (a) Appropriate written procedures, designed to prevent microbiological contamination of drug products purporting to be sterile, shall be established and followed. Such procedures shall include validation of any sterilization process.

Sterile products are manufactured using either terminal sterilization or aseptic processing. The level of sterility assurance is significantly higher with terminal sterilization; autoclaving at 121°C can easily result in a 10^{-6} microbial survivor probability, whereas aseptic processing tends to result in the order of 10^{-3}. Because of these significant differences in assurance levels, terminal sterilization should be the method of choice. Some products cannot withstand the temperature conditions of autoclaving, the ingredients may be heat labile or the package may be physically affected by the pressure changes (e.g., prefilled syringes), and aseptic processing may then be necessary. A useful compromise situation is a combination of aseptic processing with some level of heat treatment that could effectively kill off vegetative organisms without adversely affecting chemical stability or physical integrity.

The subject of aseptic processing versus heat sterilization or the compromise of aseptic plus some heat treatment can be very complex, especially for those products that have some degree of heat liability. What conditions should be used to evaluate heat treatment that is less rigorous than typical sterilization conditions? The possible permutations of temperature and time are almost limitless. Also, what are the relative benefits/disadvantages of aseptic processing with low levels of degradants and some heat treatment with higher levels of sterility assurance but also higher levels of degradants? Unless clear guidelines are agreed on or companies are allowed to make their own judgmental decisions, there will be regulatory chaos with higher development costs and delays in approvals.

Whichever process is used, the probability of having a nonsterile unit will be extremely low. Consequently, assurance of sterility cannot be demonstrated by testing a limited number of samples. For example, when sterility testing 10 units, lots with 0.1% contaminated units could be passed as sterile 99 out of 100 times. Increasing the sample size to 100 still leaves a 91% chance of passing a contaminated batch. Also, if the sample size is increased the potential for false positives also increases. This then places a greater emphasis on the need to validate the sterilization process and to ensure that the defined process is followed for every batch of product. The key parameters to be evaluated for the different types of sterilization are outlined below; whichever process is used, the same basic steps outlined previously for process validation are also to be included: product/process design, equipment qualification, services qualification, process performance, and revalidation. Validation of heat sterilization (dry heat or autoclave) includes:

1. Heat distribution within the empty sterilization chamber.
2. Heat penetration within the units of product for the various loading cycles to be used.
3. Lethality calculations based on the known numbers of resistant bacteria or spores killed, usually *Geobacillus stearothermophilus* spores placed in units that receive the least heat treatment.
4. Bioburden data showing the numbers and types of organisms, with particular reference to resistivity, likely to result from the components and the process prior to sterilization.
5. Perform studies outside the ranges of conditions that will routinely be used for sterilization cycles.

VALIDATION OF ASEPTIC PROCESSING

1. Treatments of product components and processing equipment to remove particulate matter, sterilize, and depyrogenate are critical to effective aseptic processing. This will

include ampoules, vials, stoppers, filters, intermediate storage vessels, tubing, filling equipment, gowns, masks, and gloves. The processes for each of these must be validated.

2. Environmental qualifications must include:

   a. *Air quality.* At the point of use (e.g., filling), air should be supplied by high-efficiency paniculate air (HEPA)-filtered laminar flow air at about 90 feet per minute and with a pressure differential to adjacent areas of different classification of at least 10–15 Pa with doors closed. Nonviable particle counts should be less than 100 per cubic foot equal to or larger than 0.5 μm (Class 100 or ISO 5); viable particles should be not more than one colony forming unit per 10 cubic feet. Where an aseptic processing room is adjacent to an unclassified room, an overpressure of at least 12.5 Pa from the aseptic processing room should be maintained according to the 2004 *FDA Guidance for Industry—Sterile Drug Products Produced by Aseptic Processing—Current Good Manufacturing Practice.*

   Away from the critical filling area, where product is not exposed to the environment, less stringent requirements are necessary but must still be controlled in order to minimize the bioburden load. The Class 100,000 (ISO 8) (not more than 100,000 particles per 0.5 μm or larger and not more than 25 colony forming units per 10 cubic feet) should be adequate. Air filter integrity and efficiency testing should be included.

   b. *People.* The presence of people in an area or room will have an impact on air quality. The validation study should include the maximum number of people expected to be present at any time during the process. Other people-related activities to be examined would be training programs, especially with respect to microbiological understanding, aseptic techniques, and gowning techniques. The effectiveness of these techniques can be evaluated by the use of swabs, contact plates, and touch plates.

3. *Time limitations.* Liquid preparations and wet components are prone to microbial multiplication, including the possibility of microorganisms passing through filters. Maximum time frames for key steps need to be confirmed.

4. *Product filtration.* The filtration system used to "sterilize" the drug product, usually 0.22 μm, should be challenged using a suitable small organism, usually *Brevundimonas diminuta.* The number of organisms used in the challenge will be in excess of the maximum bioburden levels measured in unfiltered solutions.

5. *Media fills.* The overall effectiveness of the aseptic process is then validated using liquid media fills.

   a. Initially, three media fills are considered desirable.

   b. According to the 2004 *FDA Guidance for Industry, Sterile Drug Products Produced by Aseptic Processing—Current Good Manufacturing Practice*, the starting point for a media fill run size is 5000 to 10,000 units. For production sizes under 5000, the number of units should be at least equal to the batch size.

   c. Each shift and each employee used for aseptic processing should be included in the validation runs.

   d. According to the guidance, the recommended criteria for assessing the state of aseptic line control are as follows:

   • When filling fewer than 5000 units, no contaminated units should be detected.
      • One (1) contaminated unit is considered the cause for revalidation, following an investigation.
   • When filling from 5000 to 10,000 units:
      • One (1) contaminated unit should result in an investigation, including consideration of a repeat media fill.
      • Two (2) contaminated units are considered the cause for revalidation, following investigation.
   • When filling more than 10,000 units:
      • One (1) contaminated unit should result in an investigation.
      • Two (2) contaminated units are considered the cause for revalidation, following investigation.

6. *Revalidation.* As with any process, revalidation should be considered whenever there is a change in the product, components, process, facility, equipment, or people. Additionally, since the aseptic process is so people dependent, regular revalidation is essential. This

routine revalidation should normally be performed every six months on each different type of process and for each shift; every operator should be included in a revalidation at least every 12 months.

The routinely collected data on bioburden levels and environmental conditions will also serve to confirm that the process is being maintained under control.

The greatest potential source of microbial contamination in a traditional aseptic environment is people. The interaction of people and process is also not consistent. One way to significantly minimize this potential microbial exposure and variability is to separate the people from the process. Newer aseptic installations and upgrades are introducing barrier technology. This technology maintains the environmental conditions around the product at Class 100 or better while allowing personnel access only by way of glove ports. Consequently, there is no direct interaction of people and process. This approach greatly enhances the potential for sterility assurance—from about 10^3 to 10^{-5} or 10^{-6}. The 2004 *FDA Guidance for Industry, Sterile Drug Products Produced by Aseptic Processing—Current Good Manufacturing Practice* discusses various aspects of maintenance, design, and monitoring of Aseptic Processing Isolators in Appendix A and Blow-Fill-Seal Technology in Appendix B.

Another benefit from the use of barrier technology is that the high-quality (expensive) air needs to be supplied only to the product operational area and not to the entire room.

VALIDATION OF ETHYLENE OXIDE STERILIZATION

This process is used for the sterilization of components but *not* for products. Because of the inherent health hazards associated with the use of ethylene oxide, its use is tending to diminish. Key parameters to be included in the validation study include:
1. Distribution of temperature, ethylene oxide, and humidity in the sterilization chamber.
2. Penetration of gas and moisture of the material to be sterilized.
3. Lethality calculations based on the known numbers of resistant bacteria or spores killed.
4. Removal of ethylene oxide and ethylene glycol residues.

VALIDATION OF RADIATION STERILIZATION

Gamma radiation using cobalt-60 is used for the sanitization and sterilization of many pharmaceutical raw materials and products. Usually, these are solids or nonaqueous preparations because water when irradiated generates free radicals, which tend to cause degradation. Gamma radiation is easy to use since time is the only variable once dosage has been established. There is also some evidence that gamma irradiation can reduce endotoxin levels.

The validation of a gamma irradiation sterilization process involves three stages.

1. Product qualification evaluates the impact of radiation on the product. Three levels of radiation may be determined: (*i*) maximum tolerated level—the highest dose that fails to induce an unacceptable change in the product. (*ii*) Maximum process dose—based on the defined sterilizing dose—to be applied and the highest level of exposure in any unit of product. (*iii*) Minimum process dose-the opposite of (*ii*). The optimum situation is for maximum and minimum process values to be close but significantly lower than the maximum tolerated level.

Assessment of impact must use real-time stability studies, since accelerated conditions may result in more rapid degeneration of free radicals and give an impression of greater stability.

2. Equipment qualification is normally performed by the operator of the facility and should address design, installation, operation, and maintenance.
3. Process qualification should include:
   a. Sterilization approach, of which there are three: (*i*) overkill, which usually involves radiation doses in excess of 25 kGy and can only be used for products that are radiation stable; (*ii*) bioburden, which relies on a lower level of radiation based on the known and constant bioburden of the product; and

(*iii*) species-specific, which uses an even lower radiation dosage and is particularly useful for products with a low, nonresistant bioburden such as pharmaceuticals.
  b. Dose distribution in the loads using well-defined loading patterns.
  c. Biological challenge using *B. pumilis.*
  d. Cycle interruption studies.

§211.115 REPROCESSING

(a) Written procedures shall be established and followed prescribing a system for reprocessing batches that do not conform to standards or specifications and the steps to be taken to ensure that the reprocessed batches will conform with all established standards, specifications, and characteristics.
(b) Reprocessing shall not be performed without the review and approval of the quality control unit.

The failure of a batch of product to meet the quality standards could be caused by special causes or variation (such as an equipment malfunction) or normal process variation. In either case, an investigation is necessary to determine the root cause of the failure and adequate CAPA. Adequate corrective action for a special cause or variation situation would be to eliminate or significantly reduce the occurrence of the event. On the other hand, a failure due to normal process variation is an unacceptable situation, and it is possible that the process had been inadequately validated or that one or more of the operating parameters are actually outside of acceptable limits. Corrective action and revalidation may be required before further lots are processed/released.

If reprocessing is a viable corrective action and permitted in the product license, QC/QA should also carefully review what testing and evaluation is to be performed on the reprocessed batch. Release specifications are usually designed to evaluate a batch produced under normal circumstances and may be inadequate for reprocessed batches. Consequently, it is essential that the evaluation of reprocessed batches is approved by quality management. Factors to be considered would include the following.

1. Whether any specification tests are not performed routinely, reliance being placed on validation data. Examples of this could include dissolution performed only at the uncoated stage of a film coated tablet; recoating may affect this, or content uniformity.
2. Whether the reprocessing might have affected product stability and its shelf life. This could happen for a liquid product requiring a reheat stage to fully dissolve some raw materials.

The practice of adding a small amount, say 10%, of the rejected batch to subsequent lots of product, based solely on the assumption that most tolerances are 10%, is nonvalid. First, there is no evidence to indicate that the rejected processed material will not change the characteristics beyond the specifications. Second, and more important, CGMPs require that the manufacturer try to attain the product specifications; tolerances are established to take into account only unavoidable processing variation and the accuracy and reproducibility of test methods. It is not a good practice to permit a lowering of target standards by adding material whose effect on the process is not known but is assumed to lower the desired quality target, even though the quality remains within specifications. If reprocessing by addition to subsequent batches is practiced, it is necessary to confirm that this does not adversely affect the target values for product quality.

EXAMPLES OF OBSERVATIONS FROM FDA 483 CITATIONS

1. Annual revalidation of the sterility test room was not performed as required by the firm's procedures.
2. The master formula does not state a time limit for holding filtered bulk drug compound prior to filling and terminal sterilization.

3. Lack of validation of the manufacturing of the various coating solutions such as gelatin solution, sub coating powder, and syrup solution.
4. Batch manufacturing instructions do not provide sufficient written detail to ensure the uniformity of the production process from batch to batch.
5. There is no final summary by management to verify that the validation data has been reviewed, that all requirements of the protocols have been met, and that the systems are considered validated.
6. The validation program for drug products is incomplete and fails to provide for physical specifications for drug substance, sampling approximately the equivalent weight of a dosage unit to demonstrate blend uniformity, in-process individual weight variation, comparison of dissolution and granulation studies between biobatches and production batches.
7. The SOP for validation or revalidation does not require that specifications and acceptance criteria be determined prior to validation.
8. There were no SOPs for the QA investigations of product failures, laboratory failure investigations, and stability investigations.
9. Validation is inadequate in that it does not include tablet thickness, hardness, weight, or dimensions.
10. There is no established time limit for sterile filling operations.
11. The firm lacks access to the source code for software.
12. Start/stop times are not routinely recorded on the batch production records (for sterile products).
13. Filter validation does not include testing of parenteral products for extractables.
14. Of the thirty-one validation studies conducted all are invalid for one or more of the following reasons: (a) lots chosen for prospective study are nonsequential, (b) lots initially identified as validation batches are eliminated from the study for various reasons, (c) study protocols do not identify critical process control points, (d) raw material and inprocess specifications such as particle size distribution, pour bulk density, tap bulk density, moisture, etc. have not been established prior to validation.

ACKNOWLEDGMENT

The editor would like to thank Vinay Bhatt for his review and input into the validation section of this chapter.

SUGGESTED READINGS

1. FDA Guidance for Industry Q8 Pharmaceutical Development, May 2006.
2. FDA. Guidance for Industry Q9 Quality Risk Management, June 2006.
3. FDA Guideline on General Principles of Process Validation, May 1987.
4. FDA Draft Guidance for Industry Powder Blends and Finished Dosage Units—Stratified In-Process Dosage Unit Sampling and Assessment, November 2003.
5. FDA Guideline on Sterile Drug Products Produced by Aseptic Processing, May 1987.
6. Agalloco JP, Carlton FJ. Validation of Pharmaceutical Processes Sterile Products. 2nd ed. 1998.
7. Hara T, Elgar E. Innovation in the Pharmaceutical Industry. The Process of Drug Discovery and Development. E. Elger Publishing Ltd, 2003.
8. J Process Anal Technol. http://www.patjournal.com.
9. Wachter AH, Nash RA. Pharmaceutical Process Validation. 3rd ed. Marcel Dekker, 2003.
10. LeBlanc DA. Cleaning Validation: Practical Compliance Solutions. 1998.
11. Rodriguez J. The challenge of implementing a centralized change control management system. J GXP Compliance 2006; 10(2).
12. Grant EL, Leavenworth RS. Statistical Quality Control. 7th ed. New York: McGraw-Hill.
13. Stoker JR, Kieffer RG, Nally JD. A different approach to quality audit. Pharm Technol 1988; 12:84.
14. Avallone H. Current regulatory issues regarding parenteral inspections. J Parent Sci Technol 1989; 43:3.
15. Shirtz JT. Sterility testing. Pharm Eng 1987; 7:35.
16. Nally J, Kieffer R. The future of validation. Pharm Technol 1992; 13(10):106.
17. Tetzlaff RF. Validation issues for new drug development part I, review of current FDA policies. Pharm Technol 1992; 16(9):44.
18. Tetzlaff RF. Validation issues for new drug development part II, systematic assessment strategies. Pharm Technol 1992; 16(10):84.

19. Tetzlaff RF. Validation issues for new drug development part III, systematic audit techniques. Pharm Technol 1993; 17(1):80.
20. Bala G. An integrated approach to process validation. Pharm Eng 1994; 14(3):57.
21. Dream RF. Qualification—validation in perspective. Pharm Technol 1994; 14(5):74.
22. Guidance for Industry for the Submission of Documentation for Sterilization Process Validation in Applications for Human and Veterinary Drug Products. FDA Center for Drug Evaluation and Research/Center for Veterinary Medicine, November 1994.
23. Reid BD. Gamma processing technology: an alternative technology for terminal sterilization of parenterals. J Pharm Sci Technol 1995; 49(2):83.
24. FDA. Supplements to New Drug Applications, Abbreviated New Drug Applications, or Abbreviated Antibiotic Applications for Nonsterile Drug Products; Draft Guide line. Federal Register 1994; 59(237):64094.
25. Nally J. Kieffer R, Stoker J. From audit to process assessment—the more effective approach. *Pharm. Technol., 19*(9), 128, 1995.
26. Akers JE, Agalloco JP, Kennedy CM. Experience in the design and use of isolator systems for sterility testing. J Pharm Sci Technol 1995; 49(3):140.

Following are some additional FDA ORA Guidance documents relevant to this chapter.

SEC. 460.800 PARAMETRIC RELEASE—TERMINALLY HEAT STERILIZED DRUG PRODUCTS (CPG 7132A. 12)

Background:

In 1985, the FDA approved supplemental new drug applications for certain large volume parenteral drug products, which substituted parametric release for routine lot by lot end-product sterility testing.

Parametric release is defined as a sterility release procedure based upon effective control, monitoring, and documentation of a validated sterilization process cycle in lieu of release based upon end-product sterility testing (21 CFR 211.167). All parameters within the procedure must be met before the lot is released.

Policy:

This policy applies only to parenteral drug products which are terminally heat sterilized. It does not apply to products sterilized by filtration or ethylene oxide. This policy does not preempt requirements of Section 505 of the FD&C Act. Approved supplements providing for parametric release are required for holders of new drug applications. (21 CFR 314.70(b))

Parametric release, in lieu of end product sterility testing, is acceptable when all of the following parameters are met and documented.

1. The sterilization process cycle has been validated to achieve microbial bioburden reduction to 10^0 with a minimum safety factor of an additional six logarithm reduction. Cycle validation includes sterilizer heat distribution studies, heat distribution studies for each load configuration, heat penetration studies of the product, bioburden studies, and a lethality study referencing a test organism of known resistance to the sterilization process. All cycle parameters must be identified by the manufacturer as critical (e.g., time, temperature, pressure) or non-critical (e.g., cooling time, heat-up time). Under parametric release, failure of more than one critical parameter must result in automatic rejection of the sterilizer load (see paragraph D concerning biological indicators). (21 CFR 211.113 (b))
2. Integrity for each container/closure system has been validated to prevent in-process and post-process contamination over the product's intended shelf-life. Validation should include chemical or microbial ingress tests utilizing units from typical products. (21 CFR 211.94)
3. Bioburden testing (covering total aerobic and total spore counts) is conducted on each batch of presterilized drug product. Resistance of any spore-forming organism found must be compared to that of the organism used to validate the sterilization cycle. The batch is deemed non-sterile if the bioburden organism is more resistant than the one used in validation. (21 CFR 211.110)

4. Chemical or biological indicators are included in each truck, tray, or pallet of each sterilizer load. For chemical indicators, time/temperature response characteristics and stability are documented and for each sterilization cycle minimum degradation values are established. Chemical indicators cannot be used to evaluate cycle lethality.

Documentation is required for biological indicators (BIs). Documentation for each BI lot shall include an organism's name, source and D-value, spore concentration per carrier, expiration date, and storage conditions. BIs can be used to evaluate cycle lethality where equipment malfunction prevents measurement of one critical cycle parameter. If more than one critical parameter is not met, the batch is considered non-sterile despite BI sterility. (21 CFR 2311.165(e) and 211.167)

Issued: 10/21/87

SUBCHAPTER 490 VALIDATION SEC. 490.100 PROCESS VALIDATION REQUIREMENTS FOR DRUG PRODUCTS SUBJECT TO PRE-MARKET APPROVAL (CPG7132C.08)

Background:

Validation of manufacturing processes is a requirement of the CGMP regulations for finished Pharmaceuticals (21 CFR Part 211). Validation is based on the documented successful evaluation of multiple full scale batches (usually at least three) to provide assurance that the processes will reliably meet predetermined specifications. Refer to the Guideline of General Principles of Process Validation (May 1987) (distributed by the CDER) for further details.

The pre-approval inspection compliance program (7346.832) also emphasizes the importance of process validation to ensure the safety, efficacy, and quality of drug products. Although the program does not require completion of multiple batch process validation before an application may be approved, completion of such process validation is a CGMP requirement that must be met before any shipments of the products are made.

In addition, for products intended to be sterile, applicants are required to submit data to the applications that demonstrate the effectiveness of the intended sterilization or aseptic processing procedures. The center evaluates the data as part of the application approval process. Such data may be derived prior to undertaking full-scale production. For example, the data may be obtained by conducting media fills to determine the effectiveness of the aseptic processing procedures or by running simulated product along with biological indicators through autoclave cycles to determine the effectiveness of steam sterilization procedures.

Applications for sterile and non-sterile drug products may be approved by a center prior to the firm's completion of full-scale process validation.

The purpose of the pre-approval inspection is to audit the completeness and accuracy of the submitted data. The inspection also evaluates the effectiveness of important CGMP facilities and systems that bear on sterility assurance including, but not limited to: HEPA, HVAC, facility conditions, and water systems.

Policy:

1. During the pre-approval inspection, any process validation data that is available should be evaluated and any process validation deficiencies reported to the firm. Based on the lack of completed process validation data, the district should monitor the firm's post-approval validation efforts, and initiate regulatory action where product has been shipped and there are deficiencies with validation that would support such action. Seizure should be considered when there are supportable deficiencies with validation or the evidence demonstrates the product does not comply with specifications. Recommendations for enforcement action due to non-compliance with the CGMP regulations should cite 501(a)(2)(B) of the Act, 21 U.S.C. 351(a)(2)(B) based on violation of 21 CFR 211.100.

    The district should recommend withholding approval of an application if any completed validation efforts include data of questionable validity or demonstrate that the process is not valid and the firm has not committed to making appropriate changes.

2. Rarely, the completion of multiple batch process validation prior to product shipment may be impractical due to public health considerations. For example, the public health benefits of expediting availability of clinically important drugs with a very limited market (e.g., certain orphan drug products) may outweigh the need to await completion of full multiple batch process validation. Under such circumstances, the firm is expected to:
   a. Document the reason that complete process validation is impractical prior to shipment of the product.
   b. Establish and commit to following an adequate protocol to complete the validation of the manufacturing process. In this regard, successful validation of each batch of the multiple batch validation study must be completed prior to shipment of the batch.
   c. Establish and follow more extensive testing and process controls to ensure batch uniformity and conformance with predetermined specifications.
      The district should consult with the center before initiating regulatory action under these circumstances.
3. Process validation requirements for bulk pharmaceutical chemicals (BPC) differ somewhat from those required for dosage form products. Refer to the BPC Inspection Guide (CDER revised, September 1991) for details. Based on recent emphasis by FDA, the industry has begun to formally validate the manufacturing processes for BPCs. The district should recommend withholding approval of an application based on the lack of process validation for the BPC where:
   a. The firm has not established or is not following an adequate plan to validate all BPCs; or,
   b. The process is not valid, as demonstrated by repeated batch failures due to manufacturing process variability not attributable to equipment malfunction or operator error.

NOTE: This compliance policy guide (CPG) also applies to pre-market approval applications submitted to the Center for Veterinary Medicine (NADAs or ANADAs). The CPG reference may be found at 7125.38 (See Sec. 638.100).

Issued: 8/30/93

SUBCHAPTER 420 COMPENDIAL/TEST REQUIREMENTS SEC. 420.100 ADULTERATION OF DRUGS UNDER SECTION 501(B) AND 501(C) OF THE ACT [DIRECT REFERENCE SEIZURE AUTHORITY FOR ADULTERATED DRUGS UNDER SECTION 501(B)] (CPG 7132A.O3)

Background:

Section 501(b) of the Food, Drug, and Cosmetic Act (the Act) deems an official drug (i.e., a drug purported to be or represented as a drug the name of which is recognized in an official compendium) to be adulterated if it fails to conform to compendial standards of quality, strength or purity. Compendial tests or assay methods are used when determining such conformance under 501(b); the standards are stated in individual monographs as well as portions of the General Notices section of the USP/NF. Standards and test methods have been established for such characteristics as potency, sterility, * dissolution *, weight variation and content uniformity.

If an official drug fails to conform to one or more compendial standards of strength, quality or purity, but plainly states on the label how it differs from the standard, then the drug is not deemed to be adulterated under Section 501 (b).

Section 501(c) of the Act deems * a drug that is not recognized in an official compendium to be adulterated if it fails to meet the strength, purity, or quality which it purports or is represented to possess. The applicable quality standards for a drug not recognized in an official compendium can be determined from such sources as the labeling of the drug (or drug product), the manufacturer's written specifications, and new drug applications. (Test methods are usually contained in the written specifications or new drug application).*

Policy:

Any official drug that, when tested by compendial methods, fails to conform to compendial standards for quality, strength, or purity is adulterated unless the differences from such standards are plainly stated on the drug's label.

Any * drug which is not recognized in an official compendium is adulterated if its strength differs from, or its purity or quality falls below that which it purports or is represented to possess, when tested by scientifically sound methods.*

Regulatory Action Guidance:

Recommendations for regulatory action will be considered in the above instances of adulteration. The regulatory action of choice will depend upon the circumstances of each case.

In cases where there is a health hazard, the first choice of action should be recall, particularly for drugs found to be non-sterile, and for narrow therapeutic range drugs that fail potency or dissolution tests. However, where the district office has advised the firm of such a defective product, and the firm fails to recall, seizure should be considered. Seizure recommendations charging adulteration under section 501(c) should be submitted to the Office of Compliance, Center for Drug Evaluation and Research (HFD-300) (CDER).

District offices are authorized to submit seizure recommendations, charging adulteration under section 501(b), directly to the Office of Enforcement without CDER review under the following circumstances, provided introduction or delivery for introduction into interstate commerce has been documented.

1. An official sample of either a compendial bulk pharmaceutical chemical or a compendial finished dosage form has been analyzed using the compendial methods without modification and found to fail both the original and check analyses.
2. The analyzing laboratory has certified in the transmittal memorandum that an unmodified compendial method was used.
   Note: No tolerance need be applied beyond that provided by the official compendium.
3. For sterile products, no check analysis is needed provided the compendial sterility test was utilized without modification, the product is one that is required to be sterile, and all relevant laboratory controls (including positive and negative) are satisfactory.

Where the analyzing laboratory deviates from the official compendial analytical method(s), a detailed description of the deviation(s) and justification for such deviation(s) can be submitted to CDER for review. In such cases, CDER will review only the deviation(s) and not the choice of regulatory action or other documentation.

For seizure actions, the charges may be drafted as follows:

That the article of drug was adulterated, when introduced into and while in interstate commerce and is adulterated while held for sale after shipment in interstate commerce within the meaning of 21 U.S.C. 351(b), in that it purports to be and is represented as a drug, the name of which is recognized in an official compendium (United States Pharmacopeia) and its strength differs from and its quality and purity falls below the standard set forth in such compendium because it fails the official (INSERT TYPE OF TEST) test.

or

That the article of drug was adulterated, when introduced into and while in interstate commerce and is adulterated while held for sale after shipment in interstate commerce within the meaning of 21 U.S.C. 351(c), in that it is a drug not subject to the provisions of 21 U.S.C. 351(b) and its strength differs from and its purity and quality falls below that which it purports or is represented to possess because (e.g., the drug contains less than the amount of (INSERT NAME OF INGREDIENT) on the label).

It should be kept in mind that the types of adulteration found under 501(b) and 501(c) may be indicative of a wider problem involving failure of the manufacturer to adhere to current good manufacturing practice that should be addressed.

Material between asterisks is new or revised
Issued: 10/1/80
Revised: 5/1/92

SEC. 420.400 PERFORMANCE OF TESTS FOR COMPENDIAL REQUIREMENTS ON COMPENDIAL PRODUCTS (CPG 7132.05)

Background:

There have been inquiries from the field and industry concerning the following four items as they apply to the manufacture of compendial (USP/NF) drug products.

1. Does a firm have to use the compendial methodology on a batch release basis, to determine whether its product meets the requirements of the monograph?
2. Does the word "specifications" as used in 21 CFR 211.165 refer to compendial specifications or those set up by the firm's quality control unit?
3. Does a firm have to test for all requirements listed in the monograph for a compendial product?
4. Are the compendial testing requirements the same for products destined for the commercial market and the military?

Policy:

1. Compendial methods need only be applied, as a batch release test, where a firm has made specific commitments to do so (as in a new drug application), or where the official method is the only appropriate test. It should be noted that neither the USP/NF nor the CGMP regulations necessarily require a firm to utilize, as a batch release test, the methods and procedures stated in the official compendia.

   What is required is that official drug products conform to the appropriate compendial standards. This conformance must be assured by suitable means, including adequate manufacturing process validation and control. Scientifically sound alternative test methods may be acceptable for the purpose of batch release testing. However, in the event of a dispute as to whether or not a drug product meets the standard, the compendial method will be applied as the referee test.

2. The term "specifications" as used in 21 CFR 211.165 refers to the criteria established by manufacturers to assure that their products have the properties they purport to possess. Typically, these specifications are identical to, or more stringent than, those contained in the compendia themselves. However, the manufacturer's specifications for standards of strength, quality, and purity may be less stringent in those cases in which the differences from the official standards are stated on the product label; such alternate standards must not adversely affect the product's safety or efficacy.

3. Where an official product purports to conform to the standards of the USP/NF the manufacturer must assure that each batch conforms to each monograph requirement. This assurance must be achieved by appropriate means, including process validation and controls and end-product testing. However, the nature and extent of end-product testing which is needed will depend upon the circumstances. Factors to consider in determining the need to test each batch for a given monograph requirement include: the adequacy of the manufacturer's process validation, adequacy of in-process manufacturing controls, and the nature of the particular product characteristic which is the subject of the specification (e.g., potency, sterility, content uniformity). Therefore, in some cases it may not be necessary for a manufacturer to test each batch for each monograph requirement.

4. Compendial testing requirements are the same for products destined for commercial and military use unless the Defense Personnel Support Center (DPSC) insists upon certain requirements as part of military contracts. For example, DPSC can insist that only compendial methods be used and that each batch be tested for every monograph specification, whereas, as explained above, the FDA considers that alternative procedures may sometimes be acceptable. Under the Government Wide Quality Assurance Program the FDA must assure that the drug manufacturer abides by the terms of the military contract, including testing requirements.

Issued: 10/1/80

SEC. 420.500 INTERFERENCE WITH COMPENDIAL TESTS (CPG7132A.O1)

Background:

The recurring question is: What is the legal status of a compendial drug in which an added substance interferes with the compendial assay of the product, even though the product may be fully potent as shown by other methods of analysis?

Section 501 (b) of the Federal Food, Drug, and Cosmetic Act states that a drug is deemed to be adulterated if it is recognized in an official compendium and its strength differs from or its quality or purity falls below the standards set forth in the compendium. Determination as to strength, quality, or purity shall be made in accordance with tests or methods of assay set forth in such compendium.

The USP XX in the section on Added Substances (p. 4) states that suitable substances such as bases, carriers, coatings, colors, flavors, preservatives, stabilizers, vehicles may be added to a pharmacopeial dosage form to enhance its stability, usefulness, or elegance, or to facilitate its preparation. The USP restrictions on the use of such added substances include "if they do not interfere with the assays and tests prescribed for determining compliance with the pharmacopeial standards."

Policy:

A compendial drug product containing an added substance which interferes with the compendial assay of the product would be adulterated under 501(b) of the Act.

Issued: 10/1/80

SUBCHAPTER 425 COMPUTERIZED DRUG PROCESSING SEC. 425.100 COMPUTERIZED DRUG PROCESSING; CGMP APPLICABILITY TO HARDWARE AND SOFTWARE (CPG 7132A.11)

Background:

The use of computers in the production and control of drug products is quickly increasing. Questions have been raised as to the applicability of various sections of the Current Good Manufacturing Practice Regulations to the physical devices (hardware) which constitute the computer systems and to the instructions (software) which make them function.

Policy:

Where a computer system is performing a function covered by the CGMP regulations, in general, hardware will be regarded as equipment and applications software will be regarded as records. The kind of record (e.g., SOP, master production record) that the software constitutes and the kind of equipment (e.g., process controller, laboratory instrument) that the hardware constitutes will be governed by how the hardware and software are used in the manufacture, processing, packing, or holding of the drug product. Their exact use will then be used to determine and apply the appropriate sections of the regulations that address equipment and records.

Issued: 10/19/84
Revised: 9/4/87

SEC. 425.200 COMPUTERIZED DRUG PROCESSING; VENDOR RESPONSIBILITY (CPG 7132A. 12)

Background:

Computer systems used in the production and control of drug products can consists of various devices (hardware) and programs (software) supplied by different vendors, or in some cases by a single vendor. It is important that such computer systems perform accurately and reliably, * and * that they are suitable for their intended use.

Questions have arisen as to the vendor's responsibility in assuring computer systems performance and suitability. When an integrated system, composed of elements from several different vendors, fails, it can be especially difficult to attribute the cause of a problem to one particular vendor.

Policy:

The end user is responsible for the suitability of computer systems (hardware and software) used in manufacture, processing or holding of a drug product.

*The vendor may also be liable, under the FD&C Act, for causing the introduction of adulterated or misbranded drug products into interstate commerce, where the causative factors for the violation are attributable to intrinsic defects in the vendor's hardware and software. In addition vendors may incur liability for validation, as well as hardware/software maintenance performed on behalf of users. *

*Material between asterisks is new or revised *
Issued: 1/18/85
Revised: 9/4/87

SEC. 460.600 CONTENT UNIFORMITY TESTING OF TABLETS AND CAPSULES (CPG 7132A. 14)

Background:

Refer to 21 CFR 211.167 Special Requirements.

There has been some misunderstanding surrounding the applicability of content uniformity testing requirements to tablets and capsules, particularly for non-official products, i.e., those not recognized as official in the USP. Added to this is the confusion created by recent changes in the USP test requirements for official products. In addition to the existing standard for the individual dosage unit assay, the USP included a specification for relative standard deviation to limit large variations in test results. However, many firms have been reluctant to incorporate the relative standard deviation specification into their SOPs.

Policy:

The following policy is applicable to tablet or capsule dosage forms.

Official Products

Any drug product recognized in the USP must comply with the USP requirement for content uniformity if such requirement is included in the monograph for the drug. The product must comply with the specifications for individual dosage unit assay and for relative standard deviation. Both requirements are applicable regardless of whether or not the product in question is subject to an NDA. If an approved NDA does not currently provide for complete content uniformity testing, or provides specifications that are in consistent with the USP monograph, the NDA holder must submit a change to provide for such testing, pursuant to 21 CFR 314.70(d)(l).

Non-official Drug Products

The Food, Drug, and Cosmetic Act requires that drug products which are not official (and therefore not subject to compendia requirements) nonetheless meet standards of strength and quality which they purport or are represented to possess. Current good manufacturing

---

[1]Applications software consists of programs written to specified user requirements for the purpose of performing a designated task such as process control, laboratory analyses, and acquisition/processing/ storage of information required by the CGMP regulations.

practice regulations (21 CFR 211.160) require the establishment of scientifically sound and adequate specifications to assure those product attributes.

Specifications for content uniformity are required, within this context, for tablets and capsules which contain less than 50 mg of any active ingredient. Requirements for content uniformity include individual dosage unit assays and establishment of specifications for relative standard deviation.

Any non-official tablet or capsule dosage form which contains <50 mg of any active ingredient and such ingredient(s) has not been tested for content uniformity is in violation of Section 501(a)(2)(B) of the Act. In evaluating the appropriateness of test specifications for a non-official product, it must be emphasized that although USP specifications are acceptable and may be adopted by a firm, they are not specifically required. Scientifically sound alternative specifications may be used.

Issued: 10/2/87

8 | Packaging and Labeling Control
Subpart G

Arlyn R. Sibille
Consultant, Harmony, Pennsylvania, U.S.A.

§211.122   MATERIALS EXAMINATION AND USAGE CRITERIA

(a)   There shall be written procedures describing in sufficient the detail of the receipt, identification, storage, handling, sampling, examination, and/or testing of labeling, and packaging materials; such written procedures shall be followed. Labeling and packaging materials shall be representatively sampled, and examined or tested upon receipt and before use in packaging or labeling of a drug product.

(b)   Any labeling or packaging materials meeting appropriate written specifications may be approved and released for use. Any labeling or packaging materials that do not meet such specifications shall be rejected to prevent their use in operations for which they are unsuitable.

(c)   Records shall be maintained for each shipment received of each different labeling and packaging material indicating receipt, examination or testing, and whether accepted or rejected.

The terms labeling and packaging in the context of this section specifically exclude containers and closures, which are covered in Subpart E. However, unlabeled packaging such as corrugated shippers and dividers are included, although the evaluation and control need not be so extensive. Homeopathic drug products are subject to the same labeling provisions of the Food, Drug, and Cosmetic Act as other drug products. However, if these products are labeled in accordance to the existing Compliance Policy Guide for homeopathic products, they will not be recommended for regulatory action.

Labels and labeling errors have been among the top ten reasons for product recall in recent years. The effective control of printed labeling (labels, inserts, cartons, foil) begins well before materials are ordered; it starts at the design and approval stage. Written procedures for the inspection, documentation, and approval/rejection of incoming labels and packaging must also be in place to effectively control them.

For new or changed labeling there must be in place a procedure or procedures that clearly defines:

1.   Who is to review and approve the copy?
   a.   This will usually include marketing, medical, legal, regulatory affairs, production, materials management, quality assurance, and editorial.
2.   What each function has to check and approve:
   a.   Individual responsibilities must be delineated. With so many people involved, individuals may be tempted to assume that someone else has checked the various points. The reviewers and approvers (item 1) should preferably be defined by job function. Since the approval process is so important, it is essential to ensure that approvers have undergone adequate training in the system. Additional individuals should not be allowed to authorize new labeling or changes without having undergone this training. A typical system might include the following detailed responsibilities:

    b.   Materials management—
        i.   Evaluate inventory situation.
        ii.   Identify an implementation date that will comply with regulatory or company requirements while minimizing stock write-offs or production interruptions.
        iii.   Identify which countries take this product so that Regulatory Affairs can evaluate any regulatory requirements and potential impacts, including potential delays in implementation.
    c.   Production—
        i.   Confirms that equipment is available, or will be, to introduce the change.
        ii.   Ensure that adequate space for batch coding and expiry dating is available.
    d.   Marketing (or other function) —
        i.   Identify whether the specific change will require modification to other materials, for example, changes to a label may require changes to inserts and cartons.
        ii.   Identify whether other pack sizes may need to incorporate the same or equivalent change.
    e.   Medical—confirm that any medical claims, warnings, dosages are correct.
    f.   Legal & Regulatory Affairs—confirm that the changes meet all legal and regulatory requirements in all countries that sell the product.
    g.   Technical Service & Quality Control—identify any stability issues (primary pack changes).
    h.   Editorial— confirm grammatical correctness and absence of typographic and print errors or omissions.
    i.   System Manager— confirm that all appropriate approvals have been given and in the event of queries to recirculate through the system, if necessary.
3.   Whether the change is mandatory or voluntary, and the date of introduction of the change. For voluntary changes this is likely to relate to depletion of inventories of existing labeling.
4.   Whether the labeling in question is interrelated with any other labeling and to note any impact. For example, a change to the dosage requirements on a label will almost certainly require amendment to inserts and cartons.
5.   The feedback loop that confirms that the change has actually been introduced.
6.   A unique numbering system, which clearly distinguishes between labeling for different products and strengths and version control to track different versions of the same labeling.
7.   Ownership of the procedure.
    a.   As with any system or procedure, especially the one that involves so many people, there should be designated responsibility for operation and monitoring of the system itself. This individual should routinely identify any potential weaknesses in the operation of the system and have them corrected.

The extent of the sampling and evaluation required by the regulations can be varied to match the importance of the material and the capabilities of the supplier. Special attention should be given to new or changed labeling, and a check against the previously approved text or artwork should be performed. As written, the regulations do not allow for the total elimination of sampling and evaluation, but the use of certified vendors will allow this to be minimal. The regulations also do not permit use of packaging and labeling materials prior to evaluation and release. This is similar to the requirement for components, but since these evaluations are usually of short duration this is unlikely to result in any practical delays.

Low levels of mix-up with labeling are difficult, if not impossible, to detect at the incoming inspection stage. This further accentuates the need to build in quality at the design and supplier stages:

1.   Avoid gang printing if possible [§211.122(f)].
2.   Design to avoid look-alike labeling especially for different strengths of the same product and if possible for different products run on the same packaging line. Different colors, sizes, and shapes can all help.
3.   In addition to the unique numbering system, include bar codes that can be used in association with on-line scanning during the labeling operation.

4. Use roll-feed labels wherever possible but ensure that any splicing done by the supplier is clearly marked.
5. Only use suppliers who have facilities and procedures with a high probability of achieving the required quality consistency; this will usually require a visit to the supplier. The evaluation suppliers of packaging and labeling have, on occasion, not received the same level of attention as the suppliers of bulk pharmaceutical chemicals Packaging Components (PCs).
   a. With the multiplicity of labels, the tendency for marketing to prefer a company image for all labeling, and the frequent changes that are implemented, the potential for errors or mix-ups is high. Consequently, supplier systems must be well designed and followed. This should include a procedure to destroy obsolete plates or other masters before commencing work on a new version. One effective way is to have the obsolete masters returned to the customer (pharmaceutical manufacturer).

The requirement in §211.122(b) to reject materials that do not meet the specification seems both obvious and, at the same time, controversial. To reject or not should depend on the specifications and inspection attributes that are included in the inspection practices for the material. If color ranges are not considered critical with respect to product quality then ranges should perhaps be expressed as "action levels" with reference to who must be consulted when material is received outside of the range. Too much variability could imply that the supplier has less control over the printing process than is required.

Records must be retained to indicate the evaluation results and disposition of each delivery of material. The data in these records should also be used to monitor supplier performance and to detect adverse trends that may require attention.

(d) Labels and other labeling materials for each different drug product, strength, dosage form, or quantity of contents shall be stored separately with suitable identification. Access to the storage shall be limited to authorized personnel.
(e) Obsolete and outdated labels, labeling, and other packaging materials shall be destroyed.

Storage of labeling must be done in a manner that will minimize the potential for mix-ups or release of unapproved materials to production. Materials for different products and for different strengths of the same product are to be kept separate—often by the use of drawers, cubicles, or cupboards. When label changes are being phased into production, the current and new versions should be kept apart. Obsolete labeling should immediately be moved to a "nonactive" location, usually a reject area, pending for destruction. The destruction of the rejected labeling/packaging materials should be documented.

Labeling storage is usually a part of the warehouse and should have restricted access. Unless the overall warehouse is operated as a secure area, the space allocated to labeling will need separate segregation. Labeling must be adequately protected from dirt and dust, which, in addition to affecting appearance, can adversely impact on adhesion characteristics. Certain labeling components, such as foil, laminates, cellophane, and self-adhesive labels, may also require storage under defined conditions of temperature and humidity.

(f) Use of gang printed labeling for different drug products or different strengths or net contents of the same drug product is prohibited unless the labeling from gang printed sheets is adequately differentiated by size, shape, or color.
(g) If cut labeling is used, packaging and labeling operations shall include one of the following special control procedures:
   (1) Dedication of labeling and packaging lines to each different strength of each different drug product;
   (2) Use of appropriate electronic or electromechanical equipment to conduct a 100% examination for correct labeling during or after completion of finishing operations; or
   (3) Use of visual inspection to conduct a 100% examination for correct labeling during or after completion of finishing operations for hand applied labeling. Such examination shall be performed by one person and independently verified by a second person.

(h) Printing devices on, or associated with, manufacturing lines used to imprint labeling upon the drug product unit label or case shall be monitored to assure that all imprinting conforms to the print specified in the batch production record.

Gang printing consists of printing of different labeling on the same sheet, and then cutting and separating the different labeling. This technique has a high potential for mix-up during the cutting and separating stages, and most pharmaceutical companies now avoid this approach. The regulations prior to 1994 acknowledged the risks but did allow gang printing on the condition that certain safeguards were employed during the cutting and separating stages. The update prohibits gang printing unless the individual items are sufficiently different.

Subsection (g) addresses the broader issue of cut labeling. While the risks at the printer are significantly lesser than for gang printing, the potential for mix-ups there and subsequently after that is still high. This is accentuated if excess labeling is returned to stock. The regulations now require additional checking, preferably by electronic or other automated means, although visual confirmation is allowed for hand labeling operations.

Some pharmaceutical operations include in-house printing of components either on-line or off-line. Examples of this include ceramic or silk-screen printing of ampoules as well as printing of tubes and plastic bottles. In these cases the printing operations should be set up, operated, and monitored in a manner equivalent to that of a supplier. Printing screens, when used, should be carefully examined for conformity to approved text, correct layout, absence of tears and holes, and absence of blocked holes or letters.

§211.125 LABELING ISSUANCE

(a) Strict control shall be exercised over labeling issued for use in drug product labeling operations.
(b) Labeling materials issued for a batch shall be carefully examined for identity and conformity to the labeling specified in the master or batch production records.
(c) Procedures shall be written describing in sufficient detail the control procedures employed for the issuance of labeling; such written procedures shall be followed.

These three subsections relate to the procedures for issue of labeling. Labeling materials should be supplied against a written order that relates to a specific packaging operation. The batch documentation should define the number of units to be issued and will normally include a small overage to allow for normal line set up and wastage. Labeling materials should be delivered to the packaging line in a secure manner to assure no potential for loss or mix-up between issue and use. Prior to acceptance onto the line, the supervisor should confirm the absence of any labeling associated with the previous packaging run and also confirm that the correct labels and numbers have been provided for the current packaging run [§211.130(c)(d)].

(d) Procedures shall be used to reconcile the quantities of labeling issued, used, and returned, and shall require evaluation of discrepancies found between the quantity of drug product finished and the quantity of labeling issued when such discrepancies are outside narrow preset limits based on historical operating data. Such discrepancies shall be investigated in accordance with §211.192. Labeling reconciliation is waived for cut or roll labeling if a 100% examination for correct labeling is performed in accordance with §211.122(g)(2).
(e) All excess labeling bearing lot or control numbers shall be destroyed.
(f) Returned labeling shall be maintained and stored in a manner to prevent mix-ups and provide proper identification.

One of the more sensitive and controversial sections of the CGMP regulations was §211.125(c). In order to reconcile issue, use, and return, it is necessary to count labels on receipt from the supplier, to count and issue precise numbers for each packaging run, to count the number of units produced, the number of labels damaged or destroyed during the packaging run, and the number returned to stock or to be destroyed. A precise count of labeling damaged or destroyed on-line is facilitated by the packaging lines that electronically monitor the labeling, and provide a count of accepted and rejected packages. Some companies require sequential numbers on labels to facilitate issuance and reconciliation. Acceptance of reconciliation limits should be based on historic achievement. If reconciliation values fall outside of the accepted range an evaluation should be initiated. This usually will include more detailed examination of the batch to identify if there is any mislabeling, also rechecking of any returned labeling and inventory records.

Over the years labeling errors have continued to be a cause for product recalls but the ranking in not as significant. The changes in §211.125(c) and §211.122(g) have addressed many of the potential causes for these errors—prohibition of gang printing and more intensive and more effective reconciliation procedures. Electronic scanning of labeling either for bar codes or total scanning should be 100% effective. The emphasis by industry now focuses on labeling design, approval, and printing.

Excess labeling may be returned to stock provided it is uncoded and undamaged. Great care must be taken to ensure return to the correct storage drawer or area. This return to stock should be controlled through written procedures. There may be an advantage to note the batch for which the returned labeling had been issued so that re-evaluation is easy in the event of a later query (before they are re-issued).

§211.130 PACKAGING AND LABELING OPERATIONS

There shall be written procedures designed to assure that correct labels, labeling, and packaging materials are used for drug products; such written procedures shall be followed. These procedures shall incorporate the following features:

(a) Prevention of mix-ups and cross-contamination by physical or spatial separation from operations on other drug products.

(b) Identification and handling of filled drug product containers that are set aside and held in unlabeled condition for future labeling operations to preclude mislabeling of individual containers, lots, or portions of lots. Identification need not be applied to each individual container but shall be sufficient to determine name, strength, quantity of contents, and lot or control number of each container.

(c) Identification of the drug product with a lot or control number that permits determination of the history of the manufacture and control of the batch.

(d) Examination of packaging and labeling materials for suitability and correctness before packaging operations, and documentation of such examination in the batch production record.

(e) Inspection of the packaging and labeling facilities immediately before use to assure that all drug products have been removed from previous operations. Inspection shall also be made to assure that packaging and labeling materials not suitable for subsequent operations have been removed. Results of inspection shall be documented in the batch production records.

Subpart F, Production and Process Controls, covers the overall manufacturing process but an equivalent section with respect to packaging does not appear to be present. Section 211.130 refers essentially to the assurance of correct labeling and packaging materials; without reference to the assurance that the correct bulk product is provided, equipment details are noted, fill volumes or quantities are monitored, or equipment is suitable, qualified, calibrated, and clean. This section, therefore, describes only the overall packaging process.

The master batch packaging formula will provide the basis for each packaged dosage form. It will contain the following data fields (where appropriate):

1. Drug product name, identification number, and strength.
2. Names, identification number, and quantities of each packaging component:

    a. Primary container: bottle, closure and liner, foil, laminate, and so on.
    b. Label
    c. Carton
    d. Insert
    e. Tamper-evident feature
    f. Child-resistant feature
    g. Shipper, dividers, other protective packaging

3. Complete description of the equipment to be utilized for the packaging operation.
4. Characteristics to be monitored during filling and packaging (e.g., temperature, fill, clarity, pH, specific gravity, color, cap tightness, seal integrity).
5. Sample requirements and frequency.
6. For the specific batch being packaged: (Items 6a–e are sometimes incorporated into a separate document, the "packaging order.")
       a. Batch number of the bulk drug product
       b. Packaging lot or control number assigned
       c. Quantities of packs expected and action level value
       d. Quantities of each packaging component issued (if different from item 2)
       e. Expiry date to be assigned

The master packaging formula and the packaging order must be written and approved by the appropriate personnel, which will usually include production and Quality Assurance (QA). No changes can be made to these master documents without reapproval.

The regulations permit spatial separation of packaging lines, but the potential for mix-ups and cross-contamination may be further reduced by physical segregation:

1. A physical barrier between adjacent lines.
2. Adequate space and physical segregation to assemble bulk product and packaging components.
3. Adequate space for assembly of finished packaged stock prior to transfer to the warehouse.
4. Accumulation tables or other space to accommodate partly packaged product in the event of temporary breakdown of part of the packaging line.
5. Dust extraction over bulk table hoppers.
6. Covers over open hoppers, open empty containers, and filled but uncapped containers.
7. Avoidance of running different, but similar looking, products or packs on adjacent lines.
8. Delay start-up until all packaging components are available.
9. During extended stoppages, such as lunch breaks and shift changes, ensure that all filled units are capped, especially for liquid products.

Before commencing a packaging operation it is essential to examine the entire line to confirm that it is documented as having been cleaned; that there is no visible evidence of bulk product, components, or finished product from the previous packaging run; that the line board notes the correct product, batch, control number, and expiry date; and that the correct materials have been delivered to the line. Results of these inspections should form part of the batch records. A useful approach, coupled with the line board, is to prominently display on the line an authorized example of the various components of the pack.

Whenever possible, coding of control number and expiry dates should take place on-line. If off-line coding is necessary, very stringent controls must be established and followed.

Subsection (b) was inserted and became effective on August 3, 1994. The intent of Subsection (b) is to address the practice of filling a product, but delaying the labeling until a future date. There are several reasons for separating these two processes, which incidentally tends to decrease efficiency. These include:

- Packaging of a batch but labeling for individual markets (with different labeling/language requirements) only when orders are received.
- Breakdown of the labeling equipment.
- Packaging of product to be shipped to another facility/country for labeling.

Obviously, the presence of unlabeled product constitutes a high potential for risk unless the packaging is unique. Companies usually label complete boxes (shippers) or shrouded pallets. The use of secured cages is also a valuable approach. Whatever procedure is used, it must be essentially fail-safe.

Food and Drug Administration (FDA) investigators have shown little interest in packaging operations other than labeling. While labeling is possibly the most critical packaging operation, the other areas should be addressed to assure the effectiveness of the overall packaging operation. These include:

- Installation qualification of equipment.
- Operational qualification of key operations such as tablet fillers (to assure that tablets will not block the feed chutes especially of slat fillers); liquid fillers (to determine range variability within and between filling heads); weight checkers (to determine the sensitivity and assure this is adequate to detect the defined deviations such as missing tablet, missing insert, etc.); cappers (for torque variability); and on-line scanners (to confirm they will progress good units and reject foreign labeling).
- Calibration checks on monitoring equipment at least at the beginning and end of each day (or packaging run if less than a day).
- Adequate in-process controls to confirm that the line is operating satisfactorily—label checks, weight/volume checks, torque checks, child resistance, and tamper checks (§211.134).

§211.132 TAMPER-RESISTANT PACKAGING REQUIREMENTS FOR OVER-THE-COUNTER HUMAN DRUG PRODUCTS

(a) General. The FDA has the authority under the Federal Food, Drug, and Cosmetic Act (the act) to establish a uniform national requirement for tamper-resistant packaging (TRP) of OTC drug products that will improve the security of OTC drug packaging and help assure the safety and effectiveness of OTC drug products. An OTC drug product (except a dermatological, dentifrice, insulin, or throat lozenge product) for retail sale that is not packaged in a tamper-resistant package or that is not properly labeled under this section is adulterated under section 501 of the act, or misbranded under section 502 of the act, or both.

(b) Requirement for tamper-resistant package. (1) Each manufacturer and packer who packages an OTC drug product (except a dermatological, dentifrice, insulin, or throat lozenge product) for retail sale, shall package the product in a tamper-resistant package, if this product is accessible to the public while held for sale. A tamper-resistant package is one that has one or more indicators or barriers to entry, which, if breached or missing, can reasonably be expected to provide visible evidence to consumers indicating that tampering has occurred. To reduce the likelihood of successful tampering and to increase the likelihood that consumers would discover whether a product has been tampered with or not, the package is required to be distinctive by design (e.g., an aerosol product container), or by the use of one or more indicators or barriers to entry that employ an identifying characteristic (e.g., an aerosol product container), or by the use of an identifying characteristic (e.g., a pattern, name, registered trademark, logo, or picture). For purposes of this section, the term "distinctive by design" means the packaging cannot be duplicated with commonly available materials or through commonly available processes. For purposes of this section, the term "aerosol product" means a product that depends upon the power of a liquified or compressed gas to expel the contents from the container. A tamper-resistant package may involve an immediate-container and closure system, or secondary-container or carton system, or any combination of systems intended to provide a visual indication of package integrity. The tamper-resistant feature shall be designed to, and shall remain intact when handled in a reasonable manner during manufacture, distribution, and retail display.

(2) In addition to the tamper-evident packaging feature described in paragraph (b) (1) of this section, any two-piece, hard gelatin capsule covered by this section must be sealed using an acceptable tamper-evident technology.

(c) Labeling. (1) In order to alert consumers to the specific tamper-evident feature(s) used, each retail package of an OTC drug product covered by this section (except ammonia inhalant in crushable glass ampoules, containers of compressed medical oxygen, or aerosol products, that depend upon the power of a liquefied or compressed gas to expel the contents from the container) is required to bear a statement that:

    i. identified all tamper-evident feature(s) and any capsule sealing technologies used to comply with paragraph (b) of this section;

    ii. is prominently placed on the package; and

    iii. is so placed that it will be unaffected if the tamper-evident feature of the package is breached or missing.

    (2) If the tamper-resistant feature chosen to meet the requirement in paragraph (b) of this section is one that uses an identifying characteristic, then that characteristic is required to be referred to in the labeling statement. For example, the labeling statement on a bottle with a shrink band could say, "For your protection, this bottle has an imprinted seal around the neck."

(d) Request for exemptions from packaging and labeling requirements. A manufacturer or packer may request an exemption from the packaging and labeling requirements of this section. A request for an exemption is required to be submitted in the form of a citizen petition under §10.30 of this chapter and should be clearly identified on the envelope as a "Request for Exemption from Tamper-resistant Rule." The petition is required to contain the following:

    (1) The name of the drug product or, if the petition seeks an exemption for a drug class, the name of the drug class, and a list of products within that class.

    (2) The reasons that the drug product's compliance with the tamper-resistant packaging or labeling requirements of this section is unnecessary or cannot be achieved.

    (3) A description of alternative steps that are available, or that the petitioner has already taken, to reduce the likelihood that the product or drug class will be the subject of malicious adulteration.

    (4) Other information justifying an exemption.

(e) OTC drug products subject to approved new drug applications. Holders of approved new drug applications for OTC drug products are required under §314.70 of this chapter to provide the agency with notification of changes in packaging and labeling to comply with the requirements of this section. Changes in packaging and labeling required by this regulation may be made before FDA approval, as provided under §314.70(c) of this chapter. Manufacturing changes by which capsules are to be sealed require prior FDA approval under §314.70(b) of this chapter.

(f) Poison Prevention Packaging Act of 1970. This section does not affect any requirements for "special packaging" as defined under §310.3(1) of this chapter and required under the Poison Prevention Packaging Act of 1970.

Product tampering is not a new occurrence. However, as a direct result of several deaths in 1982 resulting from the malicious addition of cyanide to Tylenol Capsules, this section was introduced into the CGMP regulations. The key elements of the regulation are:

1. It only applies to OTC products since these tend to be on open display with ready access to the public. It was considered that prescription products are maintained under the control of the pharmacist and consequently are less vulnerable to tampering. The exemption of insulin was for the same reason. The other excluded categories—dentifrices, lozenges, and dermatological products—were considered to be less prone to potential tampering because of their inherent nature or their use.

2. No test methodology or effectiveness criteria were established. It was considered that the development of these would be difficult, time-consuming, and probably highly controversial and would delay the introduction of tamper-resistant packaging—which an apprehensive public needed in order to retain confidence in this essential form of medication (OTC). Instead, some guidance was provided on currently available forms of tamper-resistant technology

[Sec. 450.500 Tamper-Resistant Packaging Requirements for Certain Over-the-Counter (OTC) Human Drug Products (CPG 7132a.17)]. These included film wrappers, with certain restrictions and limitation; blister or strip packs; bubble packs; heat-shrink bands or wrappers but not wet shrink, which were considered re-usable; foil, paper, or plastic pouches; bottle mouth inner seals; tape seals; breakable caps; sealed metal tubes, or plastic blind-end heat-sealed tubes; sealed cartons but not glued seals; aerosol containers; and sealed cans. The tamper-resistant feature may apply to either the primary or the secondary packaging.

Use of a tamper-resistant feature on the secondary package allows the consumer to examine the product for possible tampering before purchase. This obviously is a consumer benefit. However, any inadvertent damage to the feature during shipping or storage will result in refusal to purchase. Application to the primary container, or a bottle mouth seal, will preclude this possibility.

The Compliance Guide does not suggest that any application of the features mentioned earlier will automatically assure compliance with §211.132, but the manufacturer should be able to demonstrate effective use of the technology. Conversely, other technologies are not excluded.

3. Two-piece hard gelatin capsules have been most vulnerable to tampering since once the contents have been replaced it is unlikely that the consumer will detect the differences, especially if the contents are a white powder. Use of bead formulations, especially if colored, increases security. The regulations (the sealing requirement was effective November 4, 1999, while labeling changes had to be implemented by November 6, 2000) require that such OTC capsules have two tamper-resistant features. Two-piece capsules must have the two halves sealed and this is considered acceptable as one of the two features. Because of this inherent higher vulnerability some manufacturers have ceased to provide this form of dosage, and some companies have introduced gelatin-coated tablets that look like capsules. These provide the consumer the ability to swallow with ease the smooth elongated tablets and the strong medicine perception of a capsule.

4. The tamper-resistant feature is to be "distinctive by design or by use of an identifying characteristic." This is to preclude the possibility of removal of the feature and replacement by a commonly available material. An aerosol package is considered to be distinctive by design, as would be a sealed can. Overwraps and seals usually require a distinctive characteristic such as the company logo or the product name. A generic expression such as "Factory Sealed" may not be sufficiently specific. A further concern relates to the possibility of taking a bottle mouth seal from a wide mouth bottle, cutting it down to size and gluing it to the mouth of a narrower neck bottle. Manufacturers must use their judgment.

5. Labeling is to include specific reference to the tamper-resistant feature used and must be sufficiently explicit that a malicious replacement can be identified by the consumer.

The tamper-evident statement must be prominently placed on the drug product package to alert consumers about the product's tamper-evident features (21 CFR 211.132). The tamper-evident statement describes its feature of the product package and advises consumers that, if the feature is breached or missing when the product is purchased, tampering may have occurred. Tamper-evident packaging with an appropriate labeling statement will be more likely to protect consumers because the consumers will be in a better position to detect tampering when they have the knowledge that a tamper-evident feature has been incorporated into the product design. The Agency allows flexibility in the placement of this statement on the package and does not require that it be included within the Drug Facts section. However, if included in this section, the statement must appear under the heading "Other information" [21 CFR 201.66(c)(7)].

The Agency also noted in the final rule preamble for the Drug Facts regulation that many products are now marketed with "peel back" or "fold out" labels affixed to the product package and that these labels could be used to accommodate all of the FDA required information in the Drug Facts section (64 FR 13254 at 13268; March 17, 1999). These types of labels were not in use at the time the tamper-evident requirements became effective. Recently, interested parties have inquired whether the tamper-evident statement may be included in a Drug Facts section that appears in such "peel back" or "fold out" labels. We believe that the goals of the tamper-evident statement probably would not be achieved if the statement only appears in a "peel back" or "fold out" label and is not clearly visible without peeling back or folding out the label.

It is important that the consumer views the tamper-evident statement before purchase and use of the product so that he or she will be better aware of the tamper-evident features and any signs of tampering. Once the consumer opens the tamper-evident package, the tamper-evident features have been breached. If the consumer has failed to examine these features before opening, then the consumer will likely not know whether there were any signs of tampering. A tamper-evident statement inside a "peel back" or "fold out" label that is not visible on the outside of the package is unlikely to be viewed before breach of the tamper-evident feature. The consumer may not be aware to peel back or unfold this label to view the tamper-evident statement before opening the package. Thus, we recommend that the statement not appear within the Drug Facts box in a "peel back" or "fold out" label if the statement would not be clearly visible without peeling back or folding out the label. The FDA recommended instead in these circumstances that the tamper-evident statement be provided outside the Drug Facts box in another part of the label, where the statement is clearly visible without further manipulation of that label.

6. Tamper-resistant packaging components are to be treated identically to other components (Compliance Policy Guide 7132.14). Those coming into direct contact with the drug product are subject to the container and closure provisions of the CGMP Regulations (Subpart E). Other tamper-resistant components are subject to the appropriate provisions of Subpart G; in particular, any components with labeling information would need to comply with the provisions for labeling including accountability.

At the present time the currently available technologies do provide significant protection to the consumer, but products are not tamper-proof. Additionally, some tamper-resistant features, such as neckbands and breakable caps, may impact adversely on the ability to open the package with ease. Whereas the majority of tamper-resistant features impart a physical barrier to entry, it is anticipated that future developments will rely heavily on technologies, such as microencapsulated inks, which do not affect ease of opening. However, whatever technology is used it must be applied effectively by the manufacturer and supported by an informed and alert consumer.

Those portions of tamper-resistant packaging that contain labeling, as defined in Section 201(m) of the FD&C Act will be considered as any other labeling and, as such, are subject to the control and accountability provisions of Subpart G of the Current Good Manufacturing Practice (CGMP) Regulations.

Those portions of tamper-resistant packaging that contact the drug product are considered part of the container closure system and, as such, are subject to the control and accountability provisions of Subpart E of the CGMP regulations.

Those portions of tamper-resistant packaging that do not fall into the categories mentioned earlier will be considered as general packaging material, subject to the general controls for packaging contained in Subpart G of the CGMP regulations.

In addition, the Agency has re-evaluated the currently available tamper-resistant packaging technologies and concluded that some technologies as designed or applied are no longer capable of meeting the requirements of the TRP regulations.

In 1992, the FDA published Compliance Policy Guide 7132a.17 (Sec 450.500) Tamper-Resistant Packaging Requirements for certain OTC Human Drug Products. This guide outlined those configurations and materials that are currently considered as not acceptable to render the OTC product tamper-resistant. Sections of this guide are discussed later.

A. PACKAGING SYSTEMS

Manufacturers and packagers are free to use any packaging system as long as the tamper-resistant standard in the regulations is met. The TRP requirements are intended to assure that the product's packaging "can reasonably be expected to provide visible evidence to consumers that tampering has occurred."

Examples of packaging technologies capable of meeting the TRP requirements are listed below. The use of one of these packaging technologies does not, by itself, constitute compliance with the requirements for a tamper-resistant package. Packaging features must be properly designed and appropriately applied to be an effective TRP.

1. *Film Wrappers.* A transparent film is wrapped securely around the entire product container. The film must be cut or torn to open the container and remove the product. A tight "fit" of the film around the container must be achieved, for example, by a shrink-type process. A film wrapper sealed with overlapping end flaps must not be capable of being opened and resealed without leaving visible evidence of entry.

   The use of cellophane with overlapping end flaps is not effective as a tamper-resistant feature because of the possibility that the end flaps can be opened and resealed without leaving visible evidence of entry.

   The film wrapper must employ an identifying characteristic that cannot be readily duplicated. An identifying characteristic that is proprietary and different for each product size is recommended.

   Tinted wrappers are no longer acceptable as an identifying characteristic because of the possibility that their material or a facsimile may be available to the public.

2. *Blister or Strip Packs.* Dosage units (e.g., tablets or capsules) are individually sealed in clear plastic or plastic compartments with foil or paper backing.

   The individual compartment must be torn or broken to obtain the product. The backing materials cannot be separated from the blisters or replaced without leaving visible evidence of entry.

3. *Bubble Packs.* The product and container are sealed in plastic and mounted in or on a display card. The plastic must be torn or broken to remove the product. The backing material cannot be separated from the plastic bubble or replaced without leaving visible evidence of entry.

4. *Heat Shrink Bands or Wrappers.* A band or wrapper is securely applied to a portion of the container, usually at the juncture of the cap and container. The band or wrapper is heat shrunk to provide a tight fit. The band or wrapper must be cut or torn to open the container and remove the product and cannot be worked off and reapplied without visible damage. The use of a perforated tear strip can enhance tamper resistance. Cellulose wet shrink seals are not acceptable. The knowledge to remove and reapply these seals without evidence of tampering is widespread.

   The band or wrapper must employ an identifying characteristic that cannot be readily duplicated. An identifying characteristic that is proprietary and different for each product size is recommended.

   Tinted bands or wrappers are no longer acceptable as an identifying characteristic because of the possibility that their material or a facsimile may be available to the public.

5. *Foil, Paper, or Plastic Pouches.* The product is enclosed in an individual pouch that must be torn or broken to obtain the product. The end seams of the pouches cannot be separated and resealed without showing visible evidence of entry.

6. *Container Mouth Inner Seals.* Paper, thermal plastic, plastic film, foil, or a combination thereof, is sealed to the mouth of a container (e.g., bottle) under the cap. The seal must be torn or broken to open the container and remove the product. The seal cannot be removed and reapplied without leaving visible evidence of entry. Seals applied by heat induction to plastic containers appear to offer a higher degree of tamper-resistance than those that depend on an adhesive to create the bond.

   Polystyrene foam container mouth seals applied with pressure sensitive adhesive are no longer considered effective tamper-resistant features because they can be removed and reapplied in their original state with no visible evidence of entry.

   The Agency recognizes that technological innovations may produce foam seals that will adhere to a container mouth in a manner that cannot be circumvented without visible evidence of entry. Container mouth seals must employ an identifying characteristic that cannot be readily duplicated. An identifying characteristic that is proprietary and different for each product size is recommended.

7. *Tape Seals.* Tape seals relying on an adhesive to bond them to the package are not capable of meeting the TRP requirements because they can be removed and reapplied with no visible evidence of entry.

   However, the Agency recognizes that technological innovations may produce adhesives that do not permit the removal and reapplication of tape seals. In addition, tape seals may contain a feature that makes it readily apparent if the seals have been removed and reapplied. Tape seals must employ an identifying characteristic that cannot be readily duplicated.

8. *Breakable Caps*. The container (e.g., bottle) is sealed by a plastic or metal cap that either breaks away completely when removed from the container or leaves part of the cap attached to the container. The cap, or a portion thereof, must be broken in order to open the container and remove the product. The cap cannot be reapplied in its original state.

9. *Sealed Metal Tubes or Plastic Blind-End Heat-Sealed Tubes*. The bottom of the tube is heat sealed and the mouth or blind-end must be punctured to obtain the product. A tube with a crimped end is capable of meeting the definition of a tamper-resistant feature, if the crimped end cannot be breached by unfolding and refolding without visible evidence of entry.

10. *Sealed Cartons*. Paperboard cartons sealed by gluing the end flaps are not capable of meeting the TRP requirements. However, the Agency recognizes that technological advances may provide sealed paperboard packages that meet the requirements of the TRP regulations.

11. *Aerosol Containers*. Aerosol containers are believed to be inherently tamper-resistant because of their design. Direct printing of the label on the container (e.g., lithographing) is preferred to using a paper label, which could be removed and substituted.

12. *Cans (Both All-Metal and Composite)*. Cans may be composed of all metal or composite walls with metal tops and bottoms. The top and bottom of a composite can must be joined to the can walls in such a manner that they cannot be pulled apart and reassembled without visible evidence of entry. Rather than attaching a separate label, direct printing of the label onto the can (e.g., lithographing) is preferred.

B. CAPSULE SEALING TECHNOLOGIES

Technologies for sealing two-piece hard gelatin capsules are available that provide evidence whether the capsules have been tampered with some after filling or not. Such sealing technologies currently in use include sonic welding, banding, and sealing techniques employing solvents and/or low temperature heating. These examples are not intended to rule out the development and use of other capsule sealing technologies. Manufacturers may consult with FDA if they are considering alternative capsule sealing processes.

Sealed capsules are not TRP. They are required to be contained within a package system that utilizes a minimum of one TRP feature.

C. TAMPER-RESISTANT PACKAGES LABELING STATEMENT(S)

1. *Bottle (Container) Caps*. In the past, some manufacturers have placed the TRP labeling statement on bottle caps. This practice is unacceptable in cases where it may be a simple matter to substitute another unlabeled bottle cap for the one with the tamper-resistant warning statement. Such an act could easily be accomplished without any apparent sign of tampering.

2. *Package Inserts*. The practice of placing the TRP labeling statement solely on the product's inserts is not acceptable. While package inserts may be a useful supplement for consumer education purposes, they are not acceptable in lieu of label statements.

3. *Carton/Container (Outer and Inner)*. If the TRP feature is on an outer carton, the inner container (e.g., bottle) needs to bear a statement alerting the consumer that the bottle should be in a carton at the time of purchase. This policy applies only to situations where the inner container is so labeled that such a container might reasonably otherwise be displayed on the retail shelf without an outer carton.

4. *Identifying Characteristic*. When a TRP feature is required to have an identifying characteristic, that characteristic needs to be referenced in the labeling statement (e.g., "imprinted" neck band). It is recommended that the labeling statement specifically identify the characteristic (e.g., imprinted with XYZ on the neck band).

5. *TRP Feature(s)*. All required tamper-resistant features must be referenced in the labeling statement. When two tamper-resistant packaging features are used for unsealed two-piece hard gelatin capsules, both features must be referenced in the labeling statement. If one tamper-resistant packaging feature plus sealed capsules are used, the labeling statement must reference both the capsule seal and the tamper-resistant packaging feature.

REGULATORY ACTION GUIDANCE

The TRP requirements are part of the CGMP regulations. Regulatory actions for deviations from these requirements should be handled in the same manner as any other deviation from the GMP regulations.

§211.134 DRUG PRODUCT INSPECTION

(a) Packaged and labeled products shall be examined during finishing operations to provide assurance that containers and packages in the lot have the correct label.
(b) A representative sample of units shall be collected at the completion of finishing operations and shall be visually examined for correct labeling.
(c) Results of these examinations shall be recorded in the batch production or control records.

Compliance with the intent of this subsection requires that inspection and documentation procedures continue throughout the labeling-packaging operation to prevent incorrect components or procedures from being utilized. The majority of these on-line inspections are often performed by suitably trained production personnel, with Quality Assurance confirming compliance with the defined inspection procedures and performing periodic audit. Sampling of controlled substances carry special responsibilities as to the limitations on number taken, stored, and security thereafter.

ON-LINE INSPECTION PLAN

The on-line inspection procedure should be statistically designed and is usually based on the classification of the inspected items as "defective" or "non-defective." The aim of the system is to identify the types of defects that occur so that appropriate action can be initiated to determine and eliminate their cause and enable an objective decision to be made on the disposition of each packaged batch.

Defects or any nonconformance with specification may be classified according to their seriousness.

- *Nontolerance*: In other words, incorrect name, product, or bar code. Any one sample will reject the entire lot.
- *Critical defect*: One with a high probability of adversely impacting on the effectiveness of the product. Examples include, missing label, incorrect batch number or expiry date, incorrect carton, and incorrect dosage information on label.
- *Major defect*: One with a low probability of adversely impacting on the effectiveness of the product. Examples include low fill volume, partly legible batch number or expiry date, and missing carton.
- *Minor defect*: One unlikely to have any impact on product effectiveness. Examples include poor printing on tablets or capsules, off-center or dirty labels, missing (noncritical) print, and misspellings.

The sampling plan, inspection level, and acceptance criteria should be based on product use and performance history and should be approved by Quality Control Unit. Typical acceptance levels for each type of defect could be critical (0%), major (0.65%), or minor (2.5%). However, attempts should be made to identify the causes of any defects and to eliminate them, thereby resulting in reduction in acceptable defect levels and improvement in quality. It is interesting to note that the quality-conscious healthcare industry still operates in the percentage defects range (parts per hundred), while some other industries, such as electronics, operate in the parts per million range.

Records will note the number and types of defects found, the time, and further specify the exact defect (i.e., off-center label). All critical defects should normally be reported immediately to departmental management, who will initiate additional inspection to ascertain the extent of the problem and to eliminate critically defective items from stock. A 100% inspection of all stock packed since the previous acceptable inspection is usually regarded as essential. If further critical defects are discovered it may be necessary to suspend production until the cause has been identified and corrected. If, during the course of the packaging run, it is observed that it is likely that the number of major or minor defectives will eventually result in a rejection number, the production supervisor or manager should be informed so that appropriate action can be taken to rectify the situation. All such actions should be recorded, and will be entered into the batch record.

Repetition of deviations, especially those classified as critical or major, is an indication that the equipment involved is not operating satisfactorily—maintenance, adjustment, or even replacement may be required. Alternatively, the problems could be caused by unacceptable variability in the packaging components. This may require modification of specifications and review with suppliers.

§211.137   EXPIRATION DATING

(a) To assure that a drug product meets applicable standards of identity, strength, quality, and purity at the time of use, it shall bear an expiration date determined by appropriate stability testing described in §211.166.

(b) Expiration dates shall be related to any storage conditions stated on the labeling, as determined by stability studies described in §211.166.

(c) If the drug product is to be reconstituted at the time of dispensing, its labeling shall bear expiration information for both the reconstituted and unreconstituted drug products.

(d) Expiration dates shall appear on labeling in accordance with the requirements of §201.17 of this chapter.

(e) Homeopathic drug products shall be exempted from the requirements of this section.

(f) Allergenic extracts that are labeled "No U.S. Standard of Potency" are exempt from the requirements of this section.

(g) New drug products for investigational use are exempted from the requirements of this section, provided that they meet appropriate standards or specifications as demonstrated by stability studies during their use in clinical investigations. Where new drug products for investigational use are to be reconstituted at the time of dispensing, their labeling shall bear expiration information for the reconstituted drug product.

(h) Pending consideration of a proposed exemption, published in the Federal Register of September 29, 1978, the requirements in this section shall not be enforced for human OTC drug products if their labeling does not bear dosage limitations and they are stable for at least three years as supported by appropriate stability data.

When the CGMP regulations were introduced there was considerable comment regarding the need for expiry dating. Subsequently, some concessions were made with respect to OTC products [(h)]. The arguments against expiry dating essentially suggested that if it could be demonstrated that a product passed through the distribution chain within its acceptable life then there should be no need to display an expiry date on the product. This argument would seem to have little merit when it is known that stock rotation in the trade cannot be guaranteed and that consumers having purchased a product may not completely use it and will probably be unable to remember when it was purchased. One of the authors has personally identified products over 15 years old in the marketplace. Consequently there would seem to be very good reasons for expiry dating of all pharmaceutical products.

The need for expiration dating has consistently created problems with respect to clinical trials of new drugs being investigated. Frequently these products are supported by parallel stability studies and have only limited historical data that may be inadequate to predict their shelf life. These products are distributed to clearly identified individuals and consequently recovery is easy. Industry has indicated that it is inappropriate and unnecessary to apply an expiration date to such clinical samples provided there are ongoing stability studies and a commitment to recover products from the trials, if necessary. A further guidance specifies that when new drug products for investigational use are to be reconstituted at the time of dispensing, the drug shall bear expiration information for the reconstituted drug product. These requirements are recognized in subsection (g) of the regulations.

Products for reconstitution will show two separate shelf lives: (*i*) an expiration date after which the unconstituted product should not be used, and (*ii*) the maximum period during which the reconstituted product must be used. This requirement is included in the regulation in 21 CFR Part 211.137(c).

The exemption for homeopathic products [21 CFR 211.137(e)] is based on two facts: the inability to quantitatively evaluate the low levels of ingredients in such products and the inability to relate effectiveness to quantitative composition. While both statements are correct, it might still be considered, by reputable homeopathic producers, that some restriction on shelf life, while not proven to be essential, would certainly not be detrimental to the product user. The approaches used to establish shelf lives will be addressed in Chapter 10 (§211.166).

ALLOPATHIC DRUGS, HOMEOPATHIC DRUGS, AND DIETARY SUPPLEMENTS REGULATIONS

The information in the table given herewith summarized the regulations under which manufacturers must comply with respect to allopathic, homeopathic, and dietary supplements including Premarket Approval, GMPs, Labeling, Advertising, and Indications on the labeling.

Regulation of Allopathic Drugs, Homeopathic Drugs, and Dietary Supplements

Medication Type	Enabling Legislation	Premarket Approval	GMPs	Labeling	Advertising	Indication on Labeling
Allopathic drugs	FDCA	NDA (21 C.F.R. 300 et seq.)	21 C.F.R. 210 & 211	21 C.F.R. 201	Prescription: FDA Nonprescription: FTC	Required
Dietary supplements	DSHEA	None	Proposed	DSHEA	FTC	"Structure function" claims only
Homeopathic drugs	FDCA	HPCUS Monograph	21 C.F.R. 210 & 211	21 C.F.R. 201, FDA Compliance Policy Guide 400.400	Prescription: FDA Nonprescription: FTC	Required

Abbreviations: GMPs, good manufacturing practices; FDCA, Food, Drug, and Cosmetic Act of 1938; NDA, new-drug-application process; 21 C.F.R. Title 21 of the Code of Federal Regulations; FDA, Food and Drug Administration; FTC, Federal Trade Commission; DSHEA, Dietary Supplement Health and Education Act of 1994; HPCUS, Homeopathic Pharmacopoeia Convention of the United States.

The FDA published the Proposed Rules for "Current Good Manufacturing Practice in Manufacturing, Packing, or Holding Dietary Ingredients and Dietary Supplements" in the Federal

Register on March 13, 2003 (Volume 68, Number 49)] on pages 12157–12263 in 21 CFR chapter I, parts 111 and 112 as set forth herewith.

OVER-THE-COUNTER LABELING

Since 1982, there have been many changes in the rules and requirements to 21 CFR Part 201, Labeling. Below is a brief history of the rulemaking as published by CBER and available at http://www.fda.gov/cder/otcmonographs /Labeling/new_labeling.htm.

Rulemaking History for Labeling of Over-the-Counter Drug Products

Date	Action
7/2/1982	Notice: limitation on labeling terminology "exclusivity"
9/7/1982	Proposed Rule: proposed warning concerning use of systematically absorbed OTC drugs by pregnant or nursing women
9/21/1982	Correction: proposed warning concerning use of systematically absorbed OTC drugs by pregnant or nursing women
12/3/1982	Final Rule: warning concerning use of systematically absorbed OTC drugs by pregnant or nursing women
4/1/1983	Notice: professional labeling for reserpine drugs
8/31/1983	Notice: Final Rule: warning concerning use of systematically absorbed OTC drugs by pregnant or nursing women
11/30/1983	Notice of availability and opportunity for public comment: Final Rule: warning concerning use of systematically absorbed OTC drugs by pregnant or nursing women
4/22/1985	Proposed Rule: labeling of drug products for OTC human use ("exclusivity")
4/17/1986	Proposed Rule: proposed amendment of statement of identity requirements
5/1/1986	Final Rule: labeling of drug products for OTC human use ("exclusivity")
5/21/1986	Correction: Final Rule: labeling of drug products for OTC human use ("exclusivity")
6/3/1986	Extension of comment period: Proposed Rule: proposed amendment of statement of identity requirements
3/13/1987	Correction: Final Rule: labeling of drug products for OTC human use ("exclusivity")
7/5/1990	Final Rule: Labeling for Oral and Rectal OTC Aspirin and Aspirin-Containing Products; Pregnancy Warning;
3/6/1991	Notice (request for comments): print size and style of labeling for OTC drug products
4/9/1991	Extension of comment period: Notice (request for comments): print size and style of labeling for OTC drug products
4/25/1991	Proposed Rule: sodium labeling for OTC drugs
5/22/1991	Correction: Proposed Rule: sodium labeling for OTC drugs
6/12/1991	Extension of comment period: Proposed Rule: sodium labeling for OTC drugs
11/9/1993	Proposed Rule: labeling of drug products for OTC human use; Subject to an approved application or abbreviated application;
1/28/1994	Proposed Rule: labeling of drug products for OTC human use; similar words
8/3/1994	Proposed Rule: labeling of drug products for OTC human use
4/22/1995	Final Rule: sodium labeling for OTC drugs
8/16/1995	Notice of public hearing, Request for Comments: OTC drug labeling
10/4/1995	Proposed Rule: labeling of drug products for OTC human use
2/14/1996	Proposed Rule: OTC drugs containing phenylpropanolamine (PPA)
3/4/1996	Proposed Rule: labeling of drug products for OTC human use
4/22/1996	Proposed Rule: labeling of orally ingested OTC drug products containing calcium, magnesium, and potassium
4/22/1996	Final Rule: sodium labeling for OTC drugs
7/22/1996	Extension of comment period; Final Rule: labeling of orally ingested OTC drug products containing calcium, magnesium, and potassium
7/22/1996	Extension of comment period: Proposed Rule: sodium labeling for OTC drugs
8/19/1996	Correction: labeling of orally ingested OTC drug products containing calcium, magnesium, and potassium
2/27/1997	Proposed Rule: OTC human drugs
3/21/1997	Correction: sodium labeling for OTC drugs
4/24/1997	Partial delay of effective date: Final Rule: sodium labeling for OTC drugs
6/19/1997	Extension of comment period: Proposed Rule: OTC human drugs
3/17/1999	Final Rule: OTC human drugs
4/15/1999	Correction: OTC human drugs

(Continued)

Date	Action
12/1/1999	Notice of availability: Draft Guidance for industry on labeling of OTC human drug products using a column format
6/20/2000	Partial Extension of Compliance Date: Final Rule
4/5/2002	Partial Extension of Compliance Date: Convenience Size
3/24/2004	Technical Amendment: Final Rule: sodium labeling for OTC drugs
3/24/2004	Final Rule: labeling of orally ingested OTC drug products containing calcium, magnesium, and potassium
3/24/2004	Proposed Rule: Drug Labeling; Sodium Labeling for Over-the-Counter Drugs
4/22/2004	Proposed Rule: Toll-Free Number for Reporting Adverse Events
9/3/2004	Rule: Delay of implementation date (sunscreen Drug Facts)
11/29/2004	Drug Labeling; Sodium Labeling for Over-the-Counter-Drugs; Sodium phosphate-sodium biphosphate-containing rectal drug products; Final Rule
12/9/2004	Draft Compliance Guidance for Small Business Entities on Labeling Over-the-Counter Human Drug Products; Availability
1/4/2005	Agency Information Collection Activities; Proposed Collection; Comment Request; Format and Content Requirements for Over-the-Counter Drug Product Labeling; Notice
1/13/2005	Draft Guidance for Industry on Labeling Over-the-Counter Human Drug Products; Questions and Answers; Availability; Action: Notice
3/29/2005	Submission for Office of Management and Budget Review; Comment Request; Format and Content for Over-the-Counter Drug Product Labeling

Abbreviations: OTC, Over-the-Counter.

In addition to the revision/additions to the guidance for Labeling Drug list in the table mentioned earlier, other revisions to labeling requirement have also been implemented. In February 2002, the FDA revised section 201.16 Drugs [67 FR 4906] to include allowance of Spanish-language version of certain required statements. The guidance document states "An increasing number of medications restricted to prescription use only are being labeled solely in Spanish for distribution in the Commonwealth of Puerto Rico, where Spanish is the predominant language. Such labeling is authorized under 201.15(c). One required warning, the wording of which is fixed by law in the English language, which could be translated in various ways, from literal translation to loose interpretation. The statutory nature of this warning requires that the translation convey the meaning properly to avoid confusion and dilution of the purpose of the warning. Section 503(b)(4) of the Federal Food, Drug, and Cosmetic Act requires, to a minimum, that the label bear the statement "Rx only." The Spanish-language version of this must be "Solamente Rx."

In the Federal Register of February 26, 2004 (69 FR 9120), the FDA published a final rule requiring certain human drug and biological products to have on their labels a linear bar code that contains, at a minimum, the drug's NDC number (21 CFR 201.25). The rule also requires the use of machine-readable information on blood and blood component labels (21 CFR 606.121(c)(13)). The guidance states "Bar codes will allow health care professionals to use bar code scanning equipment to verify that the right drug (in the right dose and right route of administration) is being given to the right patient at the right time. This new system is intended to help reduce the number of medication errors that occur in hospitals and health care settings." Drugs approved on or after April 26, 2004, have 60 days from their approval date to comply with the bar code requirement (21 CFR 201.25 and response to comment 71, 69 FR at 9147). All other drugs subject to the bar code requirement, including drugs with applications approved before April 26, 2004, and drugs marketed without an application, whether prescription or OTC, must implement the requirements within two years of the effective date (i.e., no later than April 26, 2006). Prescription drug products are subject to the bar code label requirements. The bar code requirement does not apply to (a) Prescription drug samples, (b) Allergenic extracts, (c) Intrauterine contraceptive devices regulated as drugs, (d) Medical gases, (e) Radiopharmaceuticals, and (f) Low-density polyethylene form fill and seal containers that are not packaged with an overwrap.

In January 2006, the FDA published a nonbinding guidance document titled "Adverse Reactions Section of Labeling for Human Prescription Drugs and Biological Products—Content and Format." The guidance is intended to help applicants and reviewers in drafting the ADVERSE REACTIONS section of prescription drug labeling as required by 21 CFR 201.57(c)(7).

HEALTH-RELATED CLAIMS AND DIETARY SUPPLEMENT LABELING

Under the Dietary Supplement Health and Education Act of 1994 (DSHEA), the dietary supplement manufacturer is responsible for ensuring that a dietary supplement is safe before it is marketed. The FDA is responsible for taking action against any unsafe dietary supplement product after it reaches the market. Generally, manufacturers do not need to register their products with the FDA nor get FDA approval before producing or selling dietary supplements.* Manufacturers must make sure that product label information is truthful and not misleading.

The FDA's postmarketing responsibilities include monitoring safety, for example, voluntary dietary supplement adverse event reporting and product information, such as labeling, claims, package inserts, and accompanying literature. The Federal Trade Commission regulates dietary supplement advertising.

Because dietary supplements are under the "umbrella" of foods, the FDA's Center for Food Safety and Applied Nutrition (CFSAN) is responsible for the agency's oversight of these products. The FDA's efforts to monitor the marketplace for potential illegal products (i.e., products that may be unsafe or make false or misleading claims) include obtaining information from inspections of dietary supplement manufacturers and distributors, the Internet, consumer and trade complaints, occasional laboratory analyses of selected products, and adverse events associated with the use of supplements that are reported to the agency.

The Federal Trade Commission (FTC) regulates advertising, including infomercials, for dietary supplements and most other products sold to consumers. The FDA works closely with the FTC in this area. Advertising and promotional material received in the mail are also regulated under different laws and are subject to regulation by the U.S. Postal Inspection Service.

Some products encourage consumers to self-treat for a serious disease without benefit of a medical diagnosis or treatment. Products sold as dietary supplements that bear a claim to treat, mitigate, or cure a disease are drugs, and such products are subject to regulations, as such. Dietary supplements that bear unsubstantiated structure/function claims in their labeling can also defraud and harm consumers. For example, unsubstantiated claims for a dietary supplement may encourage consumers to substitute the supplement for a product or health promotion strategy whose benefit is backed up by scientifically accurate information.

The FDA currently takes the following actions to help ensure that dietary supplement labeling is truthful and nonmisleading:

1. Continue to identify and take action against dietary supplements making claims that are not supported by scientific evidence.
2. Develop and publish a draft guidance addressing what constitutes adequate scientific substantiation for structure/function claims and claims of a benefit related to a nutrient deficiency disease.
3. Identify and take enforcement action against products whose labeling fails to reveal material facts, targeting those products that pose the greatest risks to consumers.
4. Obtain and analyze samples of dietary supplements in the marketplace to verify that the contents are consistent with the labeling.
5. Review Supplement Facts panels on dietary supplement labels to determine whether the substances listed as dietary ingredients can be lawfully marketed in dietary supplements.

STRUCTURED PRODUCT LABELING

In October 2005, the FDA began requiring companies to submit their product information for approval in an XML format compliant with the Structured Product Labeling (SPL) standard. To comply, manufacturers must:

- Convert new and existing labeling content to XML
- Submit to the FDA both the narrative labeling content and structured drug listing data
- Maintain consistent product information across all outputs

The goals of Structured Product Labeling include:

- Human-readable labeling content compatible across systems
- Faster dissemination of labeling to improve risk management
- More efficient evaluation of labeling changes
- More coordinated data collection and storage
- Better support for analysis of data
- Improved interoperability with other systems
- Improved integration of clinical data
- Improved access by prescribers and consumers
- Support for retention of legacy product labeling

Labeling Over-the-Counter Human Drug Products Over-the-Counter; January 2005

The Drug Facts labeling for OTC drug products is intended to make it easier for consumers to read and understand OTC drug product labeling and use OTC drug products safely and effectively. The Drug Facts labeling regulation in § 201.66 covers all OTC drug and drug-cosmetic products, whether marketed under a new drug marketing application (NDA), abbreviated new drug application (ANDA), or OTC drug monograph (or product not yet the subject of a final OTC drug monograph).

The regulation is divided into two main parts: (1) content requirements (i.e., headings, subheadings, and the order in which certain information must be listed) and (2) format requirements (i.e., graphic specifications). This guidance primarily discusses questions received from manufacturers, packers, and distributors relating to these requirements, which are set forth in § 201.66 (c) and (d), respectively.

Section 201.66 requires that all OTC drug product labeling contain the following information about the product. This information must be organized according to the following headings and must be presented in the following order:

1. Title ["Drug Facts" or "Drug Facts (continued)"]
2. Active ingredients
3. Purpose
4. Use(s)
5. Warnings
6. Directions
7. Other information
8. Inactive ingredients
9. Questions (optional) ("Questions?" or "Questions or comments?")

This information must appear on the outside container or wrapper of the retail package, or the immediate container label if there is no outside container or wrapper. (If the Drug Facts information appears on the outside container or wrapper of the retail package, its use on the immediate container is optional.)

When the required Drug Facts content information printed in the standardized format plus any other FDA-required information for drug or drug-cosmetic products, other than information required to appear on the principle display panel, requires more than 60 percent of the total surface area available to bear labeling, the modified labeling format can be used.

"OFF-LABEL" INFORMATION AND LABELING

The FDA published the proposed rules for Dissemination of Information on Unapproved/New Uses for Marketed Drugs, Biologics, and Devices in the Federal Register / Vol. 63, No. 109 /

Monday, June 8, 1998. The proposed rule would create a new (21 CFR) part 99 entitled "Disse-mination of Information on Unapproved/New Uses for Marketed Drugs, Biologics, and Devices." The final ruling was published on November 18, 1998 under Docket No. [98 N-0222] RIN 0910-AB23

Under the proposal, firms or sponsors no longer would have to wait until the FDA approves their supplemental application before disseminating certain reliable information about unapproved uses of their products, provided the information:

- Concerns a drug or device that has been approved, licensed, or cleared for marketing by the FDA;
- Is in the form of an unabridged reprint or copy of a peer-reviewed scientific or medical journal article, or an unabridged reference publication, about a clinical investigation that is considered scientifically sound by qualified experts;
- Does not pose a significant risk to the public health;
- Is not false or misleading;
- Is not derived without permission from clinical research conducted by another manufac-turer; and
- Includes certain disclosures (e.g., that the new use has not been approved by the FDA), the official labeling, and a bibliography of other articles relating to the new use.

The manufacturer would also have to submit to the FDA, 60 days prior to dissemination, a copy of the information to be disseminated and other data specified in the proposal.

A firm that has not submitted a supplemental application for the new use could begin disseminating information if it has:

- Certified that it has completed the necessary studies and that a supplemental application will be submitted within six months;
- Provided an adequate protocol and reasonable schedule for the necessary studies and cer-tified that the application will be submitted within 36 months of the initial dissemination; or
- Received an exemption from the requirement to submit an application on the grounds that the necessary studies would be unethical or economically prohibitive.

If the FDA determines that the information is not objective and balanced, it can require the manufacturer to include additional objective and scientifically sound information or an objective statement prepared by the FDA about the safety or effectiveness of the new use.

Manufacturers would have an ongoing responsibility to provide the FDA with additional information about the disseminated new uses, and the FDA could order the cessation of the dissemination if the additional information indicated that the "off-label" use may not be effec-tive or may pose a significant risk to public health.

In the Federal Register of August 12, 1999 (64 FR 44025), the FDA published in its entirety an order entitled "Final Amended Order Granting Summary Judgment and Permanent Injunc-tion." The order was entered by the United States District Court for the District of Columbia in *Washington Legal Foundation vs. Henney,* 56 F. Supp. 2d 81 (1999). The Court of Appeals sub-sequently vacated the district court decision and injunction (and earlier decisions and injunc-tions) insofar as they declared unconstitutional (1) statutory provisions concerning the dissemination by manufacturers of certain written materials concerning new uses of approved products (21 U.S.C.360aaa et seq.), and (2) an FDA guidance document concerning certain industry-supported scientific and educational activities known generally as industry-supported continuing medical education or "CME." [*Washington Legal Foundation vs. Henney,* No. 99-5304, 2000 WL 122099, slip op. (D.C. Cir. Feb. 11, 2000)]. Consequently, these statutory provisions now constitute a "safe harbor" for manufacturers that comply with them; the CME guidance document details how the agency intends to exercise its enforcement discretion. The FDA, consistent with its longstanding interpretation of the laws it administers, may proceed, in the context of case-by-case enforcement, to determine from a manufacturer's written materials and activities how it intends that its products be used. The Court of Appeals also recognized that if the agency brings an enforcement action, a manufacturer might raise a First Amendment defense.

TABLE 1 Standard Versus Modified Labeling Format

Labeling Element	Standard Format	Modified Format
Drug Facts Box	Set off by barline	Barline may be omitted if color contrast used to set off from the rest of the labeling
Drug Facts	Larger than largest type size used in Drug Facts box or similar enclosure	Larger than largest type size used in Drug Facts box or similar enclosure
Drug Facts (continued)	No smaller than 8-point type	No smaller than 7-point type
Headings	>8-point or greater type, or 2-point type greater than point size of text	>7-point or greater type, or 1-point type greater than point size of text
Subheadings	No smaller than 6-point type	No smaller than 6-point type
Bulleted text	No smaller than 6-point type	No smaller than 6-point type
Leading	Minimum 0.5 point	Less than 0.5 point may be used, provided the ascenders and descenders do not touch
Bullets	Minimum 5-point type, Vertical alignment	Minimum 5-point type, No alignment required

Under the Federal Food, Drug, and Cosmetic Act (FD&C Act), the FDA may remove a dietary supplement from the market if it presents a significant or unreasonable risk of illness or injury when used according to its labeling or under ordinary conditions of use when there are no recommendations in the labeling. In February 2004, the FDA made such a finding for dietary supplements containing ephedrine alkaloids, and in April 2004, the agency published a final rule that prohibited their sale.

The FD&C Act also prohibits firms from marketing dietary supplements that are misbranded; that is, containing label information that is false or misleading, such as claiming to contain ingredients that are not actually present.

The governments' scrutiny of off-label drug promotion is on the rise. Investigation of pharmaceutical fraud has been included in its annual work plans of the Office of the Inspector General (OIG) since 2003. The 2005 plan specifically referenced its intent to assess the FDA's oversight and review of permissible and impermissible off-label practices, suggesting that the government's scrutiny of off-label drug marketing will continue. Of course this raises a question about the future direction of the government's enforcement efforts.

Off-label marketing will most likely remain a focus of OIG investigative efforts. Future settlements and corporate integrity agreements involving off-label marketing restrictions and obligations are likely to remain. The threat of enormous civil penalties and onerous Corporate Integrity Agreements may deter the companies from misbranding activities.

Table 1 shows the differences between the standard and modified labeling formats.

Examples of Observations from FDA Warning Letters

The FDA reported in a presentation to a group of small businesses that no GMP deficiencies were observed in the packaging and labeling area of domestic inspections for 2004 and 2005. (http://www.fda.gov/cder/about/smallbiz/Presentations/6.ppt#10). Copies of Warning letters issued by the FDA can be access on the Internet at http://www.fda.gov/foi/warning.htm. The FDA's current focus with reference to labeling is on the advertising information relating to the approved use(s) versus the advertised off-label uses of the drug, especially relating to dietary supplements and OTC products. The following examples of warning letter excerpts relate to labeling and or packaging.

1. Your firm failed to establish written procedures to cover the receipt, identification, storage, handling, sampling, examination, and/or testing of labeling and packaging materials, as required by 21 CFR 211.122(a). Specifically, you do not have written procedures to assure the quality of the incoming packaging or labeling. [Reference: Form FDA 483, Observation 8]
2. There are no procedures for an evaluation of discrepancies found between the quantity of drug product finished and the quantity of labeling issued.
3. Under the Act, dietary supplements may be legally marketed with claims that they affect the structure or function of the body (structure/function claims) if certain requirements

are met. Section 403(r)(6)(B) of the Act [21 U.S.C. 343(r)(6)(B)] requires the manufacturer of a dietary supplement containing a "structure/function" claim in the product's labeling to have substantiation that the claim is truthful and not misleading.

4. During the inspection, our investigator collected copies of labeling for some of the products distributed by your firm. This inspection, the collection and review of your labeling, and a review of your Internet site, http://www.uaslabs.com, were conducted to determine your firm's compliance with the Federal Food, Drug, and Cosmetic Act (the Act) and applicable-implementing regulations under Title 21 of the Code of Federal Regulations (21 CFR).

5. You do not identify and handle filled drug product containers set aside and held unlabeled for future labeling operations to preclude mislabeling, as required under 21 CFR 211.130(b). For example, on August 3, 2005, our investigator noted several metal trays and cardboard boxes containing unlabeled glass vials filled with drug product stored in the processing areas.

6. All of your finished injectable drug products bear an 18 month expiration date which has not been determined by appropriate stability testing to assure these products meet applicable standards of identity, strength, quality, and purity at the time of use, as required under 21 CFR 211.137

7. Failure to establish written control procedures for the issuance of labeling [21 C.F.R. § 211.125(f)]

8. The expiration date for Cromolyn Sodium lot [redacted] was extended by 18 months based solely on tests for appearance, acidity/alkalinity, water content, and clarity of solution. The supplier reported in a November 2003 correspondence that the indicated tests to check the stability of Cromolyn Sodium include: 1) appearance, 2) potentiometric assay, 3) HPLC purity, 4) water content, and 5) clarity of solution. Your firm did not perform 2 of the 5 required tests (i.e., potentiometric assay and HPLC purity test).

9. Technical grade EDTA Disodium was repackaged and labeled as EDTA Disodium Salt, Dihydrate USP grade with a statement that it was for prescription compounding. The supplier's label stated that it was not for food or drug use. While your firm performed several USP tests for the product, you failed to perform the USP test for nitrilotriacetic acid.

10. Failure to list an expiration date on the product or provide adequate stability data supporting that the product is stable for at least 3 years.

11. Failure to maintain component, drug product container, closure, and labeling controls or records.

SUGGESTED READINGS

1. FDA, Current Good Manufacturing Practice in Manufacturing, Processing, Packing or Holding of Drugs; Revision of Certain Labeling Controls; Partial Extension of Compliance Date; Reopening of Administrative Record, Fed. Reg 1994; 59 (147):39255.
2. FDA, Tamper Evident Packaging Requirements for Over-The-Counter Human Drug Products, Fed. Reg 1994; 59 (11):2542.
3. Laws Affecting Packaging—An Overview—Eric F. Greenberg, Vol. 9, No. 2 (May, 1997) Food Drug Cosmetic and Medical Device Law Digest.
4. Washington Legal Foundation v. Friedman: a new era in off-label promotion? Phelps Wu, Samp, Attinson, FDCMD Law Journal, Vol. 15. No. 3, Sept. 1998
5. Health Law Digest, HLD, v. 29, No. 2, Feb 2001, page 31 "U.S. Court District of Columbia Finds Appellate Court Wholly Vacated Injunction Barring FDA Regulation of Freedom of Speech Involving Off Label Prescription Drug Use"
6. The Dietary Supplement Health and Education Act of 1994, Young and Stolzer, 12 Food, Drug, Cosmetic and Medical Device Law Digest 16 (1995), Volume 6 (1998), Young and Wiener.
7. 21 CFR, Parts 111 and 112; PART 111–Current Good Manufacturing Practice in Manufacturing, Packing, or Holding Dietary Ingredients And Dietary Supplements, effective May 25, 2005.

Following are FDA Compliance Policy Guides relevant to this chapter.

FDA COMPLIANCE POLICY GUIDES MANUAL

The FDA Compliance Policy Guides (CPG) Manual provides a convenient and organized system for statements of FDA compliance policy, including those statements, which contain regulatory action guidance information. The CPG Manual is the repository for all agency compliance policy that has been agreed to by the center(s) and the Associate Commissioner for Regulatory Affairs. Examples of sources from which CPGs are prepared include: a) statements or correspondence by headquarters offices or centers reflecting new policy or changes in compliance policy including Office of the Commissioner memoranda, center memoranda and other informational issuances, agency correspondence with trade groups and regulated industries, and advisory opinions; b) precedent court decisions; c) multicenter agreements regarding jurisdiction over FDA regulated products; d) preambles to proposed or final regulations or other Federal Register documents; and e) individual regulatory actions.

Compliance Policy Guides (CPG) explains the FDA policy on regulatory issues related to the FDA laws or regulations. These include CGMP regulations and application commitments. They advise the field inspection and compliance staffs as to the Agency's standards and procedures to be applied when determining industry compliance. CPG may derive from a request for an advisory opinion, from a petition from outside the Agency, or from a perceived need for a policy clarification by FDA personnel.

These CPG can be accessed on the Internet at http://www.fda.gov/ora/compliance_ref/cpg/. The general guides are part of the 400 series of guides. The label and repackaging guides comprise the 430 series. There have been no revisions to the 430 Series (Labeling and Repackaging) guides since 1995. The guides in each series are listed below.

Compliance Policy Guides
___

Sub Chapter 400 – General
400.100 Drugs, Human - Failure to Register CPG 7132.07
400.200 Consistent Application of CGMP Determinations CPG 7132.12
400.210 Radiofrequency Identification Feasibility Studies and Pilot Programs
400.325 Candy "Pills" Representation as Drug CPG 7132.04
400.335 Fructose-Containing Drugs CPG 7123b.02
400.400 Conditions Under Which Homeopathic Drugs May be Marketed CPG 7132.15
400.500 Identical or Similar Product Names CPG 7132b.14
400.600 Drugs - Declaration of Quantity of Active CPG 7132.03
400.700 Drug Product Entries in Periodic Publications CPG 7132b.17
400.800 Collection and Charitable Distribution of Drugs CPG 7132.08
400.900 Class I Recalls of Prescription Drugs CPG 7132.01

Sub Chapter 430 - Labeling and Repackaging
430.100 Unit Dose Labeling for Solid and Liquid Oral Dosage Forms CPG 7132b.10
430.200 Repacking of Drug Products - Testing/Examination under CGMPs CPG 7132.13
430.300 Labeling Shipping Containers of Drugs CPG 7132b.13
430.400 Urinary Preparations - Misbranding - Lack of Rx Legend and Claims CPG 7132b.04
___

Abbreviations: CPG, Compliance Policy Guides; CGMPs, Current Good Manufacturing Practice.

POLICY ON HOMEOPATHIC DRUG LABELING FROM FDA COMPLIANCE GUIDES SEC. 400.400 CONDITIONS UNDER WHICH HOMEOPATHIC DRUGS MAY BE MARKETED (CPG 7132.15). (REFER TO CHAPTER 2 FOR THE COMPLETE TEXT OF 400.400.)

Recalls due to Labeling Information

Information on FDA recall activities can be accessed on the Internet at http://www.fda.gov/opacom/7alerts.html.

More recent recalls have been related to the labeling and insert information

- FDA and IVAX Pharmaceuticals, Inc. notified healthcare professionals of a recall of Goldline brand Extra Strength Genapap 500 mg (Acetaminophen) Caplets and Tablets and Extra Strength Genebs 500 mg (Acetaminophen) Caplets and Tablets due to a labeling error. Specifically, the product label should indicate that usage should not exceed eight tablets or caplets in a 24-hour period. The erroneous label indicates not to exceed 12 tablets or caplets in a 24-hour period.
- The FDA notified healthcare professionals and patients that cases of breathing problems, some causing death, have been reported to the FDA when the drug was used in children less than two years old. Parents and caregivers should also be careful and get a doctor's advice about giving promethazine HCl in any form to children age two and older. The labeling on all products, brand name, and generic, has been changed to reflect these strengthened warnings.
- The FDA and Centocor notified healthcare professionals of revisions to the WARNINGS, ADVERSE REACTIONS sections and PATIENT PACKAGE INSERT of the prescribing information for Remicade, indicated for the treatment of rheumatoid arthritis, Crohn's disease and ankylosing spondylitis.

SUBCHAPTER SEC. 400.500 IDENTICAL OR SIMILAR PRODUCT NAMES (CPG 7132B. 14)

Background:

* Periodically different drugs, or drugs and other products, are marketed under identical brand names similar enough to cause confusion. It is apparent that a serious danger to health could exist if a relatively mild drug or other product was dispensed in the place of a vitally needed antibiotic or vice versa. Other situations equally serious could also be imagined. We investigated a complaint where a prescription drug was dispensed in the place of the prescribed vitamin with a similar name.

Policy:

All instances of drugs of different composition including different dosage strengths being marketed under identical or similar brand names are regarded as serious violations of the Act due to the inherent potential health hazards. Regulatory action will normally be authorized.*

Regulatory Action Guidance

The following represents criteria for recommending legal action to the * Division of Drug Labeling Compliance, HFD-310*.

1. Identical Brand Names for Drugs of Different Composition. Section 502(i)(3) of the Act specifies that a drug shall be deemed to be misbranded if it is offered for sale under the name of another drug. This charge is to be used when regulatory action is recommended in this situation.
2. Similar Brand Names for Drugs of Different Composition. Section 502(a) of the Act specifies that a drug shall be deemed to be misbranded if its labeling is false or misleading in any particular. This charge is to be used when regulatory action is recommended in this situation.

The initial action of choice where no direct health hazard is involved is a * warning * letter. Recall is the initial action of choice in situations involving a hazard to health.

* Material between asterisks is new or revised *
Issued: 10/1/80
Revised: 5/22/87, 3/95

SEC. 400.600 DRUGS—DECLARATION OF QUANTITY OF ACTIVE INGREDIENT BY BOTH METRIC AND APOTHECARY SYSTEMS (CPG 7132.03)

Background:

The USP and NF allow the simultaneous use of both the metric and apothecary systems to declare the quantity of active ingredients present in drug product labeling. Prior to USP XX and NF XV, the official compendia allowed the approximate equivalent of the exact quantity to be enclosed in parenthesis; such as Quinidine Sulfate 200 mg (3 grains).

On July 1, 1980 the USP XX and NF XV became official and requires that "Where expressed in both the metric and apothecary systems, statements of quantity or strength in the labeling of drug products shall utilize the exact equivalent." (See inside back cover USP XX and NF XV.) Therefore the example mentioned earlier would now have to be modified to read Quinidine Sulfate 200 mg (3.086 grains).

Policy:

USP and NF products shipped after 7/1/80 bearing a dual declaration will be considered misbranded if the exact equivalents are not used. However, as a general rule we are not prepared to initiate regulatory action of this violation alone. It may be included as a 502(g) charge only when other violations exist or it may serve as the basis for a Notice of Adverse Findings letter.
Issued: 10/1/80

This policy has not been revised since issuance.

SEC. 430.400 URINARY PREPARATIONS—MISBRANDING—LACK OF RX LEGEND AND CLAIMS (CPG 7132B.04)

Policy:

We are not prepared to take regulatory action against the following class of products, particularly those that have been on the market for a significant period of time:

> Products offered as urinary antiseptics, urinary analgesics, acidifiers, or diuretics that are botanical mixtures, botanical with sodium biphosphate, ammonium chloride, phenazopyridine hydrochloride, or these chemicals alone or in combination.

Generally, we would prefer not to initiate regulatory action on such products based on misbranding charges (lack of Rx legend or inadequate full disclosure) until our medical position has been clarified.

If the product contains ingredients, which may cause the drug to be dangerous to health when used as directed or if its labeling makes direct claims for more serious conditions, we might want to consider action. If you encounter such products which you believe warrant action, submit full labeling and formulation to Division of Drug Labeling Compliance, * HFD-310 * for advice before collecting samples for regulatory consideration.
* Material between asterisks is new or revised *
Issued: 10/1/80
Revised: 5/22/87, 3/95

SEC. 435.100 COMPRESSED MEDICAL GASES—*WARNING LETTERS FOR SPECIFIC VIOLATIONS COVERING LIQUID AND GASEOUS OXYGEN* (CPG 7132A. 16)

Background:

This CPG provides guidance for issuing warning letters to firms processing compressed medical gases in violation of the adulteration, misbranding, and/or new drug provisions of the Federal Food, Drug, and Cosmetic Act.

Compressed medical gases, *including compressed medical oxygen and liquid oxygen, * are drug products regulated under 21 CFR 210 and 211.

*Section 201.100 requires that the labeling for prescription drugs (e.g., Oxygen U.S.P.) bear adequate directions for use. In this regard, the requirements of 201.100 would be satisfied if the article meets the labeling requirements described in the Federal Register of March 16, 1972, (37 FR 5504) entitled "Oxygen and Its Delivery Systems, Proposed Statement of Policy." Although this proposal was not finalized and is being revoked, the Agency continues to use it as a labeling guideline for medicinal oxygen.

Oxygen, U.S.P. would be misbranded if its label fails to indicate whether or not it has been produced by the air-liquefaction process as required by the United States Pharmacopeia (USP XXII).

All other compressed medicinal gases should meet the requirements of 21 CFR 201.161.

All medical gas drug labels must bear the following information: (1) name and address of the manufacturer or distributor, (2) official product name (for single-component gases), (3) contents, in units of measure commonly used, e.g., liters, cubic feet, (4) lot number, and (5) statement of ingredients (for gas mixtures).*

The Center for Drug Evaluation and Research *(CDER) "Compressed Medical Gases Guideline" (revised February, 1989) provides guidance to industry for compliance with these regulations. Many firms that fill high-pressure cylinders with medical gases do not assay the finished product for identity and strength. In some instances, the testing performed is inadequate, due to failure to: 1) establish appropriate finished product test procedures and specifications, and/or 2) maintain the test apparatus, e.g., the United States Pharmacopeia Orsat test apparatus, or calibrate the oxygen analyzer according to the analyzer manufacturer's specifications.

Of particular concern is the transfilling (cascading) of smaller high pressure cylinders (e.g., E or D) from larger high pressure cylinders (e.g., H or K). The smaller high-pressure cylinders may be transfilled individually, or more than one may be filled at a time by connecting them to a manifold with multiple outlets. The firms may fail to perform any finished product testing and to vacuum evacuate or double purge each cylinder prior to filling. Many of the firms have not been registered and are unaware of their responsibility to comply with the CGMP regulations.

Home Respiratory Care:

The revised guideline (February 1989) includes guidance for home respiratory care companies (HRC). In practice, an HRC may pick up or receive a shipment of liquid oxygen (LOX) in a cryogenic container and transport the container to a patient's home to fill a cryogenic home unit. Another common practice is for the HRC to exchange a full unit with the empty home unit. The empty units are returned to the HRC facility for filling. In both scenarios, no testing of the cryogenic home units is required, provided all of the following three elements are met:

1. The incoming LOX has been adequately tested for identity and strength by one of the methods outlined in the guideline,
2. No other liquid (gas) is being transfilled on the premises, and
3. The cryogenic home units are retained by the HRC.

However, if any other liquid (gas) is being transfilled, then ALL cryogenic home units would require full USP testing.

Furthermore, in accordance with Section 211.87, the contents of units sent out for repair/maintenance must be retested at the least for identity, prior to redistribution.

Occasionally, an HRC will use a supplier's Certificate of Analysis (CoA) to reduce the amount of testing the HRC needs to perform. In this situation, the following minimum information should be provided in the CoA:

1. Supplier's name
2. Name of product (gas)
3. Air-liquefaction statement
4. Lot number or other unique identification number

5. Actual analytical results obtained for identity and strength
6. Test method used for analysis (Note: In lieu of indicating the test method on each CoA issued, a general letter from the supplier, maintained on file with the HRC, which indicates the test method used for all supplied product is acceptable)
7. Supplier's signature/date.

If the HRC relies on a CoA to reduce the amount of testing it needs to perform, the HRC should establish the reliability of the supplier's analysis at appropriate intervals (once a year would be sufficient) by taking a recently delivered and tested vessel to a third party for full USP testing.

Most medical gas firms receive their bulk oxygen from an intrastate source. If they receive LOX from an interstate source, document the interstate movement of the bulk LOX. If the LOX is from an intrastate source, document the interstate movement of 1) the stand tank, 2) the large cryogenic vessels, 3) the high pressure cylinders, and/or 4) the home units, by determining the location of the manufacturer of the articles.*

Regulatory Action Guidance:

District offices should consider issuing warning letters under any one of the following circumstances, provided interstate movement of the gas or the various containers (cryogenic or high pressure cylinders) has been documented:

* Material between asterisks is new or revised *
Issued: 11/5/87
Revised: 8/31/92

9 | Holding and Distribution
Subpart H

Joseph D. Nally
Nallianco LLC, New Vernon, New Jersey, U.S.A.

§211.142 WAREHOUSING PROCEDURES

Written procedures describing the warehousing of drug products shall be established and followed. They shall include

(a) quarantine of drug products before release by the quality control unit,
(b) storage of drug products under appropriate conditions of temperature, humidity, and light so that the identity, strength, quality, and purity of the drug products are not affected.

As for components, containers, and closures, quarantine of drug products does not necessarily require rigid physical separation except for hazardous or toxic materials. The degree of separation necessary is dependent upon other steps to ensure that quarantined product is not used prematurely. An effective paper or computer control system is acceptable. The emphasis must be on whether the system does prevent premature distribution. If it does, the system, no matter how nonrestrictive, is acceptable; if it does not, the system, no matter how rigorous the physical separation, is not acceptable.

The following are storage definitions, as defined in the General Notices section of the United States Pharmacopeia (USP) for recommended conditions commonly specified on product labels (1).

1. **Freezer**: A place in which the temperature is maintained thermostatically between -20 and $-10°C$ (-4 to $14°F$).
2. **Cold**: Any temperature not exceeding $8°C$ ($46°F$). A refrigerator is a cold place in which the temperature is maintained thermostatically between 2 and $8°C$ (36 to $46°C$).
3. **Cool**: Any temperature between 8 and $15°C$ (46 to $59°F$). An article that requires cool storage may be stored in a refrigerator, unless otherwise specified by the individual USP monograph.
4. **Room temperature**: The temperature prevailing in the workplace.
5. **Controlled room temperature**: A temperature maintained thermostatically that encompasses the usual and customary working environment of 20 to $25°C$ (68 to $77°F$) that allows for brief deviations between 15 and $30°C$ (59 to $86°F$) that are experienced in pharmacies, hospitals, and warehouses. Articles may be labeled for storage at "controlled room temperature" or at "up to $25°C$" or other wording. An article for which storage at *Controlled room temperature* is directed, may, alternatively, be stored in a cool place unless otherwise specified in the individual monograph or on the label. (See the entire revised definition of *Controlled room temperature* in the *Ninth Supplement* to *USP XXII-NF XVII*.)
6. **Warm**: Any temperature between 30 and $40°C$ (86 to $104°F$).
7. **Excessive heat**: Any temperature above $40°C$ ($104°F$).
8. **Protection from freezing**: Where, in addition to the risk of breakage of the container, freezing subjects an article to loss of strength or potency, or to destructive alteration of its characteristics, the container label must bear an appropriate instruction to protect the article from freezing.

Aside from the special storage conditions (refrigeration, freezer etc.) there are typically two types of warehouse holding and storage conditions for room temperature storage.

1. **Ambient room temperature conditions**: For some products, an adequate shelflife can be determined that encompasses the relatively wide range of conditions that constitute "ambient." Local and national weather records are available, which provide data on temperature ranges. Maintenance of actual temperature data in a warehouse provides the assurance that the assumptions made in determining shelf-life continue to be met. Humidity and light are rarely controlled, or even monitored, since product packaging is usually designed to take these two conditions into account.
2. **Controlled room temperature**: Storage areas may be similar to the ambient conditions depending on the location and prevailing weather. These areas are typically monitored for temperature (and in some cases humidity) and most likely controlled with a heating ventilation and air-conditioning system to maintain the required temperatures. Monitoring can be daily or periodic and should represent the storage areas in-use (e.g., low to high racks).

The FDA has continued to insist that all products should be labeled with defined storage conditions. This puts the onus on the manufacturer to assure the product is stored as labeled as long as the product is in their control.

A potential problem associated with the declaration of "controlled room temperature" storage is that many warehouses do not have adequate heating and ventilation systems to maintain such conditions; on occasions the temperature may exceed 30°C for short periods. Unless warehouses expend considerable money to introduce cooling systems to accommodate these relatively infrequent occurrences, it is possible for them to be cited by the FDA for non-compliance. It is hoped that the availability of warehouse climatic data and suitable stability data will be an adequate alternative.

§211.150 DISTRIBUTION PROCEDURES

Written procedures shall be established, and followed, describing the distribution of drug products. They shall include the following:

(a) A procedure whereby the oldest approved stock of a drug product is distributed first. Deviation from this requirement is permitted if such deviation is temporary and appropriate.
(b) A system by which the distribution of each lot of drug product can be readily determined to facilitate its recall if necessary.

Distribution records must be constructed and procedures established to facilitate recall of defective product. A requisite of the system is approval and specific release of each lot of drug by the quality control function before distribution can occur. This control of finished goods for shipment allows only those drugs into commerce that have been shown by testing to conform to appropriate requirements.

The manufacturer must maintain records of all distribution transactions involving in-process or finished goods. All records should be indexed by either the manufacturing batch-lot number of the packaging control number as a means of accountability until the shipment passes from the direct control of the manufacturer. This type of indexing permits an efficient determination of the receiver of a lot to be recalled since only one shipment record need be examined. Depending on the marketing procedures of the individual company, distribution records may list shipments to consignees for packaging or labeling, or to an independent distributor, a wholesaler, a retail pharmacist, a physician, or possibly the ultimate consumer.

A variety of distribution recording systems may be utilized. Computerized tracking systems are most common but paper systems such as recording the lot or control number on the retained copies of the shipping invoices or recording the dates on which each lot commenced distribution are also used. This latter approach has disadvantages in that it does not readily accommodate the redistribution of small amounts of returned goods or the occasional need to distribute part lots out of sequence.

Many U.S. companies also distribute products to their foreign affiliates. The distribution records should also include these transactions. This can become complicated if distribution from the U.S.A. is to a central international distribution center and the U.S. operation has no records of the final distribution. In these situations, the U.S. Quality Assurance (QA) function should evaluate and audit the central international distribution center operation and confirm the adequacy of its systems and controls.

The distribution process also includes other considerations. It must be arranged so that a first in/first out movement of product occurs. This requirement is consistent with the intent of the stability and expiration dating policy. The distribution system must include provisions in order that this movement is achieved. Exceptions to this requirement that may be permitted should be described in written procedures.

All distribution records should be maintained for a minimum 3-year period after the distribution process for any control number has been completed. If expiration dating is used for a product, distribution records must be maintained at least for one year past the expiration date of the product [§211.180(b)].

COLD CHAIN DISTRIBUTION

An example of strictly controlled distribution is a "cold chain." "The cold chain can be defined as the channels of distribution, handling and storage of a temperature sensitive product (e.g., products requiring storage between 2°C and 8°C) from the time it is manufactured to the time it is consumed by the end user (2)."

For temperature sensitive products such as vaccines, biologics, and some parenteral solutions, cold chain distribution has become a practice that assures the product strength, purity, potency, and efficacy for the end user.

Approximately 10% of the total pharmaceutical worldwide sales involved biopharmaceuticals, which are typically more temperature sensitive (3).

Maintaining a complete and controlled chain of custody is no simple task and faces many challenges which include:

1. Design and qualification of the immediate packaging and packing,
2. Normal and excursion stability data,
3. Temperature monitoring and data analysis, and
4. Feedback loops to the distribution chain.

Auditing of Manufacturers/Packagers and Distributors

The design and qualification of the immediate and shipping containers must include considerations for seasonal variations in expected temperatures and reasonable extremes. Stability studies need to consider and provide data for the inevitable excursions during storage/distribution.

The technology exists for continuous temperature monitoring during distribution (data are loggers). As long as the data are within ranges there are no issues. When excursions occur, the stability data are key in the justification (or lack) of continued product quality.

Communication back to the carriers and distributors (mainly on excursions) is the most effective tool for resolution and prevention. Auditing the cold chain participants for understanding and consistency helps assure the overall system.

COUNTERFEITING

Since the fifth edition of this book was published, counterfeiting of drug products has become visible as a major issue. Current estimates state as much as 8% of the world's prescriptions are reportedly counterfeit. This has prompted the FDA to call on pharmaceutical manufacturers to start taking larger steps towards security, supply chain management, and more specifically to implement "track and trace" technology (4).

As we go to publication, drug distributors will be required to provide pedigree, or chain of custody, product documentation after December, 2006. The FDA defines a *drug pedigree* as "a statement of origin that identifies each prior sale, purchase, or trade of a drug, including the date of those transactions and the names and addresses of all parties to them. Under the pedigree requirement, each person who is engaged in the wholesale distribution of a prescription drug in interstate commerce, who is not the manufacturer or an authorized distributor of record for the drug, must provide the person who receives the drug a pedigree for that drug." "The information a drug pedigree should contain is codified in 21 CFR 203.50 but FDA does not have the statutory authority to implement a uniform requirement that would supersede state requirements (5)."

There is a draft *Compliance Policy Guide 160.900* that identifies initial proposed enforcement for:

1. High value products (price, volume, demand),
2. Prior history of counterfeiting or diversion,
3. Reasonable probability that a new drug will be counterfeited, and
4. Products in violation of PDMA or other laws.

Needless to say, the counterfeiting potential has generated a great need for additional documentation and the technology to handle the issue. The leader in track and trace technology is Radio Frequency Identification (RFID). RFID are tracking devices that send specific radio frequency signals to satellites so the exact location and trail of distribution or custody can be traced and documented. The technology is being successfully used but also has potential problems including choice of radio frequency, the evolving standards, and the physical impact to sensitive products such as biologics.

EXAMPLES OF OBSERVATIONS FROM FDA 483 CITATIONS

1. No procedures available describing distribution of oldest stock first or any record of batch numbers entering distribution.
2. Products requiring specific storage conditions, 59 to 86°F, were stored in a non-air-conditioned warehouse at 90°F.
3. No defined quarantine area for incoming finished drug products to be repacked.

SUGGESTED READINGS

1. The World Health Organization (WHO) working document QAS/04.068, Good Distribution Practices.
2. Draft *Compliance Policy Guide 160.900*.
3. CFR section 203.50.
4. PDA Technical Report 39, Cold Chain Guidance for Medical Products: Maintaining the Quality of Temperature Sensitive Medical Products through the Transportation Environment, 2005.

REFERENCES

1. USP Quality Review No. 40 Revised 6/94. http://usp.org/patientSafety/newsletter/quality/Review/qr401994-06-01c.html, accessed Aug 25, 2006.
2. Thompson Z. Validating, qualifying and ensuring cold chain packaging and distribution. American Pharmaceutical Outsourcing 2005.
3. Bishara RH. Cold chain management—an essential component of the global pharmaceutical supply chain. Am. Pharm. Rev. Jan/Feb, 2006.

4. IBM Develops RFID System for Pharmaceutical Track and Trace. Pharmaceutical Technology Aug 24, 2006. http://www.pharmtech.com/pharmtech/article/articalDetail.jsp?id=367481, accessed Aug 25, 2006.
5. Cracking down on counterfeiting. Pharmaceutical Technology Aug 2, 2006. http://www.pharmtech.com/pharmtech/article/articleDetail.jsp?=id361496, accessed Aug 25, 2006.

10 | Laboratory Controls
Subpart I

Wayne J. DeWitte
W. J. DeWitte Consulting, Suffern, New York, U.S.A.

Alex M. Hoinowski
Quantic Group Ltd., Livingston, New Jersey, U.S.A.

§211.160 GENERAL REQUIREMENTS

(a) The establishment of any specifications, standards, sampling plans, test procedures, or other laboratory control mechanisms required by this subpart, including any change in such specifications, standards, sampling plans, test procedures, or other laboratory control mechanisms, shall be drafted by the appropriate organizational unit and reviewed and approved by the quality control (QC) unit. The requirements in this subpart shall be followed and shall be documented at the time of performance. Any deviation from the written specifications, standards, sampling plans, test procedures, or other laboratory control mechanisms shall be recorded and justified.

In some instances, official compendia such as the United States Pharmacopeia (USP) dictate specifications and test methods. Although these compendial procedures are the final arbiters for drugs and components listed in them, a manufacturer is allowed to substitute other procedures or automated equipment, provided the results are consistent with those obtained using the official methods and method bias is demonstrated to be absent. In the event of a dispute or query, the official methods are to be applied.

Any compendial product must comply with the compendial specifications unless the noncompliant parameter is clearly stated on the label (Compliance Policy Guide 1732a.03). The USP monographs provide useful guidance on the typical contents of monographs for drug substances, excipients and dosage forms. Recently there has been increased emphasis from both USP and FDA to provide more details of impurities in bulk drug substances. This includes both the expected impurities from the synthesis and degradation of the bulk drug (usually limited to 2% total) and with the main impurities identified and volatile solvent residues.

Where compendial specifications and methods are not available, the manufacturers must develop their own based on current scientific practices.

Material and product specifications and test methods for new products are often generated by the Research and Development (R&D) department. They must however ultimately be approved by the quality unit before commercial implementation. FDA reviewing chemists are becoming increasingly critical of proposed specifications that are wider than the results seen in development batches. The reason for this is obvious—if the toxicology and clinical data were generated on batches with narrower specifications, there may be no justification for wider ranges. This can create a dilemma for industry, since the earlier batches (for toxicology and clinical studies) may have been small-scale batches produced by R&D chemists. Later, full-scale production may involve different equipment, operators rather than researchers and different sources of some materials. It may be impossible, due to time and financial constraints, to perform this early evaluation work on a commercial scale. Also, the processes are still under development. Consequently, if commercial specifications need to be wider than those seen during development, supporting data with justification for the wider specification will be required.

International Conference on Harmonisation (ICH) has provided guidance on specifications for impurities in new drug substances [ICH Q3A(R1), 2/2002] and new drug products [ICH Q3B(R2), 6/2006]. Impurities were classified under several headings:

1. *Organic impurities*—actual or potential impurities likely to arise during synthesis, purification, or storage. Sources include starting materials, reagents, by-products, intermediates and degradants. Impurities present at above 0.1% are to be characterized. Below 0.1%, characterization is not expected unless there is reason to expect undue toxicity.
2. *Inorganic impurities*—reagents, catalysts, heavy metals, charcoal and filter aids. These are usually evaluated by compendial methods and apply compendial limits.
3. *Solvents remaining from the process*—tests and limits are usually those included in the compendia. For other solvents, toxicity should be taken into account in defining appropriate limits.

Change control systems must be designed into the approval procedures, thereby ensuring that no changes are made without quality unit review and approval.

The requirement to document any act at the time of performance precludes the use of intermediate or temporary recording of data such as weighing into notebooks that are discarded after transcription of the information into the formal system. Such intermediate records are acceptable if retained. However, wherever possible, data should be recorded directly into the final format, eliminating the possibility for transcription errors.

Any deviations from written procedures must be recorded and be properly evaluated. The reason for the deviation should be identified and it should be determined whether it will have any adverse impact on the drug product. Approval for any deviation should be by a suitably qualified individual—usually a supervisor or manager.

(b) Laboratory controls shall include the establishment of scientifically sound and appropriate specifications, standards, sampling plans, and test procedures designed to assure that components, drug product containers, closures, in-process materials, labeling, and drug products conform to appropriate standards of identity, strength, quality, and purity. Laboratory controls shall include:
   (1) Determination of conformance to appropriate written specifications for the acceptance of each lot within each shipment of components, drug product containers, closures, and labeling used in the manufacture, processing, packing, or holding of drug products. The specifications shall include a description of the sampling and testing procedures used. Samples shall be representative and adequately identified. Such procedures shall also require appropriate retesting of any component, drug product container, or closure that is subject to deterioration.
   (2) Determination of conformance to written specifications and a description of sampling and testing procedures for in-process materials. Such samples shall be representative and properly identified.
   (3) Determination of conformance to written descriptions of sampling procedures and appropriate specifications for drug products. Such samples shall be representative and properly identified.

Whereas subsection (a) deals with the drafting and approval of any specifications, standards, sampling plans and test procedures, subsection (b) applies to the application of these to components, containers, closures, in-process materials, labeling and drug products.

The procedures are to be scientifically sound and appropriate. Where possible, established specifications and test methodology such as USP will be applied. Otherwise a knowledge of the composition, potential impurities (synthesis intermediates, solvent residues, heavy metals, etc.) and degradation products should be taken into account. The specifications should be designed to control any such impurities within acceptable levels and to monitor trends. As indicated previously, the application of "action levels" that are based on historical data and are more rigorous than the specifications is a useful and practical way of highlighting adverse trends and bringing them to the attention of QC management.

A common practice is to set the action levels such that 95% of all acceptable results will fall within these levels; the exceptional 5% will then be highlighted. In-process control action levels for physical parameters such as tablet weight or fill volumes are more usually approached by way of control charts.

For a new product, with no available historic data, control levels may be calculated from the USP criteria for "Uniformity of Dosage Units." In-process testing involves weighing a composite of 10 tablets at a determined frequency, often every 15 minutes.

The test methods used may vary for different applications. For example, in-process test methods performed in production areas by production personnel may need to be more robust than those performed by QC laboratory personnel; nonavailability of equipment or servicing may result in the use of different methods in different countries.

The effectiveness of test methodology is further dependent on two additional factors. First, the methods must be written in sufficient detail that no interpretation is necessary. If there is any doubt, then query should be raised with QC management and the procedure should be rewritten. Second, only trained individuals must be allowed to perform testing.

Sampling requirements may also vary with component or product history. Comments on reduced testing and supplier validation were included earlier (Chap. 6, §211.84).

The regulations also require retesting of components, closures, or containers that may be prone to deterioration. Testing should be restricted to evaluation of parameters known or expected to change during storage. This subject was addressed in Chapter 6 (§211.87).

Confirmation of conformance to specifications usually involves two groups within the quality unit. The laboratory function (QC) is responsible for sampling and testing, whereas a quality assurance unit reviews the resulting data and conclusions. This review by quality assurance forms part of the overall batch review procedure.

> (4) The calibration of instruments, apparatus, gauges, and recording devices at suitable intervals in accordance with an established written program containing specific directions, schedules, limits for accuracy and precision, and provisions for remedial action in the event accuracy and/or precision limits are not met. Instruments, apparatus, gauges, and recording devices not meeting established specifications shall not be used.

Equipment qualification and calibration are critical aspects of test methodology. Laboratory equipment and software must be qualified to demonstrate that the equipment performs as expected and required. Calibration programs should define who is responsible for the calibration, the frequency, how the calibration is to be performed and action to be taken if the equipment is found to be outside acceptable ranges. Refer to the chapter 5 "Maintenance and Calibration" section.

In some instances calibration programs are contracted out to third parties. The responsibility for calibration must still reside within the manufacturer's organization. The manufacturer must approve the calibration procedure, the acceptance criteria and the frequency. Calibration results must be recorded; it is not sufficient to report that the equipment is acceptable.

Frequency is usually determined on the basis of the equipment manufacturer's recommendation, experience and past performance. Any equipment found to be outside of acceptable operational ranges must be taken out of service until it is returned to normal performance. Additionally, the potential impact of such equipment on testing performed since the previous calibration needs to be evaluated. The potential implications of this are extensive. For example, an analytical balance that is serviced and calibrated every three months, that is found to be significantly inaccurate, might place in jeopardy some of the analytical results generated since the previous calibration. The rechecking of all analytical results would involve a significant amount of work. However, the recording of the specific pieces of equipment used for production and testing (§211.105) will narrow the field. To avoid or at least minimize the possibility outlined in the example above, it is usual to perform more frequent performance checks or verifications (such as daily performance checks on balances). Although less comprehensive, and not adequate as calibrations, these do provide a high degree of assurance that the equipment is performing satisfactorily. Additional calibrations should be initiated if there is reason to suspect that equipment may not be performing satisfactorily.

§211.165 TESTING AND RELEASE FOR DISTRIBUTION

(a) For each batch of drug product, there shall be appropriate laboratory determination of satisfactory conformance to final specifications for the drug product, including the identity and strength of each active ingredient, prior to release. Where sterility and/ or pyrogen testing are conducted on specific batches of short-lived radiopharmaceuticals, such batches may be released prior to completion of sterility and/or pyrogen testing, provided such testing is completed as soon as possible.

(b) There shall be appropriate laboratory testing, as necessary, of each batch of drug product required to be free of objectionable microorganisms.

As a result of a court ruling against Barr Laboratories in 1993, FDA investigators increased their level of attention to laboratory operations. The main points relating to laboratory operations noted from the Barr case included:

- Inadequate evaluation of the cause of out-of-specification results,
- Use of an outlier test to discount failing results,
- Lack of a defined procedure to evaluate out-of-specification results,
- Long delays in completion of the evaluation of failures—a maximum of 30 days was proposed,
- Frequency of failures (product history),
- Use of different samples to reevaluate a failing result, especially for content uniformity and dissolution,
- Averaging of results, for example, three assay values of 89, 90 and 91 cannot be averaged to allow release of a product with a 90% minimum assay specification,
- Samples taken from blends were too large, they should not be more than three times the active ingredient dosage size,
- Sampling blends after they have been transferred to drums is not an acceptable alternative to sampling from the blender,
- For retrospective validation, all batches produced in the designated time frame must be included unless there was a reason to exclude based on non-process-related error,
- Ideally, 20 to 30 batches should be evaluated for retrospective validation,
- All retrospective validation batches must be made by the same process,
- Concurrent and prospective validation requires at least three batches, and
- Particle size distribution specifications should be included in validation studies.

The regulations require confirmation of conformance of drug products to specifications prior to release. Identity testing and assay of active ingredients by the QC laboratory is specifically required. In-process data from production personnel may be acceptable for most other parameters, provided operators have been properly trained, have adequate equipment and performance is audited. The availability of process validation data and process control data does not eliminate the need for finished product testing. This is somewhat at variance with the acceptance of parametric release (ICH Q6A) as an alternative to sterility testing for terminally sterilized products (details presented in §211.167). However, for sterility testing to be statistically valid, the sample size would be impracticable with a high probability of obtaining false-positive results.

An exception is made to the testing before release requirement for the sterility and pyrogen testing of short-lived radiopharmaceuticals. Since the test time may be a significant part of the product shelf-life, release prior to completion of testing is allowed. Obviously, in such instances, the process should be thoroughly validated and controlled to minimize the chance of a sterility or pyrogen failure.

The need to test each batch of product required to be free of objectionable microorganisms applies to both sterile products and to those products where specific organisms are to be absent (e.g., absence of *P. aeruginosa* and *S. aureus* in topical products). Products covered by this requirement include terminally sterilized products and aseptically processed products such as injections, and products produced under clean and hygienic conditions to exclude specific organisms and/or to minimize the level of microorganisms. The processing conditions for these products have been described earlier (§211.113).

The effective microbiological control of nonsterile products, where required, will usually include evaluation of levels of total microbial content, absence of specified organisms, presence of adequate levels of any added antimicrobial agent or preservative and review of the environmental data generated during the process. The subject of sterility testing is addressed in §211.167.

(c) Any sampling and testing plans shall be described in written procedures that shall include the method of sampling and the number of units per batch to be tested; such written procedure shall be followed.

(d) Acceptance criteria for the sampling and testing conducted by the QC unit shall be adequate to assure that batches of drug products meet each appropriate specification and appropriate statistical QC criteria as a condition for their approval and release. The statistical QC criteria shall include appropriate acceptance levels and/or appropriate rejection levels.

The need for written sampling and testing plans and definition of acceptance criteria is basic. Appropriate "action levels" should be built into the acceptance criteria. As emphasized elsewhere, it is important that any atypical situations are brought to the attention of sufficiently senior people so that appropriate actions can be initiated.

Where testing is delegated to production personnel, there should be adequate supporting data to demonstrate that the personnel were adequately trained, that equipment is suitable and properly maintained and calibrated and that the results obtained are equivalent to those obtained by QC. Audit programs should be in place to confirm these points.

(e) The accuracy, sensitivity, specificity, and reproducibility of test methods employed by the firm shall be established and documented. Such validation and documentation may be accomplished in accordance with §211.194(a)(2).

Analytical testing plays a key role in the overall control of product quality, and consequently it is imperative that the methodology used should provide accurate and reliable results. Although this subsection specifically applies to drug products, it is obvious that method validation should be applied to all methods, wherever used—components, in-process, process validation, finished product release and stability.

Analytical Validation

The subject of analytical validation has been covered in numerous publications over the years. However, this section will primarily focus on the approach described in ICH Q2A, ICH Q2B, FDA Draft Guidance (8/2000) and current USP/NF.

The first step in analytical validation is to ensure that the analytical method is defined in detail and includes any specific instructions or precautions (including safety) that are required to enable different trained analysts to perform the method reproducibly.

Four different analytical applications were reviewed—identification tests, quantification of impurities, limit tests for impurities and assay of actives or other key components of drug products. It was acknowledged that there are other important analytical procedures, including dissolution testing for drug products and particle size characterization of materials, but they were not addressed at this time.

Identification tests normally compare the sample under evaluation with a known reference standard. The methods are frequently spectrophotometric (IR/UV) or chromatographic, but some older methods involve chemical tests such as functional group analysis procedures

TABLE 1 ICH Validation Guideline

Type of analytical procedure; characteristics	Identification	Testing for impurities		Assay; content/potency (dissolution; measurement only)
		Quantitation	Limit	
Accuracy	−	+	−	+
Precision				
Repeatability	−	+	−	+
Intermediate precision	−	+[a]	−	+[a]
Reproducibility[b,c]	−	−	−	−
Specificity[d]	+	+	+	+
Detection limit	−	+[b]	+	−
Quantitation limit	−	+	−	−
Linearity	−	+	−	+
Range	−	+	−	+

Note: − signifies that this parameter is not normally evaluated; + signifies that this parameter is normally evaluated.
[a]In cases where reproducibility has been performed, intermediate precision is not needed.
[b]May be needed in some cases.
[c]Not listed in ICH Q2A Table.
[d]Lack of specificity of one analytical procedure could be compensated by other supporting analytical procedure(s).

(e.g., derivatization). Validation of identification tests is essentially confirmation of specificity (Table 1).

Impurity tests may be either quantitative or limit tests, and different validation requirements apply. For limit tests, validation of specificity and the limit of detection (LOD) may only be required. For quantification, the requirements are similar to those for assay methods (including those used for dissolution and content uniformity) except that for assay methods, the limit of quantitation does not need to be established since the methods are operating well in excess of these limits. Some additional details on specific characteristics are provided here.

- *Accuracy*—defines the agreement between the true value and the value found in the testing.
- *Precision*—defines the degree of variability in a series of measurements from multiple testing of the same homogeneous sample. Precision is usually expressed as a standard deviation or coefficient of variation (relative standard deviation). Three levels of evaluation of precision are defined:

(a) Repeatability—relating to testing performed over a short time interval on the same sample(s) by the same analyst on the same instrumentation.
(b) Intermediate precision—evaluations performed on different days with different analysts and possibly different equipment.
(c) Reproducibility—relates to collaborative studies between laboratories. This evaluation is a measure of the ruggedness of the method since many variables are involved—different facilities, different equipment, different analysts, and different reagents. This is a key element in analytical method transfer and confirmation that a new laboratory (e.g., QC laboratory) will obtain equivalent results to the originator laboratory (e.g., R&D).

- *Specificity*—confirms the ability of the method to evaluate the desired analyte in the presence of other known components: degradants, impurities, potential contaminants and excipients. Frequently this is assessed by comparing results from "normal" material with those from stressed samples (heat, light, moisture, acid and/or base).
- *Detection limit*—the lowest amount of analyte in a sample that can be detected but not necessarily quantitated as an exact value.
- *Quantitation limit*—relates to the lowest level of analyte that can be determined quantitatively in a sample with suitable accuracy and precision.
- *Linearity*—applies only to methods involving quantification and involves the demonstration of a linear response bracketing the range being evaluated. For example, an assay method

may be evaluated only over the range of 50 to 150% of the specification since any results outside of these values would be out of specification. Linearity is performed using standards in the absence of the sample and in this aspect differs from the Range.

■ *Range*—is established by confirming the analytical procedure and provides an acceptable degree of linearity, accuracy and precision when applied to samples containing analyte within or at the extremes of the specification. The following minimum specified ranges bracketing the specification should be considered as defined by ICH Q2B:
  ■ For assay of a drug substance or a finished drug product: 80 to 120%.
  ■ For content uniformity: 70 to 130%.
  ■ All other, consult Q2B and bracket the specification.

■ *Robustness*—though not a specific requirement in the ICH approach, it is an additional measure of the reliability of the method when normal variabilities in the product or method are experienced. Product variables can include excipient levels, pH ranges for liquids, hardness of tablets (potential impact on dissolution). Analytical method variables could include extraction process, sample preparation, HPLC flow rate, wavelength and mobile phase composition. The potential impact of these variables may be examined using a matrix design approach.

ICH Q2(R1), "Validation of Analytical Procedures: Text and Methodology" provides guidance on how to perform the actual testing of the validation parameters (e.g., linearity, specificity, LOD, etc.).

The validation and method transfer (technology transfer) protocols should include acceptance criteria and be approved. Any discrepancies from the agreed acceptance criteria need to be evaluated and explained.

Revalidation should be considered when there are changes to the analytical methods, equipment or to the product formulation and/or manufacturing process. For all analytical methods, new analysts need to be trained appropriately.

Validation and verification data must be reviewed and approved by responsible persons. In the case of new methods, it is advisable to have sign-off by both the method development unit (R&D) and the method receiving unit (QC), as well as a final approval by Quality Assurance. This acknowledges that the method has been validated and transferred effectively.

> (f) Drug products failing to meet established standards or specifications and any other relevant QC criteria shall be rejected. Reprocessing may be performed. Prior to acceptance and use, reprocessed material must meet appropriate standards, specifications, and any other relevant criteria.

Failure to meet specifications or noncompliance with the approved process should result immediately in quarantine of the material until the cause of the event is ascertained. Since products are in a quarantine status until released or rejected, it may be appropriate to retain this quarantine status, particularly if it is expected that the problem can be resolved quickly. The approach to be used in evaluating an out-of-specification result should be clearly defined in an SOP. As indicated earlier, this area was one of the main issues in the Barr Laboratories case.

In October 2006 the FDA formalized the 1998 Draft guidance on Investigating Out-of-Specification (OOS) Test Results for Pharmaceutical Products and issued Guidance for Industry Investigating Out-of-Specification (OOS) Test Results for Pharmaceutical Products. Differences between the 1998 Draft and the 2006 Final Guidance can be summarized as:

■ Scope is expanded to include:
  ■ Raw materials, in-process materials and stability testing rather than just API and finished drug testing
  ■ Contract laboratories
■ Introduction of new terminology: phase I and phase II investigations:
  ■ Phase I: Laboratory Investigation with delineation of the responsibilities of the analyst and the laboratory supervisor

- Phase II: Full-Scale OOS Investigation with guidance on:
  1. The review of production and sampling procedures
  2. Additional laboratory testing (retesting and resampling)
- Clearer definition of and emphasis on the responsibilities of the Quality Control Unit for full investigation and evaluation of investigation results
- Additional detailed examples for averaging of test results and resampling
- Description of the steps of the procedure:
  - The failure investigation
  - Identifying root cause
  - Identifying impact (severity and impact on other batches including those already distributed)
  - Evaluating the need for corrective and preventive actions
  - Implementing the corrective action(s) and preventive action(s), as appropriate
  - Verifying effectiveness of the corrective action(s) and preventive action(s)
- Discussions on Field Alerts for impacted marketed products

With the above in mind, for laboratory testing noncompliance situations, the following guidance is provided.

There are typically three aspects to a noncompliance situation:

The first thing to evaluate is the validity of the actual analytical result. During the testing process an obvious, recognized error may occur that invalidates the results. Such errors are caused by the analyst or by instrument failure and do not require the performance of a formal laboratory investigation, whether or not data have been generated. When recognized, testing should be stopped, the supervisor notified, and the data invalidated.

Analysts detecting an out-of-specification result must report that result promptly to the supervisor. The analyst and supervisor then immediately conduct a review for obvious errors. If no obvious error is revealed, the out-of-specification result is reported to the quality assurance unit with information on the impacted or potentially impacted lots. The analyst and the supervisor then initiate and conduct phase 1 of the laboratory investigation (see CDER guidance document) to determine whether or not the out-of-specification result is assignable to the testing laboratory. Checklists of possible sources of error are typically used to aid this part of the investigation.

Second, how did the problem arise and what was the underlying cause. If the cause of the out-of-specification result can be assigned to the testing laboratory, then the original sample is tested again. If the result of the repeat test meets the acceptance criterion, then the original out-of-specification result is invalidated. The root cause/most probable cause for the out-of-specification result is documented, and the effect of the finding on other data is determined. Corrective actions and preventive actions (CAPA) that may be applicable should be determined and taken to prevent recurrence of this source of error. The investigation then is closed, but the effectiveness of the CAPA monitored.

If the phase I investigation determines that the out-of-specification result cannot be assigned to the testing laboratory, then a phase II full-scale OOS investigation with review of production and sampling procedures by the Quality Control Unit is initiated (as applicable) with additional laboratory testing, retesting and resampling if justified and applicable (see CDER 2006 guidance document).

The rationale for additional planned laboratory investigation should be documented. This should include the tests that will be conducted, justification for retesting or resampling if planned, the number of samples to be tested and the criteria for evaluation of the data. The Quality Contol Unit (quality assurance) should be involved in defining the retest/resample protocol and in determining whether or not retesting is justified and when and how it is conducted and concluded.

The third aspect is the future of the specific batch in question. The 2006 CDER guidance states that the QCU is responsible for interpreting the results of the investigation. An initial OOS result does not necessarily mean the subject batch fails and must be rejected. The findings of the investigation, including retest results, should be interpreted to evaluate the batch and reach a decision regarding release or rejection (§211.165).

For confirmed OOS, the batch does not meet established standards or specifications and should result in the batch's rejection, in accordance with §211.165(f). This also indicates a batch failure and the investigation must be extended to other batches or products that may have been associated with the specific failure (§211.192). The CDER guidance discusses actions for inconclusive investigations.

If reprocessing is viable, it must be done according to written and approved instructions. Where NDA/ANDA products are involved, the reprocessing should be in conformance with the approved NDA/ANDA methods. The reprocessed product must meet all of the product specifications. Additional data may also be required to confirm that the product will behave in a similar manner to a typical batch. Such additional data could include accelerated stability and ingredient degradation evaluation, which might be included in a stability monograph but not in a release monograph.

Analyst training is not specifically referenced in §211.165. However, its importance is obvious. Training, which must be recorded, should include:

■ basic analytical techniques,
■ specific methods where these are complex,
■ proper laboratory practices and relevant SOPs,
■ laboratory safety,
■ new methods transferred from R&D (method transfer), and
■ retraining for analysts whose results are atypical.

§211.166 STABILITY TESTING

(a) There shall be a written testing program designed to assess the stability characteristics of drug products. The results of such stability testing shall be used in determining appropriate storage conditions and expiration dates. The written program shall be followed and shall include:
   (1) Sample size and test intervals based on statistical criteria for each attribute examined to assure valid estimates of stability.
   (2) Storage conditions for samples retained for testing.
   (3) Reliable, meaningful, and specific test methods.
   (4) Testing of the drug product in the same container–closure system as that in which the drug product is marketed.
   (5) Testing of drug products for reconstitution at the time of dispensing (as directed in the labeling) as well as after they are reconstituted.
(b) An adequate number of batches of each drug product shall be tested to determine an appropriate expiration date and a record of such data shall be maintained. Accelerated studies, combined with basic stability information on the components, drug products, and container–closure system, may be used to support tentative expiration dates provided full shelf-life studies are not available and are being conducted. Where data from accelerated studies are used to project a tentative expiration date that is beyond a date supported by actual shelf-life studies, there must be stability studies conducted, including drug product testing at appropriate intervals, until the tentative expiration date is verified or the appropriate expiration date determined.
(c) For homeopathic drug products, the requirements of this section are as follows:
   (1) There shall be a written assessment of stability based at least on testing or examination of the drug product for compatibility of the ingredients, and based on marketing experience with the drug product to indicate that there is no degradation of the product for the normal or expected period of use.
   (2) Evaluation of stability shall be based on the same container–closure system in which the drug product is being marketed.
(d) Allergenic extracts that are labeled "No U.S. Standard of Potency" are exempted from the requirements of this section.

The purposes of stability studies are to predict and confirm product shelf-life under the climatic conditions expected during trade storage, shipping, house storage and use. Refer to Appendix B for the latest list of ICH Guidelines involving stability testing.

Before commencement of a stability evaluation, the stability protocol must be written and approved—usually by technical services and QA. The key elements of a stability protocol include:

1. *Product name and packaging details.* The information should be sufficiently detailed to clearly identify the specific formulation to be evaluated, the specific container–closure type (and source), and the batch size.

2. *Storage conditions.* For several years, there was a lack of uniformity by the FDA on this subject. The Division of Generic Drugs usually required "ambient" conditions to be 30°C, whereas other divisions accepted 25–30°C. This complicated the situation for companies, which then needed two sets of stability chambers. The rationale for the 30°C requirement appeared to be related to the then-current USP definition of controlled room temperature (CRT) as 15–30°C. It was stated that if the product labeling indicated storage at CRT, then the stability should be performed at 30°C. Industry argued that the product, even when labeled for storage at CRT, would not be exposed to 30°C throughout its shelf-life. Eventually the issue was resolved via the ICH and the redefinition of CRT by the USP.

ICH, which used the climatic zone concept as part of the basis for its decisions, applied only to new chemical entities and also only to Europe, Japan and the U.S.A. The decisions were published in the Federal Register in September 1994.

The key points included:

- Stability storage conditions will normally involve long-term studies at $25 \pm 2°C$ with 60% RH $\pm$ 5% with at least 12 months of data before filing; accelerated studies at $40 \pm 2°C$ and 75% RH $\pm$ 5% with at least 6 months of data.
- Where "significant change" occurs during the 40°C accelerated study, an additional intermediate station should be used, such as $30 \pm 2°C/60\%$ RH $\pm$ 5%. "Significant change" was defined as a 5% loss of potency, any degradant exceeding its specification limit, exceeding pH limits, dissolution failures using 12 units, and failures of physical specifications (hardness, color, etc.).
- For less stable products, the storage conditions may be reduced but the accelerated conditions should still be at least 15°C above those used for long-term evaluation.
- For products where water loss may be important, such as liquids or semisolids in plastic containers, it may be more appropriate to replace the high-RH conditions by lower RH such as 10–20%.
- The same storage conditions are to be applied for the evaluation of bulk drug substances. However, retest dates may be used instead of expiration dates.

The ICH guideline does not specifically address the position of samples during storage. This is especially important for liquid products where leakage and product closure interaction need to be evaluated. One approach is to store samples both upright and inverted but only to test the inverted samples. The upright samples may be used as controls in the event that problems are identified with the inverted samples. For products with closures at two ends, such as pre-filled syringes and semisolids in tubes, horizontal storage is more appropriate.

As indicated earlier, these ICH guidelines relate only to new chemical entities and the products made from these materials. However, the FDA expects the same conditions for stability studies (with less data at the time of filing) for ANDAs and for supplements. The FDA has published several Scale Up and Post Approval Changes (SUPAC) guidelines for different types of dosage forms. These guidelines provide suggested stability studies that are expected to support various levels of change to previously approved processes. An approach significantly different from that indicated in the respective SUPAC guideline should be discussed in advance with the FDA to avoid problems with ultimate NDA supplement approval.

The subject of light sensitivity/stability is addressed in ICH Q1B.

3. *Number of batches to be evaluated.* Normally, a minimum of three batches is required to provide a sufficient basis for shelf life prediction for a new or significantly reformulated

product. Development and stability batches may be used provided they are of the same formulations as the commercial product and they were processed in an equivalent manner. The ICH guideline requires stability data on three batches, two of which should at least be pilot scale (not less than 1/10th commercial scale and the same process or 100,000 tablets or capsules, whichever is the larger), and the third batch can be smaller. Stability results from laboratory-scale batches may be used only as supporting data. The first three commercial-scale batches are also to be included in the stability evaluation program.

For some drug products there can be a number of variants—different package sizes, different strengths (some with the same formulation) and different packaging/closure arrangements. In such circumstances, the extent of the stability evaluation can become enormous. To accommodate these situations, bracketing and "matrixing" approaches were introduced to reduce the amount of testing required.

Bracketing involves making conclusions about all levels of a parameter based on the evaluation of the extremes. Suggested applications include:

- Same formulation and container–closure system involving different container sizes and/or different fill volumes.
- Different strengths of the same formulation (e.g., different capsule sizes or different tablet weights from the same granulation).

Matrixing involves a statistical experimental design that allows only a fraction of the total number of samples to be tested at each sampling point. Since fewer tests are performed, there is usually more variability in the data and a shorter predicted shelf-life may result. However, this can be "corrected" when more data eventually become available. Matrix designs should be applicable:

- for the same formulation in different strengths (same granulation),
- for different but closely related formulations, and
- for different sources of bulk drug substance.

An example of a matrix design would be a tablet product produced in three strengths (same granulation, different compression weights) and packaged into three different bottle sizes. Three batches of granulation are produced, each of which is compressed into three sublots with the different compression weights. Each of these nine sublots of tablets is then packaged into the three different bottle sizes—27 sets of stability samples. Testing is to be performed at 0, 3, 6, 9, 12, 18, 24 and 36 months.

A complete evaluation would therefore involve 27×8 (216) sets of testing. Three alternative matrix designs could be applied. In each of these, all of the different combinations are tested at 0 and 36 months. In a complete one-third design, one-third of the samples are tested at each intermediate point and in the complete one/two-thirds design one-third of the samples are tested at some points and two-thirds at others. The complete one-third design is depicted in Figures 1 and 2.

This matrix results in the testing of 108 samples—half of the total if all combinations had been tested. The number of samples tested could be further reduced by testing only one sample of each granulation batch (three samples rather than 27) or one sample from each strength of each granulation batch (nine samples rather than 27).

A matrix approach may also be applied to the actual tests performed on the samples—each test need not be performed at each test interval.

It is recommended that bracketing and matrixing proposals should be reviewed and agreed on with the FDA prior to introduction. With time and experience (FDA and industry) the need for this prior agreement should reduce.

4. *Test methodology.* The stability-testing monograph need not include all of the criteria defined in the product release monograph. Only those parameters that are potentially susceptible to change during storage and that may impact on quality, safety, or efficacy need to be evaluated. The characteristics evaluated will include actives, degradants, antimicrobial agents, antioxidants and key physical characteristics such as dissolution, fragility, color,

```
Granulation batches: G1, G2, G3

Compression weights: 100, 200, 300

Packages:           P1, P2, P3

Test sample groups  T1, T2, T3
```

	MATRIX			
Granulation Batch	Strength	Packages in group		
		T1	T2	T3
	100	P1	P2	P3
G1	200	P2	P3	P1
	300	P3	P1	P2
	100	P3	P1	P2
G2	200	P1	P2	P3
	300	P2	P3	P1
	100	P2	P3	P1
G3	200	P3	P1	P2
	300	P1	P2	P3

FIGURE 1   Matrix.

moisture and volume. Closure integrity may also be required; however, evaporation or leakage will normally show up in other tests. Functional effectiveness should also be evaluated: child-resistant closures, tamper-evident packaging, syringability for prefilled syringes and openability of containers. For tamper-evident packaging there are currently no generally accepted methods of evaluation and in-house methods should be developed.

For parenteral or other sterile products, testing should include sterility assurance. This may be achieved by normal sterility testing, but because of the large number of samples evaluated during the stability protocol, there would be some chance of a "false-positive" result, which could create difficulties in interpretation. A validated closure integrity test may therefore offer a better approach and can be used with a larger number of samples if required. Recently, FDA microbiologists appear to be favoring this approach. However, there are currently no universal closure integrity test methods that have both industry and FDA support.

5. *Test frequency.* This should be adequate to demonstrate any degradation and to provide enough data points for statistical evaluation. For the scale-up batches and the first three

	Testing					
Test Group	Testing Interval (months)					
T1	0	3		12		36
T2	0		6		18	36
T3	0			9	24	36

FIGURE 2   Matrix design.

commercial batches testing is expected initially, at three-month intervals during the first year, six-month intervals in the second year, and yearly thereafter. Some companies do not evaluate beyond 36 months. A different frequency may be more appropriate for ongoing stability evaluation (see later).

Stability studies can be classified into three types:

1. Studies, usually under accelerated conditions to predict a tentative shelf life for a new or modified product or process. For a new drug substance, these studies usually commence with a preformulation evaluation. The effect of stress conditions, such as temperature, humidity, light, acidity and oxygen, can provide much useful information to the formulator. The potential interactive effects of the bulk drug and the anticipated dosage form excipients may also be evaluated. It should also be noted that any ingredient that interferes with the official assay of a USP product automatically makes the product noncompliant with that monograph—regardless of whether an alternate assay has been developed. Where degradation is observed, attempts should be made to identify the decomposition products, since this information could be of value later in developing analytical methodology for product stability studies.

   The accelerated studies at elevated temperature on the dosage form should allow some extrapolation to provide a tentative shelf life. The ICH guidelines allow extrapolation of six-month data under accelerated conditions with 12 months data at 25°C/60% RH to predict a shelf-life of up to 24 months. Shelf life in excess of 24 months should rarely be extrapolated from accelerated data. There are also some parameters such as dissolution, whose shelf-life performance cannot be predicted from accelerated study data. Consequently, any significant change in dissolution during accelerated studies should be a signal for caution until adequate real-condition data are available.

   For changes in container–closure, formulation, or material supplier, the FDA usually requires accelerated data comparing the revised product with the existing product plus a commitment to continue the stability study. The previously designated shelf-life may be retained if there are no observed differences. A similar approach should be used when reprocessed material is incorporated into a batch.

   Where there is a change of manufacturing facility for the dosage form, but using the same process and similar equipment, three months accelerated data may suffice, again with the commitment to monitor the first three commercial batches.

2. Studies under conditions appropriate to the market are used to provide real-time data for confirmation of the predicted tentative shelf life. These studies are usually performed using controlled environmental cabinets. A typical warehouse may be an acceptable alternative provided temperature and humidity are recorded. For certain physical parameters such as dissolution, tablet fragility and parenteral sterility, accelerated conditions may not provide useful data for extrapolation.

   Where such studies demonstrate that the predicted tentative shelf life was too optimistic, it would be necessary to consider recall of released batches.

   Real-time studies are also used to extend the defined shelf life when sufficient satisfactory real-time data have been obtained, as defined in the approved ANDA.

3. Stability studies on current production. Once the shelflife is established, it is necessary to evaluate some ongoing batches to confirm that current production is behaving in a similar manner. This is to detect the possible impact of any subtle or unknown changes to the components or process. In the event that a change is observed, it will be necessary to perform a root cause analysis.

At this stage, there should be a considerable amount of available stability data that identify the shelf life limiting factors. This may allow elimination of some tests. The frequency of testing should also relate to the shelf life.

The FDA is prepared to recommend action, such as a warning letter or seizure, if there is inadequate evidence to support the shelf life (Compliance Policy Guide 7132a.04). Specific concerns include lack of sterility assurance; lack of, or noncompliance with, a stability program;

absence of an expiry date; inadequate test methodology; lack of ongoing stability; lack of assurance of preservative effectiveness; and distribution after expiration date.

The stability requirements for homeopathic products are less demanding than for other drug products. The levels of "active ingredients" are frequently so low that determination of degradation products, or even assay of the active itself, may not be practicable. The requirements allow examination for compatibility as an alternative to testing.

The immediate container and closure play an important role in the product shelf-life. They may accelerate degradation reactions, be an additive to or an absorbant of the drug substance, or be ineffective in protecting the contents from environmental conditions. Four types of containers–closures are commonly analyzed for pharmaceutical preparations: glass, plastic, rubber (natural and synthetic), and metal. Each has characteristic properties that should be recognized.

Glass

Glass, because of its many variations and resistance to chemical and physical change, is still used as a container material. Several inherent limitations exist with glass:

1. Its alkaline surface may raise the pH of the pharmaceutical and induce chemical reaction.
2. Ionic radicals present in the drug may precipitate insoluble crystals from the glass (such as barium sulfate).
3. The clarity of the glass permits the transmission of high-energy wavelengths of light, which may accelerate physical or chemical reactions in the drug.

To overcome the first two deficiencies, alternate types of commercial glass, each possessing different reactive characteristics, are available. Borosilicate (USP type I) glass contains fewer reactive alkali ions than the other three types of USP-recognized glass. Treatment of glass with heat and/or various chemicals, as well as the use of buffers, can eliminate many ionic problems normally encountered. Amber glass transmits light only at wavelengths above 470 nm, thereby reducing light-induced reactions. When light sensitivity is a stability issue, the secondary packaging, with appropriate labeling, may provide adequate protection.

Plastics

These packaging materials include a wide range of polymers of varying density and molecular weight, each possessing different physical and chemical characteristics. Various additives to the polymeric material are often required to provide suitable characteristics for molding, to minimize impact damage or for color. As a result, each must be considered in relation to the pharmaceutical product that will be in contact with it to determine that no undesirable interaction occurs. Several problems are encountered with plastic:

1. Migration of the drug through the plastic into the environment.
2. Transfer of environmental moisture, oxygen and other elements into the pharmaceutical formulation.
3. Leaching of container ingredients into the drug.
4. Adsorption or absorption of the active drug or excipients by the plastic.

Since each plastic possesses intrinsic properties, varying conditions and drug formulations must be tested to optimize stability of the final product by selecting the appropriate container. Again, chemical treatment of the material prior to use may reduce reactivity, migration characteristics and transmitted light. It must be remembered that neither the drug nor the container should undergo physical or chemical changes that affect the safety and efficacy of the product. The use of light transmission by plastics as a measure of light protection is complicated by the fact that plastics are only semitransparent. Light that is admitted to the container is reflected and diffused back into the product so that light energy available to degradation processes is much higher than that which might be indicated by transmission characteristics. The proper test is a diffuse reflectance measurement. Appropriate testing procedures and specifications are given in the USP.

Metals

Various alloys and aluminum tubes frequently are utilized as containers for emulsions, ointments, creams and pastes. These materials are generally inert to their contents, although instances of corrosion and precipitation have been noted with products at extreme pH values or those containing metallic ions. Coating the tubes with polymers, epoxy, or other material may reduce these tendencies but impose new stability problems on the pharmaceutical product. The availability of new, less expensive polymers has sharply reduced the use of metal packaging components during the last few years (except for metal screw cap closures).

Rubber

The problems of extraction of drug ingredients and leaching of container ingredients described for plastics also exist with rubber components. The use of neoprene, butyl or natural rubber, in combination with certain epoxy, Teflon®, or varnish coatings, substantially reduces drug–container interactions. The pretreatment of rubber vial stoppers and closures with water and steam removes surface blooms and also reduces potential leaching that might affect chemical analysis, toxicity, or pyrogenicity of the drug formulation. The impact of additional treatments, such as siliconization to enhance movement of elastomeric components during handling in production or for plunger action in syringes, must also be evaluated.

§211.167 SPECIAL TESTING REQUIREMENTS

(a) For each batch of drug product purporting to be sterile and/or pyrogen-free, there shall be appropriate laboratory testing to determine conformance to such requirements. The test procedures shall be in writing and shall be followed.
(b) For each batch of ophthalmic ointment, there shall be appropriate testing to determine conformance to specifications regarding the presence of foreign particles and harsh or abrasive substances. The test procedures shall be in writing and shall be followed.
(c) For each batch of controlled release dosage form, there shall be appropriate laboratory testing to determine conformance to the specifications for the rate of release of each active ingredient. The test procedures shall be in writing and shall be followed.

Specific testing requirements for sterile products, ophthalmic ointments and controlled release products are delineated in this section.

As written, §211.167(a) requires testing to confirm sterility and where appropriate pyrogen testing on each batch of sterile or pyrogen-free product. The necessity for such testing would seem superfluous for terminally sterilized products prepared by the application of validated processes. The sterility test in these circumstances is more a challenge of the technique in the microbiology laboratory than an assurance of sterility. The FDA, which in 1985 approved the replacement of the sterility test by parametric release for certain large-volume parenterals, recognized this. This was followed in 1987 by the issuance of a Compliance Policy Guide (7132a. 13). This guide defined the criteria for parametric release.

1. Only terminally sterilized products may be considered.
2. The sterilizer validation should include
   a. chamber heat distribution,
   b. heat distribution for each load configuration,
   c. heat penetration studies for the products,
   d. lethality study using organisms with known resistance, usually *Bacillus stearothermophilus*,
   e. presterilization bioburden—number of organisms and their resistivity to the cycle,
   f. recording of all key cycle parameters—time, temperature, pressure, and

g.   demonstration of bioburden reduction to 10% and a minimum safety factor of 6 log-arithm reduction.
3.   Closure integrity validation should be performed on each container–closure system to ensure no ingress of organisms during the shelf life.
4.   Bioburden is required on each batch of product prior to sterilization. The resistance of any spore-forming organisms is to be measured and compared with those found during the validation study. It is also indicated that if such organisms had a higher resistance, the batch would be considered nonsterile.
5.   Biological or chemical indicators are to be used in each sterilizer load.

The detailed requirements of the Compliance Policy Guide essentially replace traditional sterility testing with an alternative sterility testing procedure; validation of the sterilization process forms only a part of this alternative procedure. The main points of concern include:

1.   The F_0 concept does not appear to be accepted since the Guide states that "failure of more than one critical parameter must result in automatic rejection of the sterilizer load." Critical parameters include time, temperature and pressure, whereas heat-up and cool-down times are considered noncritical.
2.   Evaluation of container–closure integrity would seem to be more appropriate to stability considerations and has no direct correlation with parametric release.
3.   Without considerable validation evaluation, the use of chemical indicators in each load would not be reliable. The alternative, biological indicators, appears essentially to replace the traditional sterility test by an alternative version. With a validated process, shown to be operating under control, there would seem to be no need for biological indicators in each batch. Although less prone to false-positives than sterility tests, their use prevents the opportunity for early batch release since an incubation period is still involved.

The special requirements for ophthalmic ointments relate to the potential presence of abrasive particulate matter. This is of obvious concern in such preparations and especially since metal tubes are frequently used for their packaging. The USP ⟨751⟩ includes specifications and methodology for the presence of metal particles in ophthalmic ointments. Although metal particles are considered to be the biggest risk, especially from metal tubes, it should be noted that §211.167(b) refers more generally to "foreign particles and harsh or abrasive substances." For products packaged in other configurations, such as plastic tubes, it would seem appropriate to apply the USP metal particle limits and to establish appropriate methodology to allow visualization of other particulate matter.

Subsection (c) refers to controlled release products and is somewhat generic in nature—"there shall be appropriate laboratory testing to determine conformance to the specifications for the rate of release of each active ingredient." Products with the same active ingredients may be formulated by different manufacturers to have different release patterns. This creates no problems with respect to drug registration but it does for the USP and for the consumer with respect to OTC products. The USP is moving toward generic-style monographs, which define ranges for release rates at three or four time intervals: 0.125, 0.250, 0.500 and 1.00D, where D represents the dosing interval (e.g., 8 hours). Where possible, the criteria defined in the USP for Drug Release ⟨724⟩ will be applied. This allows for different release patterns from different products. The release pattern would be presented on the product label, which allows the knowledgeable consumer some choice.

§211.170 RESERVE SAMPLES

(a)   An appropriately identified reserve sample that is representative of each lot in each shipment of each active ingredient shall be retained. The reserve sample consists of at least twice the quantity necessary for all tests required to determine whether the active ingredient meets its established specifications, except for sterility and pyrogen testing, The retention time is as follows:

(1) For an active ingredient in a drug product other than those described in paragraphs (a) (2) and (3) of this section, the reserve sample shall be retained for 1 year after the expiration date of the last lot of the drug product containing the active ingredient.

(2) For an active ingredient in a radioactive drug product, except for nonradioactive reagent kits, the reserve sample shall be retained for:

    (i) Three months after the expiration date of the last lot of the drug product containing the active ingredient if the expiration dating period of the drug product is 30 days or less; or

    (ii) Six months after the expiration date of the last lot of the drug product containing the active ingredient if the expiration dating period of the drug product is more than 30 days.

(3) For an active ingredient in an OTC drug product that is exempt from bearing an expiration date under §211.137, the reserve sample shall be retained for 3 years after distribution of the last lot of the drug product containing the active ingredient.

The regulations require retention of active ingredients but not of inactive ingredients. This relaxation for inactive ingredients was in response to comments that some materials are hazardous or unstable. However, samples of hazardous or unstable active ingredients are to be retained. It should be noted, however, that European Union (EU) countries require the retention of excipient samples, with varying time periods for their retention.

The rationale for retaining samples is to allow evaluation in the event of a complaint or query. Consequently, it is prudent to retain samples of all ingredients, active and inactive.

If a batch of ingredient is delivered on more than one occasion, samples from each delivery are to be retained. This is in line with the evaluation of such deliveries.

(b) An appropriately identified reserve sample that is representative of each lot or batch of drug product shall be retained and stored under conditions consistent with product labeling. The reserve sample shall be stored in the same immediate container–closure system in which the drug product is marketed or in one that has essentially the same characteristics. The reserve sample consists of at least twice the quantity necessary to perform all the required tests, except those for sterility and pyrogens. Except for those drug products described in paragraph (b)(2) of this section, reserve samples from representative sample lots or batches selected by acceptable statistical procedures shall be examined visually at least once a year for evidence of deterioration unless visual examination would affect the integrity of the reserve samples. Any evidence of reserve sample deterioration shall be investigated in accordance with §211.192. The results of the examination shall be recorded and maintained with other stability data on the drug product. Reserve samples of compressed medical gases need not be retained. The retention time is as follows:

(1) For a drug product other than those described in paragraphs (b) (2) and (3) of this section, the reserve sample shall be retained for 1 year after the expiration date of the drug product.

(2) For a radioactive drug product, except for nonradioactive reagent kits, the reserve sample shall be retained for

    (i) three months after the expiration date of the drug product if the expiration dating period of the drug product is 30 days or less; or

    (ii) six months after the expiration date of the drug product if the expiration dating period of the drug product is 30 days.

(3) For an OTC drug product that is exempt for bearing an expiration date under §211.137, the reserve sample must be retained for 3 years after the lot or batch of drug product is distributed.

The retention of batch samples of product allows evaluation in the event of complaints or queries. For large package sizes, where product costs and storage space could be a problem, it is acceptable to retain the samples in a smaller version of the immediate container–closure system. The actual storage conditions for retained samples are not defined. It would seem appropriate to use conditions that are reasonably related to those likely to be experienced by the commercial product. This would probably equate to warehouse conditions for products with no special storage requirements. However, this would not be appropriate for a product required to be stored in a refrigerator (see also §211.166).

The FDA acknowledged that the evaluation of all retained batches was a time-consuming exercise. As a consequence, (b) was revised in 1994 to allow evaluation of a statistically selected number of batches only.

The FDA allows the annual review to be omitted if in so doing the integrity of the sample would be affected. For example, if a product is stored in a colored or translucent container that must be kept closed, then visual examination of the dosage form may be impractical. However, examination of the exterior of the container and the label should still be performed. If any problems are noted with these aspects, the quality of the dosage form enclosed in the package may be suspect.

The results of any visual examination may be held with other stability data and need not be entered into individual batch records.

§211.173 LABORATORY ANIMALS

Animals used in testing components, in-process materials, or drug products for compliance with established specifications shall be maintained and controlled in a manner that assures their suitability for their intended use. They shall be identified and adequate records shall be maintained showing the history of their use.

Minimum standards for the care and health of research and test animals are described in the following sources:

1.   Animal Welfare Act (7 USC §§2131–2156).
2.   Title 9, Code of Federal Regulations, §§ 1.1–11.41.
3.   Good Laboratory Practice for Nonclinical Laboratory Studies, 21 CFR Part 58.
4.   Guide for the Care and Use of Laboratory Animals, DHEW Publication No. (NIH) 78-23, revised 1978, U.S. Government Printing Office, Washington, D.C., 017-040-00427-3, 1978.

In addition to these requirements, current interpretation of Good Manufacturing Practices would regard animals as sources of product contamination. Considerations such as separate facilities, constructed away from manufacturing areas, with closed water, waste removal, air conditioning, and other systems would, therefore, be ideal. If these are not possible due to construction or other limitations, animal areas should be segregated as far as possible from all production activities with closed air, water and waste systems, as well as limited personnel access. The same standards of cleanliness prescribed for other work areas are also applicable to these spaces.

Record requirements for animals are necessary to maintain control of their use in experimentation, testing, or assay procedures. Data fields for individual animals should include

1.   Identification number or letter assigned to each animal or group of animals,
2.   Characteristics and description of animal,
3.   Source of animals (breeder, vendor),
4.   Date of arrival,
5.   Age at arrival,
6.   How used, and
7.   Date used.

If the animal is to be used for repeated assay procedures, i.e., pyrogen testing, a time period sufficient to permit complete clearance of the drug and recovery of the test animal is required.

§211.176 PENICILLIN CONTAMINATION

If a reasonable possibility exists that a nonpenicillin drug product has been exposed to cross-contamination with penicillin, the nonpenicillin drug product shall be tested for the presence of penicillin. Such drug product shall not be marketed if detectable levels are found when tested according to procedures specified in "Procedures for Detecting and Measuring Penicillin Contamination in Drugs," which is incorporated by reference. Copies are available from the Division of Research and Testing (HFD-470), Center for Drug Evaluation and Research, FDA, 200 C St. SW, Washington, D.C. 20204, U.S.A. or available for inspection at the Office of the Federal Register, 800 North Capitol Street, NW, Suite 700, Washington, D.C. 20408, U.S.A.

This regulation permits low-level contamination of drug products with penicillin. The permitted tolerance is at the "undetectable" level, using specific methodology.

If there is possibility of contamination of raw materials by penicillin because of its place of production or warehousing, it is appropriate for a manufacturer to require that the supplier test and certify that the material is not contaminated. If the possibility of contamination arises from the conditions of shipping, those conditions should be changed, or the manufacturer must test for the absence of contamination.

EXAMPLES OF OBSERVATIONS FROM FDA 483 CITATIONS AND WARNING LETTERS

1. Failure to establish scientifically sound and appropriate specifications, standards, sampling plans and test procedures designed to assure that drug products conform to appropriate standards of identity, strength, quality and purity. WL 04-NWJ-18 9/15/2004.
2. SOP uses a statistical outlier test to invalidate out-of-specification results; statistical outlier tests are inappropriate for use with validated methods.
3. Data acceptance/rejection was done selectively.
4. Stability testing SOPs contained no provision for increased testing of either additional lots or additional intervals or shortened intervals after confirmed stability failures.
5. There is neither statistical analysis nor graphical representation of the firm's stability data in the annual product reviews.
6. There are no data to show that the methods used to analyze stability samples were validated as stability indicating with respect to acid and base hydrolysis, oxidation, thermal degradation and photolysis.
7. There is no system in place that assures that senior management are made aware of problems that may affect product quality.
8. Failure to validate the software, which is used to collect raw data from the HPLC units, to integrate peaks and to perform analytical calculations for assaying products.
9. The firm used the service of an outside microbiology laboratory for microorganism quantitation and identification. The laboratory had never been audited by the firm.
10. Chromatograms are run for an extended length of time without additional standard solution injections being made to check on the stability of the chromatographic system.
11. There are no criteria established for out-of-specification results defining at what points testing ends, product is evaluated and rejected if results are not satisfactory.
12. Stability test failures not reported to the FDA.

ACKNOWLEDGMENTS

The authors wish to acknowledge the very valuable comments and suggestions provided by Dr. Henry C. Stober during the editing of this chapter.

SUGGESTED READINGS

1. Fusari SA and Hostetler GL. Reference thermal exposures and performance of room temperature stability studies, Pharm Technol, 1984.
2. Haynes JD. J Pharm Sci 1971; 60: 927–931.
3. Riggs TH. Bull, Parenteral Drug Assoc, 1971; 25: 116–123.
4. Biebart C. Drug Dev Ind Pharm, 1979; 5: 349–363.
5. Carstensen J. J Pharm Sci, 1974; 63: 1–14.
6. Davis J. "Expiry Dating Proposals," FDA reprint of a speech presented at Purdue University, Sep 12–15, 1978.
7. Davis J. Pharm Tech, 1979; 3: 65–67.
8. Ferguson L. FDA By-Lines, 1978; 6: 281–320.
9. Mollica J et al. J Pharm Sci, 1978; 67: 443–465.
10. Dovich RA. Modified control chart limits. Quality 1984; 36.
11. American National Standard (ANSI 21.3-1958(r-1975)). "Control Chart Method for Controlling Quality During Production."
12. International Conference on Harmonization. Stability Testing of New Drug Substances and Products, Guideline, Fed Reg, 1994; 59(183): 48754.
13. Draft FDA Interim Guidance. "Immediate Release Solid Oral Dosage Forms. Pre-and Post-Approval Changes: Chemistry, Manufacturing and Controls, In Vitro Dissolution Testing and In Vivo Bioequivalence Documentation," Nov 29, 1994.
14. International Conference on Harmonization. Guideline on Impurities in New Drug Substances, (ICH Q3AR) and Guideline on Impurities in New Drug Products (ICH, Q3BR), Impurities: Guidelines for Residual Solvents (ICH Q3C) (available at www.ich.org). See Appendix B.
15. Tetzlaff RF. GMP Documentation Requirements for Automated Systems: Part III, FDA Inspections of Computerized Laboratory Systems, Pharm Technol, 1992; 16(4): 60.
16. FDA. "Guide to Inspection of Pharmaceutical Quality Control Laboratories," Jul 1993.
17. FDA. Center for Drug Evaluation and Research, "Review Guidance, Validation of Chromotographic Methods," Nov 1994.
18. FDA. Changes to an Approved NDA or ANDA, Revision 1, Apr 2004.
19. Hokanson GC. A life cycle approach to the validation of analytical methods during pharmaceutical product development, Part II. Changes and the need for additional validation. Pharm Technol, 1994; 18(10): 92.
20. Hokanson GC. A life cycle approach to the validation of analytical methods during pharmaceutical product development, Part I. The initial method validation process. Pharm Technol, 1994; 18(9): 118.
21. PMA. Measuring quality performance. Pharm Technol, 1994; 18(3): 140.
22. International Conference on Harmonization. Guideline on Validation of Analytical procedures: Definitions and Terminology, ICH Q2A and Q2B. See Appendix B.
23. Chrai S, Hefferan G, Myers T. Glass vial container–closure integrity testing—an overview, Pharm Technol, 1994; 18(9): 162.
24. United States of America v. Barr Laboratories Inc. et al. Defendants, No. CIV A 92-1744, United States District Court, D. New Jersey, Feb 5, 1993. See Chapter 21.
25. International Conference on Harmonization. Extension of the ICH Text "Validation of Analytical Procedures," ICH Q2B. See Appendix B.
26. Kieffer RG, Stoker JR. Quality performance measurement and reporting, Pharm Technol, 1986; 10(6): 54.
27. International Conference on Harmonization. Impurities Testing Guideline: Residual Solvents, ICH Q3C. See Appendix B.
28. Nally J, Kieffer RG, Stoker JR. From audit to process assessment—the more effective approach, Pharm Technol, 1995; 19(9): 128.
29. Guidance for Industry, Analytical Procedures and Methods Validation, Draft Guidance, 8/2000.
30. Mark Green J. A practical guide to analytical method validation, Anal Chem, 1996; 68: 305A.
31. Larry Paul W. USP perspectives on analytical methods validation, Pharm Technol, 1991.
32. USP29/NF24 <1225> for discussion of method verification as well as other sections referenced in the chapter.
33. Guidance for Industry, ANDAs: Impurities in Drug Substances, 11/1999.
34. Huber L. Validation and qualification in analytical laboratories, Oct 1998.
35. Stefan RI, Van Staden J, Singer D. Laboratory Auditing for Quality and Regulatory Compliance, Marcel Dekker, Jul 2005.
36. Guidance of Industry, Investigating Out-of-Specification (OOS) Test Results for Pharmaceutical Production October 2006.
37. Investigation of Out-of-Specification Results, Alex M. Hoinowski, Sol Motola, Richard J. Davis, and James V. McArdle, Pharmaceutical Technology, January 2002.

11 | Records and Reports
Subpart J

Arlyn R. Sibille
Consultant, Harmony, Pennsylvania, U.S.A.

Steven Ostrove
Ostrove Associates Inc., Elizabeth, New Jersey, U.S.A.

Joseph D. Nally and Laura L. Nally
Nallianco LLC, New Vernon, New Jersey, U.S.A.

Before reviewing Subpart J and CFR sections 211.180 to 211.198 it is worthwhile to review CFR Part 11 on Electronic Records and Electronic Signatures. Electronic records are playing an ever increasing role in pharmaceutical operations and while few firms have only electronic records, many have hybrid documentation systems utilizing both electronic and manual or hard copy records.

Electronic signatures/initials frequently involve a personal password and a personal magnetic card with a secure system to manage allocation and review. For some time the FDA disagreed with this approach and stated that signatures and initials must be handwritten. After an extensive review, involving other areas of government and industry, the FDA in 1994 issued a Proposed Rule that stated that electronic recording was an acceptable alternative (see Suggested Reading, 1). The Final Rule, CFR Part 11 "considers electronic records, electronic signatures and handwritten signatures executed to electronic records, to be trustworthy, reliable, and generally equivalent to paper records and handwritten signatures executed on paper." This acceptance of electronic alternatives has application at various points in the CGMP regulations. It allows the use of electronic record-keeping systems in complying with regulations. Part 11 (also known as "Electronic Records; Electronic Signatures" or ERES) works in tandem with a predicate rule, which refers to any FDA regulation that requires organizations to maintain records. With this guidance came the obligation for each manufacturer to ensure that the computer systems and software used in these operations were validated and revisions histories of software(s) maintained. The validation efforts proved burdensome on manufacturers and confusion relating to "what constitutes validation of computer system and software" was rampant. The FDA embarked on a re-examination of part 11 as it applies to all FDA regulated Products in the *Federal Register* of February 4, 2003 (68 FR 5645), they announced the withdrawal of the draft guidance entitled "Guidance for Industry, 21 CFR Part 11; Electronic Records; Electronic Signatures, Electronic Copies of Electronic Records" because the FDA wished to limit the time spent by industry reviewing and commenting on the guidance, which might not have been representative of FDA's approach under the CGMP initiative. As a result the scope of application for electronic records and computer systems in general has been narrowed.

CFR PART 11
Steven Ostrove

Plated to electronic equipment in section §211.68 is the CFR Part 11 Electronic Records and Electronic Signatures regulations. In the Federal Register of March 20, 1997, at 62 FR 13429, the FDA issued a notice of final rulemaking for 21 CFR, Part 11, Electronic Records; Electronic Signatures. The rule went into effect on August 20, 1997.

There are three subparts to the Part 11 regulations:

- Subpart A is for General Provisions including 11.1 Scope; 11.2 Implementation; 11.3 Definitions.
- Subpart B is for Electronic Records and includes 11.10 Controls for closed systems; 11.30 Controls for open systems; 11.50 Signature manifestations; 11.70 Signature/record linking.
- Subpart C is for electronic signatures and includes 11.100 General requirements, 11.200 Electronic signatures components and controls; 11.300 Controls for identification codes/passwords.

Subpart A—General Provisions
11.1 Scope
The scope of Part 11 is all records or signatures that are kept in electronic format and are "created, modified, maintained, archived, retrieved, or transmitted." The purpose is to assure that the electronic information is as trustworthy as the written records and signatures. Part 11 does not apply to any written record that is transmitted by electronic means but only for those records that fall into one or more of the above conditions. If Part 11 is met, the electronic records "may be used in lieu of paper records."

11.2 Implementation
This section discusses the use of electronic records and signature in lieu of written records on paper. Only those records in a recognized format as specified by the agency will be accepted by the agency (FDA). The regulations specify that the agency should be contacted prior to submitting an electronic record to be sure that the records comply in all ways including format, technical composition, and means of transmission.

11.3 Definitions
This section sets forth the definitions for closed and open systems, digital signatures, electronic records and signatures, handwritten signatures, and biometrics.

Subpart B—Electronic Records
11.10 Controls for Closed Systems
This section establishes the conditions needed to control a closed system. The system must be able to assure that the signer cannot dispute the signed record. The system must be validated and its ability to distinguish between valid and invalid data must be proven. In addition, accurate copies of the records in usable forms need to be assured. The use of audit trails, limited system access, and other general data security are addressed.

The section goes on to discuss system checks, access to input and output devices connected to the system, and the validity of the data source. As with other CGMP functions, the personnel who have access to the system must have the necessary training, education, or experience to perform their assigned tasks. There must be written procedures to "hold individuals accountable and responsible for actions initiated under their electronic signatures."

11.30 Controls for Open Systems
All of the controls specified in 11.10, as appropriate, apply to this section also. Control of who, what, and how data is entered into the system must be in place in order to ensure the integrity of the system. In addition, electronic record confidentiality, as appropriate, must be assured from the time of their creation until they are received.

11.50 Signature Manifestations
Information in the signed electronic record must clearly indicate the following:

- Printed name of the signer
- Date and time signature was made
- The purpose of the signature (reason it was made)

All controls for electronic records "shall be included as part of any human electronic records and shall be included as part of the human readable form."

11.70 Signature/Record Linking

"Electronic signatures and handwritten signatures executed to electronic records shall be linked to their respective electronic records." Thus, the signatures cannot be copied or removed from the files by ordinary means. This is to prevent falsification of the records, both written and electronic.

Subpart C—Electronic Signatures
11.100 General Requirements

The section starts with the specification that each electronic signature shall be unique and attributable to only one person. The section goes on to describe the requirements of verifying personnel using electronic signatures as well as how the verification is to be made. It specifies that before a person is given access to the system using electronic records or signatures, they must prove their identity and provide written verification of their handwritten signatures to the agency. The agency reserves the right to request additional certification of identity and that the person has agreed that the electronic signature is legally binding.

11.200 Electronic Signature Components and Controls

There are two types of controls allowed for electronic signature and records. These are biometric and non-biometric. Examples of biometric controls are fingerprint identification, voice recognition, or retinal scans. Examples of non-biometric identification are passwords or identification codes. If non-biometric identification is used, then there must be at least two unique identifiers. These identifiers should be used each time an entry is made. However, if several entries are made during one continuous session, then the two should be used for the first identification and one may be used for each successive entry. If a biometric identification is used, only one form needs to be applied for each use. In either case, assurance need to be made that only each individual has his or her own unique identifier and that the identifiers are secure and non-transferable. In addition, the identifiers cannot be used by anyone else than the designated owner.

11.300 Controls for Identification Codes/Passwords

If passwords and identification codes are used, controls should:

- Maintain their uniqueness
- Be periodically checked (e.g., changed on a regular basis)

If a password is lost or stolen, or possibly so, management needs to be able to inactivate the users identification and to issue a replacement using strict controls as before. All transactions need to be safeguarded from fraud or unauthorized attempts to use another's identification or codes. Any device used to generate codes, such as tokens, need to be checked on a periodic basis to assure proper functionality and that no alterations have taken place.

§211.180   GENERAL REQUIREMENTS

(a) Any production, control, or distribution record that is required to be maintained in compliance with this part and is specifically associated with a batch of a drug product shall be retained for at least one year after the expiration date of the batch or, in the case of certain OTC drug products lacking expiration dating because they meet the criteria for exemption under §211.137, three years after distribution of the batch.

(b) Records shall be maintained for all components, drug product containers, closures, and labeling for at least one year after expiration date or, in the case of certain OTC drug products lacking the expiration dating because they meet the criteria for exemption under §211.137, three years after distribution of the last lot of drug product incorporating the component or using the container, closure, or labeling.

(c) All records required under this part, or copies of such records, shall be readily available for authorized inspection during the retention period at the establishment where the activities described in such records occurred. These records or copies thereof shall be subject to photocopying or other means of reproduction as part of such inspection. Records that can be immediately retrieved from another location by computer or other electronic means shall be considered as meeting the requirements of this paragraph.

(d) Records required under this part may be retained either as original records or as true copies such as photocopies, microfilm, microfiche, or other accurate reproductions of the original records. Where reduction techniques such as microfilming are used, suitable reader and photocopying equipment shall be readily available.

(e) Written records required by this part shall be maintained so that data therein can be used for evaluating, at least annually, the quality standards of each drug product to determine the need for changes in drug product specifications or manufacturing or control procedures. Written procedures shall be established and followed for such evaluations and shall include provisions for:

(1) A review of a representative number of batches, whether approved or rejected, and, where applicable, records associated with the batch.

(2) A review of complaints, recalls, returned or salvaged drug products, and investigations conducted under §211.192 for each drug product.

Procedures shall be established to assure that the responsible officials of the firm, if they are not personally involved in or immediately aware of such actions, are notified in writing of any investigations conducted under §211.198, 211.204, or 211.208 of these regulations, any recalls, reports of inspectional observations issued by the Food and Drug Administration, or any regulatory actions relating to good manufacturing practices brought by the Food and Drug Administration (FDA).

This section was last amended at 60 FR 4091, January 20, 1995.

The intent of subsections (a) and (b) is to have available records for review for a reasonable time after the expiration date of the product. The period chosen was based on FDA experience as to the time when the records are most likely to be needed. The prudent manufacturer will keep these records until the statute of limitations runs out for liability to the consumer.

In general, the regulations do not require retention of original records; retention of suitable true copies such as on microfilm or portable document format (PDF) files are permissible, provided equipment for reading of the microfilm record and the making of hard copies is available. The records must be available at the location at which they were generated, but not necessarily stored there in hard form, if the record can be readily accessed on demand, such as from a data processor terminal or by electronic transmission.

It should be noted that the FDA does not have the authority to inspect records regarding the manufacture of non-prescription drugs that are not "new drugs" as defined in Section 201(p) of the Federal Food, Drug and Cosmetic Act. The FDA reviews records under its mandatory inspection authority contained in Section 704 of the act and further extended in the 1953 amendments contained in Public Law 82-217, which established section 704(a), and in the Drug

Amendments of 1962 contained in Public Law 87-781 (1961). Note, however, that manufacture of any drug product without compliance to good manufacturing practices makes the product adulterated under Section 301(b) of the Act and a federal crime under Section 303 of the Act. The FDA, therefore, if it had reliable information that a non-prescription or old drug product was being manufactured in violation of current good manufacturing practices, could obtain a search warrant that would authorize the inspection of the records to seek evidence regarding the alleged criminal offense.

In short, while the FDA does not have authority to inspect the records required under this subpart, or even inquire into the existence of such records for non-prescription drugs that are "not new," the records may be obtained or their absence determined under a search warrant obtained on reliable information, such as provided by an FDA inspection, that there was violation of current good manufacturing practices in the production of the product.

Section (c) authorizes the FDA to copy the records and, by implication, to remove the copies from the premises. These copies are required to be released under the Freedom of Information Act, unless they contain information of such a nature that a request for release could be denied. Denial would usually be based on the prohibition of disclosure of "trade secret" information in FDA files in 21 CFR 20, particularly in Section 20.61.

Subsection (e) requires a review, at least once a year, of the quality standards for each drug product in order to determine if there are any needs for changes in specifications or controls. It also requires that written procedures shall exist for how the evaluation is to be made. It is suggested that a review schedule should be set within the quality control unit (QCU) in order that the burden of review is spread throughout the year.

The requirements for annual reporting are contained in 21 CFR 314.70 (d) and 314.81 (b)(2). The FDA has been concerned at the variability of submitted reports and in some instances the absence of data.

The annual review of data should not be considered as a bureaucratic exercise for the benefit of the FDA. It should be used by the management team as basic data to drive improvement. In addition to the data required by the FDA, management should review deviations, reworks, rejections, and complaints (and other customer feedback related to quality). The annual report should be the basis for a life-cycle approach to product quality. It provides important input for in-house pharmacoepidemiologic analysis.

Prior to 1994, §211.180(e)(l) required review of every batch, but the regulation was changed. With the current application of computerized systems such as LIMS, it is possible to continuously evaluate data for trends.

In August 2003, the FDA issued a draft guidance for Providing Regulatory Submissions in Electronic Format for Annual Reports for New Drug Applications (NDAs) and abbreviated NDAs (ANDAs). This guidance is one in a series of guidance documents intended to assist applicants making regulatory submissions in electronic format to the FDA. The document discusses issues related to the electronic submission of annual reports for NDAs and ANDAs. Documents that qualify for electronic submission are listed in public docket number 92S-0251, as required by §11.2 (21 CFR 11.2). The most recent version of the guidance is available at the CDER guidance page at http://www.fda.gov/cder/guidance/index.htm.

The documents for the annual report should be divided into different types based on the regulations. Reports for non-clinical [§314.81(b)(2)(v)] and clinical studies [§314.81(b)(2)(vi)] and information for CMC [§314.81(b)(2)(iv)] should be organized as described in the guidance for industry on Providing Regulatory Submissions in Electronic Format-NDAs or as described in the guidance Providing Regulatory Submissions in Electronic Format-ANDAs. All other documents for the annual report should be placed in a folder named US FDA Form 2252 should be provided as a single PDF file.

Subsection (f) speaks to FDA experience that corporate officials were not advised of potential or real adverse conditions uncovered by the firm's own quality assurance system or by the FDA. Corrective actions that might have been taken therefore were not implemented. Correspondence by the FDA regarding findings on inspection and recall are directed to the corporate officials, but not all items in Section 211.180(f) are necessarily the subjects of FDA inspection and recall. In a professionally managed QA/QC operation, there should be a clearly written communication procedure that clearly identifies what quality issues are to be reported, to whom, and at what frequency—immediate, monthly, quarterly, annually. This is essential if senior management is to be involved in the drive for quality compliance and improvement. The subsection also enhances the ability of the corporate officers to carry out their very rigid legal duties to take action on conditions leading to drug adulteration as shown in *United States vs. Park* (421 U.S. 658, 1975) and *United States vs. Dotterweich* (320 U.S. 277, 1943).

§211.182   EQUIPMENT CLEANING AND USE LOG

A written record of major equipment cleaning, maintenance (except routine maintenance such as lubrication and adjustments), and use shall be included in individual equipment logs that show the date, time, product, and lot number of each batch processed. If equipment is dedicated to manufacture of one product, then individual equipment logs are not required, provided that lots or batches of such product follow in numerical order and are manufactured in numerical sequence. In cases where dedicated equipment is employed, the records of cleaning, maintenance, and use shall be part of the batch record. The persons performing and double-checking the cleaning and maintenance shall date and sign or initial the log indicating that the work was performed. Entries in the log shall be in chronological order.

This section requires written designation of which equipment is "major." The intent of the regulations is not to include small items such as ladles, scoops, stirrers, and spatulas. The exclusion of "non-major" items from the record-keeping requirement does not, however, exclude them from the requirements that they be properly cleaned.

Because the log is for a repetitive operation, the record may be initialed rather than signed. Note that a separate log, which may be a completely separate bound volume, or consecutive pages in a bound or loose-leaf format, or a number of individual records or logs is required for each piece of major equipment that is not dedicated to the manufacture of a single product. The issue of signatures and initials has involved considerable industry–FDA interaction. As new computerized technology became available it has been possible to move to paperless control of manufacturing processes. These computerized controls had several advantages over manual systems:

- More consistent control.
- Only approved (trained) personnel could perform a process.
- Processing could be prevented until any prior steps or checks were performed.
- Precise recording of the times of operations were possible.

Although routine maintenance and adjustment are not specifically required to be logged, it is strongly suggested, because it is just this information which will be needed to show compliance with §211.67. The data referring to cleaning should be incorporated into the appropriate batch record so that this information is readily available for review prior to release. For units that are fully dedicated, the information may be part of the batch record. Even though there is this exemption for dedicated equipment, it is suggested that the log be kept since it is easier to retrieve a history of cleaning and maintenance from a record dedicated to logging such information than it is from perusal of individual batch records. An annual review of equipment function is implicit in §211.180(e)(l).

§211.184   COMPONENT, DRUG PRODUCT CONTAINER, CLOSURE, AND LABELING RECORDS

These records shall include the following:

(a) The identity and quantity of each shipment of each lot of components, drug product containers, closures, and labeling; the name of the supplier; the supplier's lot number(s) if known; the receiving code as specified in §211.80; and the date of receipt. The name and location of the prime manufacturer, if different from the supplier, shall be listed if known.

(b) The results of any test or examination performed [including those performed as required by §211.82(a), §211.84(d), or §211.122(a)] and the conclusions derived therefrom.

(c) An individual inventory record of each component, drug product container and closure and, for each component, a reconciliation of the use of each lot of such component. The inventory record shall contain sufficient information to allow determination of any batch or lot of drug product associated with the use of each component, drug product container and closure.

(d) Documentation of the examination and review of labels and labeling for conformity with established specifications in accord with §211.122(c) and §211.130(c).

(f) The disposition of rejected components, drug product containers, closure, and labeling.

The regulations require identification and recording of the name of the producer of components, product containers, closures, and labeling. Where the name of the producer is not known, the supplier must be identified. This provides for the use of agents who may be unwilling to divulge their source of supply. However, without knowledge of the actual producer, it is not possible to evaluate the producer's facility or integrity and the agent may be tempted to switch producers for various business reasons. This could have significant impact especially with respect to raw materials, which may have different impurity profiles, different processibility, and different stability. Consequently, purchasing through such agents should be discouraged.

As with every aspect of the regulations, documentation is required—in this instance, reporting of test results and conclusions reached and dispositions.

Subsection (c) requires that individual inventory records for product containers and closures (but not other packaging materials) be maintained to allow for the identification of the specific lots used in each batch of drug product. Components are to be treated similarly but there is an additional requirement for reconciliation of usage. These procedures are important in the effective evaluation of production problems and consumer queries. Action levels should be established for component usage, based on historical data, and any usage falling outside of these levels should be brought to the attention of the management and be investigated.

Subsection (d) requires that records for the disposition of rejected components, drug product containers, closures, and labeling be maintained.

§211.186 MASTER PRODUCTION AND CONTROL RECORDS

(a) To assure uniformity from batch to batch, master production and control records for each drug product, including each batch size thereof, shall be prepared, dated, and signed (full signature, handwritten) by one person and independently checked, dated, and signed by a second person. The preparation of master production and control records shall be described in a written procedure and such written procedure shall be followed.

The master production and control records for each drug product describe all aspects of its manufacture, packaging, and control (inspection and testing). The preparation by one competent individual and independent verification of its correctness with endorsement and dating by both parties is a basic concept of good manufacturing practices. Competence infers the possession of sufficient knowledge, through academic training and experience, to allow proper compilation and checking. The two individuals involved with each master record are required specifically to sign, not initial, the document. The preamble to the CGMPs infers that it is not always possible to decipher an initial. The same is true for some signatures and the current expectation of the agency is that master production and control records contain personnel identification and signature lists within the documentation. Most companies also maintain a master signature list to ensure interpretation of signatures over time.

Specific reference is made to the requirement for master production and control records for each batch size. For manufacturing and packaging operations this is important in order to eliminate the need for recalculation of quantities of components, packaging materials, and in-process samples. This is not relevant for control documentation where specifications and finished product testing are not usually related to batch size.

This is the only section of the CGMPs that specifically uses the term "handwritten" with respect to signature or initial. The proposed revision to 21 CFR Part 11 retains this requirement but allows the use of more recent technology. "The act of signing with a writing or marking instrument such as a pen, or stylus is preserved. However, the scripted name, while conventionally applied to paper may also be applied to other devices which capture the written name." The second check, signature is not required to be handwritten and the electronic alternative proposed in Part 11.3 would be acceptable.

The FDA considers part 11 to be applicable to the following records or signatures in electronic format (part 11 records or signatures): electronic signatures that are intended to be the equivalent of handwritten signatures, initials, and other general signings required by predicate rules. Part 11 signatures include electronic signatures that are used, for example, to document the fact that certain events or actions occurred in accordance with the predicate rule (e.g. approved, reviewed, and verified).

Surprisingly, with so much emphasis on signatures and initials, largely as a basis for potential legal cases, there is no specific mention, in either the original CGMPs or the proposed revision to Part 11, that any handwritten form should be indelible. However, industry has long accepted that pencil is inappropriate and the FDA has repeatedly cited companies when white correction fluid was used to obliterate an incorrect entry. The agency intends to exercise enforcement discretion regarding specific part 11 requirements related to computer-generated, time-stamped audit trails [§11.10 (e), (k)(2), and any corresponding requirement in §11.30]; this will in effect link the documented information to the person documenting the activity.

(a) Master production and control records shall include:
    (1) The name and strength of the product and a description of the dosage form.

The product name is usually the manufacturer's trade or proprietary name and should be used consistently in all documentation. Dosage form refers to tablet, capsule, injection, and so on. Since many pharmaceutical products are manufactured with more than one strength this should be clearly obvious in the master documentation.

(2) The name and weight or measure of each active ingredient per dosage unit or per unit of weight or measure of the drug product and a statement of the total weight or measure of any dosage unit.

Master records usually record weights and measures of both active and inactive compounds per dosage unit for solid dosage forms such as tablets and capsules, whereas for liquid dosage forms percentage is more common.

(3) A complete list of components designed by names or codes sufficiently specific to indicate any special quality characteristic.

Components are usually specified by name and by an internally generated alphanumerical code. This double identification, although frequently used primarily for accounting purposes, helps to reduce the potential for usage of incorrect components—particularly if the chemical name is complex or similar to other materials. This is particularly relevant with regard to different varieties of the same component such as hydrated/anhydrous (e.g., citric acid) or crystalline/powder.

(4) An accurate statement of the weight or measure of each component, using the same weight system (metric, avoirdupois, or apothecary) for each component. Reasonable variations may be permitted, however, in the amount of components necessary for the preparation in the dosage form, provided they are justified in the master production and control records.
(5) A statement concerning any calculated excess of component.

In order to prevent errors, all component weights for a product should be stated in the same system (metric, avoirdupois, or apothecary). Ideally, only one unit such as kilos or pounds, should be used throughout with a consistent policy with respect to zeros and decimal points. However, where there are significant variations such as kilos and milligrams, the consistent use of one unit may be confusing to operators (e.g., 100 mg would be 0.0001 kilo). In these instances, it may be better to use both designations, for example, 0.0001 kilo (100 mg). There is also advantage in using the same weight system throughout a production facility to reduce the potential for misunderstanding when operators transfer between processes.

Theoretical variations in the amount of components are permitted, provided they are justified in the master records. Where variations are routine, such as a standard overage to accommodate processing losses, they should be included in the master production record so that individual calculations are not required. Other routine variations include adjustment of the amounts of components in response to assay variations. In these instances the amounts required will vary between different batches of components and it is not possible to include a standard overage in the master production record. The actual quantities are to be approved by QC, but inclusion of a "generic" calculation in the master batch record will minimize the potential for calculation error. These assay-related adjustments should not be applied to allow usage of out-of-specification components; batches of components that are within specification but which show atypical assay results should also not be used without adequate review and resolution of the causes of the atypical situation.

Other component variables include acids and bases used to adjust the pH of solution/suspension formulations. In these instances, the actual quantity to be used is not usually included in the master documentation but is recorded in the operational batch record at the time of use. Some FDA reviewing chemists have been requesting that the master record indicate the maximum amount of acid and/or base that should be used. The rationale for this was reputedly to provide a warning flag if there was something atypical with the batch and to prevent frequent readjustment that could increase the level of sodium chloride in the solution (if hydrochloric acid and sodium hydroxide are used for pH adjustment). It is not known whether these concerns are based on real situations or only represent inherent FDA distrust of industry. However, there would seem to be some merit in including maximum or typical amounts as "action levels," which, if exceeded, would require evaluation and managerial involvement.

(6) A statement of theoretical weight or measure at appropriate phases of processing.
(7) A statement of theoretical yield, including the maximum and minimum percentages of theoretical yield beyond which investigation according to §211.192 is required.

The definition of theoretical and acceptable actual yields at the main phases of processing makes it easier to evaluate and identify causes of discrepancies and to correct them. Acceptable actual yields (action levels) are calculated from historical data and, like all action levels, should be regularly reviewed and revised where appropriate. Production management personnel are usually more concerned about overall variance than individual yield variances on each batch. However, a wide range of yield variations tends to indicate that the process is not fully under control. In such cases some technical evaluation may be required to identify and correct the cause of variability. Refer to Chapter 7 for additional examples of yield calculations.

(8) A description of the drug product containers, closures, and packaging materials, including a specimen or copy of each label and all other labeling signed and dated by the person or persons responsible for approval of such labeling.

The label is the written, printed, or graphic descriptive information placed on the product or on the immediate container. Section 502 of the Act should be consulted for the requirements of label content. Labeling is all other descriptive material, packaging for the product container, and inserts that accompany the product as a part of integrated shipment into interstate commerce (see 21 CFR 321 k and 1).

In *Kordell vs. U.S.* (164 F. 2d 913 and 335 U.S. 345), it was determined that labeling may have a variety of meanings. Any printed or verbal claims relating to a drug product's efficacy or use and that are available to potential customers may be defined as labeling. Labeling does not have to directly accompany the drug product in interstate commerce or be shipped simultaneously with the container. An integrated or related transaction with the function of promotion is sufficient to constitute labeling. The company must, therefore, carefully monitor all types of advertising and correspondence, which may allude to the product in order to maintain control over labeling and to be in conformance with CGMP and the Act. See also Chapter 8.

Labels and labeling copy, that are authentic specimens of those used in production must be attached to the master formula to which the signature and date of approval are affixed to the specimens. Photocopies are insufficient in this instance, since colors and paper quality are not apparent. The master batch records do not need to include advertising-related items of labeling.

Master formula labels and labeling serve as originals against which all incoming copy, designated for production, are compared prior to release. As such, they must be kept current, reflecting all changes in dosage levels, indications, contraindications, administration, warnings, and other information. The use of sequential revision number prefixes or suffixes on the basic label or labeling identification code number is strongly recommended to achieve the desired control.

Current specimens of labels and labeling that are attached to the master formula must be signed and dated by the person responsible for their approval. Most companies have a formal procedure for the approval of new and modified labeling. The review process includes several disciplines or functions, each with defined responsibilities. These usually include medical, legal, drug regulatory affairs, production, marketing, and quality assurance. It would seem unnecessary to require signatures from each of these functions on the master labeling, and usually only the person with responsibility for the overall process will sign. The records maintained by this person will, however, include the sign-off by each function.

Means for determining whether the labels and labeling being used in production exactly match the sample specimen attached to the master must be established. Alpha or numerical code designators and machine-identifying bars are two alternatives.

(9) Complete manufacturing and control instructions, sampling and testing procedures, specifications, special notations and precautions are to be followed.

The master manufacturing and control records should provide sufficient details to ensure that different, but properly trained, people will perform the process similarly. These documents are critical with respect to assuring consistent quality. If a document is too detailed it may not be read or followed; individuals may try to work from memory or establish personal informal and unapproved procedures. If details are insufficient, there may be too much opportunity for individual judgment.

The master manufacturing records should clearly identify:

1. Name of product, product type, strength;
2. Ingredients to be added: name, alphanumeric code, amounts or dosage unit, or percentage;
3. Amount of each ingredient for a batch;
4. Sequence of adding ingredients;
5. Equipment to be utilized designated by name and, where appropriate, by number;
6. Processing steps with details of conditions such as time, temperature, speed;
7. In-process samples, testing, acceptance criteria;
8. Special precautions and hazardous conditions which exist and the necessary safety equipment to be used;

9. Theoretical yields and actual yields (action levels);
10. Space for signature and date of operator/supervisor performing or checking each significant step.

A second document, designed essentially on the same basis, is provided for the packaging operation (master packaging record). The master packaging record should include:

1. Name of product, product type, strength;
2. Product specifications;
3. Test methodology.
4. Sampling requirements. In-process sampling, testing, and acceptance criteria may be included in a separate SOP rather than a part of the master packaging documentation.
5. Reduced testing criteria, if appropriate.
6. Action levels beyond which QC management are to be alerted for review, comment, and action.

Pharmaceutical master records are generally very detailed and in some cases quite long (collectively hundreds of pages for one complete batch). Many firms have tried to simplify the format, instructions, signatures and so on. while still retaining the necessary detail and required documentation. Some techniques for simplification and error prevention include:

- Standardizing the columns on each page such as a five-column approach (instruction, control value or range, actual result, performed by, verified by).
- Instructions are short statements highlighting the verb to indicate the desired action (add the ingredient, mix the solution, verify the connection etc.)
- Grey out boxes, columns that need not be completed.
- Using color (text, boxes, key instructions) is useful to differentiate or call attention to:
- Using visuals—pictures and icons are popular. In addition to safety figures, icons can be used to denote specific types of checks or operations (e.g., a picture of a two-person figure where a double check is required). Icons are also used to designate critical process parameters (CPPs), regulatory commitments, and corrective and preventive action (CAPA) (things that cannot be changed).

§211.188   BATCH PRODUCTION AND CONTROL RECORDS

Batch production and control records shall be prepared for each batch of drug product produced and shall include complete information relating to the production and control of each batch.

The batch production and control record follows every production batch through the plant. It provides a detailed description of all processing operations and controls, when they are performed, by whom, and where.

Production and control operations occur at different locations within the plant. The batch records that accompany material through processing provide information for operators and also serve as a means for documenting which ingredients were added, which control measures were exercised in process and final assay of the drug product, and the huge amount of information produced during the manufacturing cycle. Because this flow of information accompanies the product through all operations, the medium of transmission must be durable and provide protection for the forms which it encloses. Since it is advisable to keep the manufacturing and packaging portions of the batch record together during these operations, many manufacturers keep batch records for a single production cycle consolidated in a polyethylene bag. In order to minimize handling and possibility of loss, laboratory records for the batch may be added just prior to release review by the control section. In addition to the information that is attached to the batch production record, the departments contributing to the manufacturing cycle must retain accurate records and comments about operations within the department.

It is insufficient if the production and control records for a released batch are correctly completed but are not available for rapid retrieval from archives. This section suggests that all completed records associated with a single batch of production be consolidated and filed in order to ensure ready access. Since this section implies that an identifying control number be assigned to each individual batch, this number could serve as a means of indexing these record files.

There is no single correct method for assigning control or lot numbers to production batches. Many larger companies utilize a production planning function that coordinates market requirements, inventory levels, and projected manufacturing necessities. This function may assign control numbers sequentially to the batch formulas as released to production, or each product may have a block of control numbers assigned to it over a specific time period. The former method permits a general idea of when the product was manufactured; the latter indicates how many batches of a specific product have been processed. Both are compatible with the first-in/first-out method of inventory control.

These records shall include:

(a) An accurate reproduction of the appropriate master production or control record, checked for accuracy, dated, and signed.

These instructions, procedures, controls, and specifications established in the master batch formula must be duplicated to serve as a guide for the actual production operations. Accurate duplication may be achieved by copying the master by hand, mimeographing, photocopying, or computer printout methods. Copying the master for each production batch manually is generally less accurate and less economical; therefore, the last three methods are recommended. The volume of operations and equipment are the ultimate determinants.

Although the necessity for the checking of automated copying has been questioned, errors can occur when the copy is blurred, contains extraneous specks, or is partly cut-off. Each page of each copy, therefore, must be checked, dated, and endorsed by an individual able to detect any error, either centrally or by a department supervisor, before manufacturing operations may commence. This may be done within each production department or by the control function by direct comparison with a verified copy of the master, or by the data processor itself if it has programming for verification.

(b) Documentation that each significant step in the manufacture, processing, packing, or holding of the batch was accomplished, including:
  (1) Dates;
  (2) Identity of individual major equipment and lines used;
  (3) Specific identification of each batch of component or in-process material used;
  (4) Weights and measures of components used in the course of processing;
  (5) In-process and laboratory control results;
  (6) Inspection of the packaging and labeling area before and after use;
  (7) A statement of the actual yield and statement of the percentage of theoretical yield at appropriate phases of processing;
  (8) Complete labeling control records, including specimens or copies of all labeling used;
  (9) Description of drug product containers and closures;
  (10) Any sampling performed;
  (11) Identification of the persons performing and directly supervising or checking each significant step in the operation;
  (12) Any investigation made according to §211.192;
  (13) Results of examinations made in accordance with §211.134.

The documentation requirements record that the steps referenced in §211.186(b)(9) have been performed. The regulations require two signatures for significant steps: that of the individual performing the step and the supervisor for confirmation. This places an important, yet difficult, responsibility on the supervisors. If the supervisor is actually present when the operation is performed then the signature for confirmation creates no problems. However, in real life a supervisor has many activities with several processes and people to supervise and it is not always practicable to be present when a specific operation is being performed. The supervisor must then exercise some judgment. Some examples illustrate the point.

1. To confirm that specific materials have been added to a batch, it may be acceptable to check the labels from the dispensed materials. However, if confirmation is required of the weight of material added, then the supervisors would need to be present during the weighing operation.
2. To confirm that a material has been dried for a specified period of time and at a specific temperature, checking of the oven chart should be sufficient. But to confirm the pH of a solution, unless the instrument provides a printout, the supervisor would need to be present.
3. To confirm that in-process weight checks have been performed properly and at the required frequencies, the supervisor may be able to rely on review of the operator data possibly coupled with an occasional weight check himself.

The need to clearly reference the specific piece of equipment used has previously been noted. In the event of a problem it may be important to identify all key parameters associated with the process stage in question: materials, process, operator, supervisor, equipment, environment. Without this information a comprehensive evaluation of the problem to identify the root cause may be unsuccessful.

Section 211.188(b)(6) requires that there shall be an inspection of the packaging and labeling area before and after use. This can be met by incorporating into the records a line clearance form which:

1. Notes the previous product packaged on the equipment.
2. Confirms that the equipment has been properly cleared of components from the previous run. If the new packaging run is the same product, then confirmation of the removal of precoded labeling components and of bulk and finished product may be adequate.
3. Confirms that all components from the previous run, except as noted in (2), have been removed from the vicinity.
4. Confirms the presence of the required components for the run about to commence.

A similar procedure is recommended for manufacturing operations. This necessity to confirm readiness for use does benefit from the availability of equipment and facility status labeling; obviously, it is easier for a supervisor to sign for readiness if in addition to the check there is documentation to confirm cleaning and clearing of the previous batch of product.

The batch record must incorporate the complete labeling controls record and must contain a specimen or copy of all labels. A specimen is preferred since this gives a more accurate picture, especially with respect to color. A copy is acceptable if a specimen cannot be conveniently prepared, such as from preprinted tubes or ampoules.

A description of drug product containers and closures should include all elements of packaging that can impact on product quality. This will include, in addition to the primary pack, closure, and label, other packaging components such as cap liners used to provide product protection for stability or as tamper evidency, secondary packaging such as cartons which carry product labeling information, tamper-evident and child-resistant features, secondary closures such as the metal coverings holding in place the elastomeric closures of vials, and also tertiary packaging such as shippers and dividers, especially where these play an important role in product protection during storage and transportation.

Subsections (12) and (13) require supervisory review, comment, and satisfactory resolution of any discrepancy or deviation from the standards prescribed. Further, any previous lot or batch of the product which might have been subjected to a similar variation from the specified procedures must be identified and reconciled before the product may be released into interstate commerce for marketing purposes. The recorded commentary should include:

1. Description of the problem.
   a. The specific parameter that exhibits variance from the norm.
   b. Identification of the variance: when, how, and by whom.
   c. Potential extension to other batches of the same product or to different products.
   d. Confirming that the problem is real and not due to atypical reporting or analytical error.

2. Viability of rework or reprocessing of the batch and identification and approval of the method. This is to include reference to any NDA or ANDA and also the need for additional testing, inspection, or stability. Unless the rework procedure is included in the NDA/ANDA, approval from the FDA is required before the reworked product can be shipped. Also, the need for frequent reworking would tend to indicate that the production process needs to be re-evaluated.

3. Identification of the root cause of the problem and initiation of appropriate corrective actions to prevent or minimize the potential for reoccurrence.

4. Appropriate supervisory or managerial review of the entire problem with signature and date.

If a single lot of bulk manufactured product is utilized in more than one packaging order, a record should exist which shows:

1. Each packaging order to which the bulk was assigned.
2. Packaging control numbers.
3. Quantity utilized in each order.
4. Date of each packaging operation.

Conversely, completion of large quantity packaging orders may require product from more than one production batch. If this condition exists, the records must list each batch used. A method must also be defined to indicate when each different production batch entered the packaging sequence and to permit accountability determinations of the total amounts used. Three systems tend to be used:

1. Separate packaging control numbers are used each time a bulk product from a different batch is introduced into the packaging program. This method segments the packaging run with two or more control numbers. The main disadvantage is that the line needs to be stopped, cleared of all bulk and finished product and precoded labeling components, and set up with the new control numbers. This can obviously result in significant disruption of packaging with consequent labor and machine utilization inefficiencies.

2. A single control number can be assigned to the entire packaging order. This obviates the problem referred earlier. A compromise situation may be applicable where multiple batches are involved. When a new batch of bulk product is introduced onto the packaging line, the control number is changed but the line is not cleared from the previous bulk or labeled components; these are simply allowed to be "flushed" through the line by the new batch.

The intent of the above procedures is to ensure that in the event of a recall all products involved can be identified and removed from the market expeditiously. This requires that the batch records for both manufacturing and packaging show, and cross-reference, which batch numbers were assigned to each packaging control number and which packaging control numbers were filled by bulk product from any specified batch. The first alternative described above clearly maintains a discrete correlation between individual bulk batches and individual packaging runs, thereby making any recall specific to the bulk batch or packaging run in question. The use of a single packaging control number associated with several batches of bulk product would require the recall of the entire packaging order even if only one bulk batch was suspect. The productivity benefits of this approach could outweigh the disadvantages for a production process which has been fully validated and whose performance has been confirmed by historical data. However, a further disadvantage relates to the evaluation of complaints. It would not be possible to properly evaluate a complaint on the drug product itself, since there is no way to identify the specific batch of bulk product involved. The third alternative does provide a compromise. In the event of a recall, it might be necessary to withdraw additional batches on either side of the affected batch, since discrete segregation was not possible. But for complaint review most of the distributed units are identified by packaging numbers that do relate to specific bulk batches.

NUMERICAL MATERIAL IDENTIFICATION SYSTEMS

Throughout the text there are references to material identification numbers. At this point it may be useful to more fully describe a pharmaceutical alphanumeric identification system.

1. Raw material, components, and other supplies: Each material should be assigned a specific number that clearly identifies the material. Different physical or chemical forms of the material should be provided with different numbers. When package-labeling text is changed, it is usual to apply a suffix to the existing number.
2. Receivables of raw materials and components: Each receivable should be allocated a sequential stock or receivable number. When more than one supplier lot is included in the receivable, each lot should be given a separate number.
3. Manufacturing batch number: Each scheduled manufacturing batch should be given a sequential number. Often batches manufactured in different departments (e.g., tablets and liquids) are given a different letter prefix. Sometimes a different number, a control number, is assigned to a bulk batch after it has been released by quality assurance. This serves as an additional check that unreleased bulk product cannot be used in the packaging cycle.
4. Product formulations are assigned unique identification, thereby allowing differentiation between products by both name and number.
5. Packaging control numbers are designated to each packaging order to provide a means of correlating packaging and bulk product and also act as the number to be used in the event of customer complaint or recalls.

§211.192   PRODUCTION RECORD REVIEW

All drug product production and control records, including those for packaging and labeling, shall be reviewed and approved by the QCU to determine compliance with all established, approved written procedures before a batch is released or distributed. Any unexplained discrepancy (including a percentage of theoretical yield exceeding the maximum or minimum percentages established in master production and control records) or the failure of a batch or any of its components to meet any of its specifications shall be thoroughly investigated, whether or not the batch has already been distributed. The investigation shall extend to other batches of the same drug product and other drug products that may have been associated with the specific failure or discrepancy. A written record of the investigation shall be made and shall include the conclusions and follow-up.

This section requires that a product be released only after review of the entire batch record for compliance with approved written procedures. That the intent of the section has been carried out can be shown by the use of a checklist which defines the specific documents which should be in the batch record and what is to be checked on each document. If the release criteria are not stated on the batch record they should be included on the checklist. Items to be entered on the checklists would include:

1. Batch record is current and approved as an accurate copy.
2. Correct, released, components were used in manufacturing.
3. Correct quantities of components were used in manufacturing.

4. All components were within the retest dating period.
5. Manufacturing control document is properly completed.
6. Correct product was packaged.
7. Correct packaging components were used.
8. Labeling bears the correct control number, expiry date.
9. Yields and accountability are within action levels.
10. Packaging control document was properly completed.
11. Test data, in-process, and control laboratory are within specifications.
12. Retained samples have been taken.
13. Written investigation of any deviation from procedure, with any approvals and data to support remedial action.

A full and comprehensive review of every aspect of the manufacturing, packaging, and control documentation is very time consuming and occasionally identifies more than the absence of signatures or the misplacement of a document. Consequently, the emphasis must be on the operations themselves, ensuring that all employees understand the importance of the procedures and the need to follow them or document atypical or non-complying situations; and that supervisors and managers pay enough attention to this during their routine activities.

When production deviations occur they must be documented, investigated, and appropriate levels of management must be involved in the review of the data and in any decision-making. It is particularly important to decide what actions are required to minimize the potential for reoccurrence—retraining, process improvement, revalidation.

FAILURE INVESTIGATIONS
Joseph D. Nally and Laura L. Nally

Over the last decade, written failure investigations have evolved into a very comprehensive and complete document. In practice, they are primarily used to help justify batch release but are more important in determining root causes and eliminating deviations, variations, and problems. Figure 1 is an example of an investigation process.

All investigations start with a failure event or deviation. The first step is to determine whether the event is reportable and requires an investigation. Some firms have developed a three-tier approach to reportable events:

1. Minor reportable event—no product impact. May require limited immediate action but does not require an investigation. Minor reportable events are tracked for repeat occurrences or trends. Multiple recurrences would generate a trend investigation.
2. Major reportable event—potential product impact, requires a full investigation, root cause analysis, and CAPA.
3. Critical reportable event—potential adverse product impact or failure, requires a full investigation, root cause analysis, CAPA, and senior management notification.

As an aid to making the criticality decision, the use of FMEA and other tools are recommended. See Investigation Tools list table.

The quality assurance organization needs to be intricately involved in the investigation process. In some organizations they conduct the investigation; in others they approve the decisions/go forward steps. At a minimum they need to be involved in the decision-making for event classification, root cause determination, and overall report approval. The practical application is that quality assurance or the QCU responsible person will likely have to answer the FDA investigators' questions about the investigation scope, content, decisions, and CAPA and it helps if that person was involved in the investigation. Failure investigations have a 90% plus probability of being requested/reviewed on any FDA inspection (see Chapter 21).

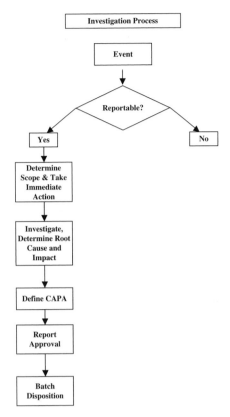

FIGURE 1 An investigations process.

An example of a comprehensive, stand-alone investigation report is presented in the following table.

Section	Content
Description	Complete but concise presentation of event resulting in problem identification.
Immediate action	Complete description of the who, when, and how of event notification including any decisions made and actions taken.
Background	Information needed to put the event into context. Section should be designed to give an independent reader enough information to fully understand the event and related facts.
Chronology	Outlines the sequence of events (days/hrs/mins) before, during, and after the event.
History/trends	Based on database queries and searches, a complete history of prior or related events is presented along with a conclusion of how it relates to the current investigation.
Investigation text: PEMME or logical analysis of events/conclusions	Each cause and effect branch (PEMME-Personnel, Equipment, Materials, Methods, and Environment) is brainstormed for potential impact/involvement/cause and investigated. The facts and relevance of each branch is documented in the investigation.
Impact	GMP and material or batch impact is evaluated based on the investigation scope and facts.
Root cause	Using the PEMME template, the impact or cause conclusion is listed for each branch and by process of elimination the root cause or most probable cause is identified.
CAPA	Actions that in total provide a correction, a corrective action (includes elimination of recurrence), and if applicable a preventive action (e.g., trends indication a potential failure).

A stand-alone report means that the report can be read and understood by an independent reviewer without looking for other documents and relying on interviews to recreate the event, investigation, or root cause analysis.

The investigation process can be simple to quite detailed depending on the available facts and knowledge/skills of the investigator and investigation team. Typically, for major and critical events/investigations there are two predominant tools used: cause and effect analysis known as the PEMME Fishbone analysis or the five "Whys." Figure 2 shows an example of the PEMME approach.

Each leg of the fishbone is brainstormed with possible causes of the event. It is very important not to jump to an immediate conclusion (usually based on some expert's idea of the cause) and go through a logical process of elimination in determining root cause.

The five "Whys" is a fairly simple process by starting with the event and asking why the event happened and then why the first cause happened until one arrives at the apparent cause (after about three or more "Whys") and the root cause (after five or more "Whys"). The trick to the five "Whys" technique is asking the right why.

A real example of the use of the five "Whys" is presented:

- *Event*: Process step in bulk manufacturing process was not completed within the required timeframe per batch record requirements.
- *Investigation*: Valve sequencing was performed incorrectly during the manufacturing process.

Why # 1: Root cause (per original investigation): Operator failed to follow the SOP describing the valve sequencing process.

Corrective action: Retrain operator on the appropriate SOP.
    Investigation was questioned by quality assurance and was reopened.

Why # 2 (after investigation originally closed out as per above): Why did operator not follow the SOP (per operator interview)?

Answer: To follow the SOP exactly as written would have created a potential safety hazard by opening a steam valve prematurely.

Why # 3 Why did not the operator point this out to supervision and request a revision of the SOP or batch record?

Deviation Investigation Model

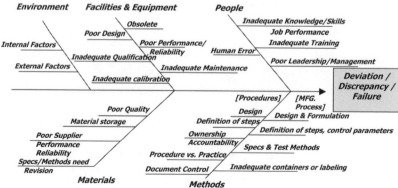

FIGURE 2   The PEMME approach.

Answer: Operator did point this out to supervision and redrafted SOP with correct valve sequencing.

Why # 4 Why was not the new SOP put into effect?

Answer: It was bogged down in the SOP review/approval process (for months)

Why # 5 Why was it still not through the approval process?

Answer: Because other reviewers wanted to make additional changes, SOP was not given a priority status and no one was taking ownership in oversight of the SOP revision/review process. No metrics were in place to evaluate cycle time for SOP revisions and no effort was made to expedite those revisions where safety was at risk.

Thus, the ultimate root cause was the failure in the overall document control/approval process and not in the training system that was originally indicted after Why # 1. It could also implicate the change control process as making a change in an operational procedure should have required a change control review.

There are a number of tools that can be used during an investigation to prioritize the event, perform the data analysis, arrive at the correct root cause and CAPA that actually eliminates the event from recurring. The following list of tools are presented:

1. Problem understanding and collecting information:

   - Interviews—obtain direct information from involved parties.
   - Surveys—collect data/information from a select group.
   - Sampling—representative data from a targeted population.
   - Flowcharts—identifies sequence of events in a process.

2. Determining risk and potential impact

   - Severity and probability matrix—identifies degree of GMP risk
   - FMEA and FMECA [failure modes effects (critically) analysis]—identifies potential failure points.
   - FTA (fault tree analysis)—identifies elements of risk.

3. Data (numbers) collection and analysis

   - Pareto Chart—finds the few elements causing the most effects.
   - Check Sheet—registers data in a systematic fashion.
   - Histograms—portray data graphically.
   - Run Charts—detect trends or patterns.
   - Scatter Diagrams—find relationships between two variables.
   - Control Charts—detect variability, shifts, and trends.
   - Process Capability (Cp, Cpk, Cpu, Cpl)—determines capability of process to meet predetermined specifications.

4. Working with problem and cause information/ideas:

   - Brainstorming/Brainwriting—generate as many ideas as possible.
   - Cause and Effect Diagram (Fishbone PEMME)—generate and group problem causes.
   - Why/Why Diagram and Analysis—identifies chain of cause and effect.
   - Nominal Group Technique—prioritizes ideas.
   - Affinity Diagrams—finds relationships otherwise not easily seen.
   - Relations Diagram—find relationship among many elements.
   - Matrix Diagram—analyze causal relationships.
   - Prioritization Matrix—systematic approach to narrowing down options by comparing choices.
   - Interrelationship Digraph—systematically identify, analyze, and classify cause and effect relationships.

5. CAPA implementation

-   Force Field Analysis—identifies factors that support or work against a solution.
-   Tree Diagram—maps tasks for implementation.
-   Gantt Charts—details tasks and implementation timelines.

The quality of the most probable or root cause analysis part of the investigation is critical in achieving real CAPA. Often the process is stopped before the real root cause is revealed. This happens because of time limitations, cost of permanent fixes, and often political considerations. In the editor's experience, if the predominant root cause is people, then the analysis has not gone far enough. The next "Why" is "Why did the people/personnel fail?" Typically, this will raise a number of additional questions such as definition/adequacy of methods/instructions, adequacy of equipment, adequacy of management/supervision etc. A generic root cause list is presented. Although some of these causes may be controversial, it is based on the editor's experience.

Root Cause List
People/Personnel:
1. Knowledge/skills/understanding deficient due to:
    a. Inadequate job description, instruction, or training
    b. Person not capable
2. Human error:
    a. Careless
    b. Human mistake
3. Poor quality leadership/management due to:
    a. Lack of planning
    b. Inadequate resource allocation
    c. Inadequate communication
    d. Lack of verification and follow-up
    e. Lack of accountability

Equipment and Facilities:
1. Poor design:
    a. Equipment design inadequate or obsolete for use (capacity, tolerance, speed)
    b. Facility/room/area design not adequate for use (size, environment, finishes)
2. Inadequate qualification:
    a. Facilities or equipment not qualified, capability is unknown or not documented.
3. Inadequate performance or reliability:
    a. Facility/room/area fails to maintain specifications (also see design)
    b. Equipment breakdowns (unpredictable); capability or reliability unknown
    c. Equipment not calibrated (also see methods)
4. Inadequate maintenance (also see methods)
    a. Lack of or inadequate facility maintenance (unscheduled/reactive, routine, preventive, or predictive maintenance)
    b. Lack of or inadequate equipment maintenance (unscheduled/reactive, routine, preventive, or predictive)

Materials:
1. Poor quality
    a. Unacceptable variation (in or out of specification)
    b. Wrong design/item for use
    c. Wrong specification for use
    d. Wrong test method (does not evaluate critical material parameters or functionality)
    e. Outdated material (also see methods)
    f. Inadequate container or storage
2. Poor supplier/manufacturer reliability

    a.   Unacceptable consistency (in or out of specification)
    b.   Unacceptable supplier/manufacturer performance (quality, delivery)

Methods:

1. Inadequate manufacturing methods/process/procedures
    a.   Inadequate design of formulation (stability, functionality)
    b.   Inadequate design of manufacturing process (sequence, timing, complexity)
    c.   Wrong or inadequate equipment (also see equipment)
    d.   Inadequate definition of steps, critical parameters in batch records
    e.   Process science not understood (also see people)
    f.   Process is not capable of consistent performance to meet specifications (periodic failures, also see design and validation)
    g.   Process is not adequately validated; critical parameters unknown
    h.   Improper process/product test methods and/or specifications (see support methods)
2. Inadequate support methods/processes/procedures:
    a.   No procedure
    b.   Inadequate design for use (too complicated, too many patches, does not handle exceptions, not fail safe where needed, no feedback or communication loops)
    c.   Inadequate definition or unclear/understandable instructions (critical steps to reproduce the task consistently are not defined in the SOP)
    d.   Ownership (individual) of the tasks and results are not defined
    e.   Accountability for results not accepted (also see people, management)
    f.   Inadequate communication of procedure or results (also see design and management)
    g.   Results of the procedure/process are not measured/trended/communicated (also see management)
3. Inadequate documentation
    a.   Forms missing information, does not reflect task
    b.   Format confusing and not user-friendly
    c.   Obsolete or uncontrolled editions

Environment:

1. External factors
    a.   Weather extremes (temperature, humidity, rain, wind etc.)
    b.   Inadequate on non-existent utilities supply (electricity, steam, gas etc.)
    c.   Inadequate supplier or contractor service (also see methods)
2. Internal factors
    a.   See facilities and equipment
    b.   Inadequate site-supplied services (utilities, support services)

The last and probably most important phase of an investigation is determining the CAPA. The FDA draft guidance on Quality Systems Approach to CGMP describes CAPA as involving three concepts:

- Remedial corrections;
- Root cause analysis with corrective action to prevent recurrence;
- Prevention action to prevent initial occurrence.

This concept is close to the current ISO 8402 definitions of:

- Correction—a re-cleaning, repair, rework etc. in order to dispose of an existing deviation or discrepancy.
- Corrective action—eliminating the cause of a deviation/discrepancy so that it will not occur again
- Preventive action—action taken to eliminate the cause(s) of a potential non-conformity, defect or other undesirable situation in order to prevent occurrence.

An example of the above would be:

- Event—process water pipe developed a leak.
- Correction—pipe was patched to stop the leak.
- Corrective action—pipe segment was replaced.
- Preventive action—because of the age and prior history of sporadic pipe failures, the whole distribution system will be replaced during the summer shutdown.

After the investigation process and report is complete, QA has to evaluate the contents, root cause analysis, and CAPA effectiveness independent of and as part of any involved batch release. No easy task for any complicated failure investigation. This is another reason why the QA responsible personnel must have a high level of knowledge and skills (refer to Chapter 3) to be able to make those decisions and too often to fend off the operational personnel who want to get the batch out of the door.

§211.194 LABORATORY RECORDS

(a) Laboratory records shall include complete data derived from all tests necessary to assure compliance with established specifications and standards, including examinations and assays as follows:

   (1) A description of the sample received for testing with identification of source (that is, location from where sample was obtained), quantity, lot number or other distinctive code, date sample was taken, and date sample was received for testing.

The description of the sample simply requires that there be an adequate identification, such as the name or identification code of the material; it does not mean visual appearance. The lot number specific to the material sampled must also be noted.

Where samples are taken from a representative number of containers, it is necessary to cross-reference the samples to the containers from which they were taken. In the event of any query it is then possible to precisely pinpoint the sample source.

The amount of sample taken needs to be recorded to allow effective reconciliation. For materials in bulk (raw materials, granules, tablets) it may not be necessary to actually weigh or measure each sample. The use of sampling equipment or sample containers which have been "roughly calibrated" may provide data of sufficient accuracy.

The need to record the date of sampling and the date of receipt for testing seem to be of no value. Recording the sampling date is useful since some materials may undergo change, such as moisture pick-up, between sampling and testing. The date of receipt for testing would seem to have little relevance. However, most QC departments do record the date of sample receipt as part of the managerial evaluation of laboratory effectiveness.

(2) A statement of each method used in the testing of the sample. The statement shall indicate the locations of data, which establish that the methods used in the testing of the sample meet proper standards of accuracy and reliability as applied to the product tested. (If the method employed is in the current revision of the United States Pharmacopeia, National Formulary, Association of Official Analytical Chemists, Book of Methods, or in other recognized standard references, or is detailed in an approved NDA and the referenced method is not modified, a statement indicating the method and reference will suffice.) The suitability of all testing methods used shall be verified under actual conditions of use.

Test methodology may be modified with time and it is important that the retained records clearly indicate which methodology was actually used. This is usually done by indicating the monograph reference number or issue date. Obviously, copies of superceded monographs that relate the retained records must be retained.

Analytical methodology must be validated and the validation data must be retained. As written, the regulations would require that for each method used the records for each sample tested should reference the location of the validation data. This would seem to be onerous and unnecessary provided the monograph, or some other procedure, indicates the location. It is assumed that "official" methods have been validated and reference to the official source is considered to be adequate. However, for both official methods and for methods validated in another laboratory, it is necessary to verify suitability in the individual laboratory [§211.165(e)]. The degree of work to verify suitability may vary with the complexity of the method. It may be sufficient to perform the method with samples of known composition or that have previously been analyzed.

(3) A statement of the weight or measure of sample used for each test where appropriate.
(4) A complete record of all data secured in the course of each test, including all graphs, charts, and spectra from laboratory instrumentation properly identified to show the specific component, drug product container, closure, in-process material, or drug product, and lot tested.

It is difficult to understand why it was necessary to specify the recording of "weight or measure . . . where appropriate" since without such data it would not be possible to evaluate quantitative results. The retention of raw data does bulk-up the size of batch records. However, their retention does allow re-evaluation in the event of a future query.

The requirement for a "complete record of all data secured in the course of each test" took on new emphasis when the FDA identified that some companies were reporting only "good" results in submissions. All results must be reported (see also §211.165) and any out-of-specification results fully evaluated and explained.

(5) A record of all calculations performed in connection with the test, including units of measure, conversion factors, and equivalency factors.
(6) A statement of the results of tests and how the results compare with established standards of identity, strength, quality, and purity for the component, drug product container, closure, in-process material, or drug product tested.

As indicated earlier (§211.68), the use of automatic calculation procedures is acceptable; in fact, these are preferable since they minimize the potential for individual random calculation errors.

The results obtained from testing must be compared with "established standards." It is common practice to establish action levels based on historical data. Confirmation that results lie within the action levels may be delegated to a responsible analyst. Results outside of the action level, but inside specification, are usually referred to a more senior individual and require evaluation for possible cause before making a decision on status. The written methodology or procedures should clearly identify these action levels and the review process.

(7) The initials or signature of the person who performs each test and the date(s) the tests were performed.
(8) The initials or signature of a second person showing that the original records have been reviewed for accuracy, completeness, and compliance with established standards.

Records must indicate who performed each test. Either initials or signature are considered adequate. Since the initials and signatures of many people are indecipherable, it is recommended that printed or typed names should also be shown. Alternatively, a master record could be maintained of all departmental initials and signatures. This approach would have benefit anywhere in the production process where signatures or initials are required.

The depth of checking necessary to confirm the accuracy and completeness of the records will vary with the degrees of automation and complexity in the procedures. However, replacement of routine checking by random or periodic audit is not acceptable.

> (b) Complete records shall be maintained of any modification of an established method employed in testing. Such records shall include the reason for the modification and data to verify that the modification produced results that are at least as accurate and reliable for the material being tested as the established method.

It is important to remember that the applicability of test data is dependent on confirmation of the validity of the methodology. Because of the work involved in validating a modified method, it is desirable to make changes only when essential.

> (c) Complete records shall be maintained of any testing and standardization of laboratory reference standards, reagents, and standard solutions.
> (d) Complete records shall be maintained of the periodic calibration of laboratory instruments, apparatus, gauges, and recording devices required by §211.160(b)(4).
> (e) Complete records shall be maintained of all stability testing performed in accordance with §211.166.

Standardized reagents and reference standards play key roles in analytical methodology. Procedures must define how standardization is to be performed, the frequency of re-standardization, and the operational limits. These limits will identify the required degree of precision between replicate results and also the maximum allowable values from nominal. If individual values vary too much from nominal, revalidation may be necessary. Where secondary reference standards are used their suitability must be confirmed by cross-calibration with a defined primary standard.

The subjects of calibration and stability have been addressed previously in §211.160(b)(4) and §211.166, respectively.

This section was last amended in 70 Federal Register 40880, July 15, 2005.

§211.196 DISTRIBUTION RECORDS

> Distribution records shall contain the name and strength of the product and description of the dosage form, name and address of the consignee, date and quantity shipped, and lot or control number of the drug product. For compressed medical gas products, distribution records are not required to contain lot or control numbers.

The primary purpose of this section is to ensure that adequate data are available to access trade customers should a recall be initiated. The recording of lot number to each order will certainly accomplish this purpose; other approaches can achieve the same result. The recording of dates on which a specific lot of product commenced and ceased distribution may be used. All customers receiving the product between these dates could then be contacted. Obviously, on the first and last days of distribution, some of the customers may have received product from the end of the previous lot or the beginning of the next lot. This overlap should in no way adversely impact on the effectiveness of a recall.

Whatever system is used, it must accommodate the re-introduction of returned goods into the distribution chain.

Distribution records include a wide range of documentation such as invoices, bills of lading, customers' receipts, and internal warehouse storage and inventory records. The information required need not be on every document. Also customer codes and product codes may be used as alternates to customer names and addresses and product names.

Compendial articles are to be in compliance with the current compendium when shipped. This can create some difficulties with respect to inventories of components, labeling or products when there are compendial changes. Compliance Policy Guide 7132.02 does suggest that regulatory action should not be taken if a company is using up existing stock of labels in a reasonable time and that the product or material is otherwise in compliance with the new monograph.

§211.198 COMPLAINT FILES

(a) Written procedures describing the handling of all written and oral complaints regarding a drug product shall be established and followed. Such procedures shall include provisions for review by the QCU, of any complaint involving the possible failure of a drug product to meet any of its specifications and, for such drug products, a determination as to the need for an investigation in accordance with §211.192. Such procedures shall include provisions for review to determine whether the complaint represents a serious and unexpected adverse drug experience which is required to be reported to the FDA in accordance with §310.305 and §514.50 of this chapter.

(b) A written record of each complaint shall be maintained in a file designated for drug product complaints. The file regarding such drug product complaints shall be maintained at the establishment where the drug product involved was manufactured, processed, or packed, or such file may be maintained at another facility if the written records in such files are readily available for inspection at that other facility. Written records involving a drug product shall be maintained until at least one year after the expiration date of the drug product, or one year after the date that the complaint was received, whichever is longer. In the case of certain OTC drug products, lacking expiration dating because they meet the criteria for exemption under §211.137, such written records shall be maintained for three years after distribution of the drug product.

    (1) The written record shall include the following information where known: the name and strength of the drug product, lot number, name of complaint, and reply to complainant.

    (2) Where an investigation under §211.192 is conducted, the written record shall include the findings of the investigation and follow-up. The record or copy of the record of the investigation shall be maintained at the establishment where the investigation occurred in accordance with §211.180(c).

    (3) Where an investigation under §211.192 is not conducted, the written record shall include the reason that an investigation was found not to be necessary and the name of the responsible person making such a determination.

Complaints received from consumers, professionals, and the trade serves as a primary means of obtaining feedback about product quality after distribution. It is necessary, therefore, that each complaint or inquiry be evaluated by knowledgeable and responsible personnel.

The records of production, packaging, and distribution of drug and the retained samples provide the basis for assessing the validity and seriousness of the alleged deviations that precipitated the complaint. It is important, therefore, that the records for each production lot are readily available. The complaint file itself also plays an important role in determining whether any other similar complaints have been received on the lot in question, or on any other lots of the same product.

The evaluation of complaints serves several valuable purposes. First, there is the urgent need to confirm whether consumers are potentially at risk and to initiate any appropriate action. A second value is the review of the product and its production process to establish whether any modifications are required. Third is the need to rapidly respond to the customer, thereby attempting to maintain confidence in the product and company.

Various surveys have shown that only a proportion of those people receiving substandard product will actually complain. This needs to be remembered when evaluating the extent of a problem. Some companies have attempted to increase the amount of feedback, in order to gain more information, by making it easier for the consumer to contact the company. The use of toll-free telephone numbers on products has been successful. A secondary benefit arising from such approaches has been an increase in the number of positive suggestions for improvements from consumers.

It should be noted that for NDA and ANDA products, the FDA requires a "field alert report" to be submitted within three days of a company being made aware of any mislabeling, bacterial contamination, any significant deterioration, or failure to meet the registered specifications for a distributed product (§314.81).

This section does not address specific adverse drug reactions. However, reports of adverse drug reactions, if considered atypical with respect to the reaction itself or the frequency, will also require evaluation of the product.

The recording of complaint data should allow examination by product, by lot number, by complaint type, and by pack type and size. These data, along with sales volumes, makes it possible to pinpoint the source of the problem and to monitor trends.

Although complaint data provide useful quality data, this must not be seen as an end in itself. Where complaints continue, it may be worth considering a field evaluation of the product to obtain more extensive information on the potential problem.

This section must also be considered in the context of drug products liability where it has served as a fertile field for exploration by plaintiffs' lawyers. It is, therefore, another instance of where occasional, or as-needed, review in-house by internal local counsel should be undertaken.

EXAMPLES OF RECENT OBSERVATIONS FROM FDA 483 CITATIONS OR WARNING LETTERS

1. There is a failure to follow written procedures for the cleaning and maintenance of equipment used in the manufacture, processing, packing, and holding of drug products [21 CFR §211.67(b)].
2. There is a failure to follow written procedures describing the handling of all written and oral complaints regarding a drug product [21 CFR §211.198].

    For example, "Your firm failed to investigate a complaint to determine the origin of black flakes in a drug product. Your preventative action plan only states that the product is undergoing reformulation to address the issue. Your firm did not determine the impact of these black flakes on the identity, strength, quality, and purity of the drug product."
3. Failure to maintain records of the identity and quantity of each shipment of each lot of components, drug product containers, closures and labeling, and of the name of the supplier [21 CFR §211.184(a)]. For example, your firm does not maintain records of the origins of the used cylinders used in your oxygen, USP manufacture that would allow identification of the original supplier of a cylinder.
4. Failure of the QCU to conduct an adequate investigation of any unexplained discrepancy whether or not the batch has been distributed as required by 21 CFR 211.192. The current inspection revealed that your firm received three complaints: one of a burning sensation from the gas (received November 30, 2004); one that the oxygen in the tanks gave the consumer headaches (received February 2004); and, one complaint of a diesel smell in four oxygen tanks received (received January 2004). The inspection revealed that in all three cases your firm reportedly conducted USP testing on the cylinders, but failed to perform testing for impurities.
5. Written records are not always made of investigations into unexplained discrepancies, nor did investigations of unexplained discrepancies extend to other batches of the same drug product or other drug products that may have been associated with the specific failure or discrepancy [21 CFR §211.192]. Specifically:
    a. Product samples tested in conjunction with a complaint regarding loose caps on nitroglycerin tablets, lot 201105, produced out-of-specification results for assay and content uniformity. There was no examination of product retains or a review of the batch

record. The sample results were invalidated due to product damage from environmental exposure although there was no provision for this in your firm's written procedures.

b. Product samples tested in conjunction with complaints regarding tablets not dissolving for nitroglycerin tablets, lots 201105 and 214605, were tested for all release specifications except assay. There was no justification for omission of this test, nor was there any evidence of an examination of stability lots, retained samples, or a review of the batch records.

c. The reference standard injection following assay and content uniformity testing of hyoscyamine sulfate tablets, lot P3274, failed to show any peaks due to a leaking column. There was no documented investigation of this deviation, there was no assessment of the impact of the leaking column on the hyoscyamine sulfate analysis or any other analysis conducted with the same column, and the hyoscyamine samples were not re-injected.

6. Laboratory records fail to include the initials or signature of the person who performs each laboratory test [21 CFR §211.194(a)(7)]. Specifically, laboratory analysis records for analyses performed on HPLCs and do not indicate performed the injections.

7. Failure to establish master production and control records for each drug product including a statement of the maximum and minimum percentages of theoretical yield beyond which investigation is required [21 CFR §211.186(b)(7)]. Specifically, actual and theoretical yields are calculated after tablet compression, but your firm has not established appropriate specifications for acceptable yields.

8. Complaint records are deficient in that they do not include the findings of the follow-up [21 CFR 211.198(b)(2)]. The batch production and control records do not include complete labeling control records, including specimens or copies of all labeling used [21 CFR 211.188(b)(8)].

9. The batch records lack records of the cleaning for dedicated equipment [21 CFR 211.1821]. Laboratory records do not include a complete record of all data secured in the course of each test, including all graphs, charts, and spectra from laboratory instrumentation, properly identified to show the specific drug product and lot tested [21 CFR 211.194(a)(4)].

10. Failure to clean and maintain the packaging equipment at appropriate intervals to prevent contamination that would alter the safety, identity, strength, and quality of the drug products. Failure to maintain accurate, complete, and current records relating to your participation in an investigation [21 CFR 812.140(a)]. Responsibilities of clinical investigators include maintaining accurate, complete, and current records relating to the investigator's participation in an investigation.

SUGGESTED READINGS

1. Guidance for Industry Part 11, Electronic Records; Electronic Signatures—Scope and Application; U.S. Department of Health and Human Services Food and Drug Administration Office of Regulatory Affairs (Ora), August 2003.
2. Andersen B, Fagerhaug T. Root Cause Analysis Simplified Tools and Techniques. Milwaukee, Wisconsin, Marcel Dekker, NY: ASQ Quality Press, 2000.
3. Swarbrick J. Encyclopedia of Pharmaceutical Technology. 2nd ed. Marcel Dekker, 2002.
4. Turner SG. Change Control. Inter Pharm Press, 1999.
5. Sharp J. Good Pharmaceutical Manufacturing Practice. CRC Press, 2004.

Following are some specific guides provided by the FDA to their field staff and others that are pertinent to this chapter.

SEC. 470.100 ORDERS FOR POST-APPROVAL RECORD REVIEWS (CPG 7132C.07)
Background:

This document states the FDA's policy and procedures for the issuance of orders to conduct record reviews for approved new drug products for human and animal use. During the generic drug investigations, the agency encountered some problems that could not be addressed with traditional legal tools. One such problem involved situations where omissions, inconsistencies, untrue statements of material facts, or outright fraud were found in records (e.g., biobatch manufacturing records) submitted as part of some ANDAs (for human use). Another problem involved departures from approved manufacturing procedures. In these instances, where only a few applications are implicated, the FDA can readily initiate action against specific products found to have been approved on the basis of false or incomplete information, or which are not made in accordance with approved procedures. However, where many applications are implicated, the sheer volume of records which must be reviewed makes it difficult to determine how many products are involved. In the 1989 generic drug cases, the affected firms cooperated by engaging qualified outside consultants to review all records, and report results to the FDA. In the absence of such voluntary cooperation, the agency may have to issue orders requiring firms to conduct such reviews.

The agency has concluded that it has a legal basis for requiring drug manufacturers to conduct and report post-approval record reviews under authority of Sections 505(k), 505(e), 512(e), 512(l), 512(m)(4), 512(m)(5), 701(a), 704(a) of the Federal Food, Drug, and Cosmetic Act (the Act) and the Current Good Manufacturing Practice Regulations for drugs that are enforceable under Section 501(a)(2)(B) of the Act. Sections 505(k), 512(l), and 512(m) of the Act sanction such orders on the basis of a finding that such records and reports are necessary in order to determine, or facilitate a determination, whether there is or may be ground for invoking Sections 505(e), 512(e), and/or 512(m)(4).

Policy:

The FDA may issue an order, requiring a records review and report, where there are questions about the safety or effectiveness of an approved drug, or about the truth or falsity of information submitted in support of the original application, in order to determine whether or not such questions are serious enough to warrant withdrawal of the application approval. Such questions may arise, for example, from findings of non-compliance with approved manufacturing procedures, untrue statements of material facts, fraud, or application omissions and inconsistencies. Such orders shall afford the applicant an opportunity to respond informally to the basis for the order.

Regulatory Action Guidance:

Recommendations for a post-approval record review order may be made by field district offices or scientific review divisions in accordance with the above policy. Recommendations initiated by field district offices should be forwarded to the CDER or CVM compliance office. Recommendations initiated by review divisions within the CDER Office of Generic Drugs, the Office(s) of Drug Evaluation (I or II), or the CVM Office of New Animal Drug Evaluation shall be made in consultation with their respective compliance office and forwarded to the director of the office which approved the application.

The proposed order for post-approval record review should include the following paragraphs [substitute Section 512(l) or 512(m)(5); and 512(e), and/or 512(m)(4) as required]: In accordance with Section 505(k) of the Act, this order requiring a post-approval record review is based on a finding that such records and reports are necessary to determine, or facilitate a determination, whether there is or may be ground to withdraw approval of the drug application(s) covered by this order. Non-compliance with this order would be ground to withdraw approval of the application(s) under Section 505(e).

The order for a records review and report shall afford the applicant an opportunity to respond informally to the basis for the order and may permit the applicant to engage outside consultants to perform all, or part of, the review. The order shall issue over the signature of the director of the office which approved the application. A copy of the order should be sent to the director of the field district office.

Where the applicant fails to comply with the order for records review, or where the results of such a review indicate that withdrawal of application approval may be warranted, existing procedures for initiating withdrawal of application approval apply.

The guidance was originally issued: 6/25/92 and has not been updated.

12 | Returned and Salvaged Drug Products
Subpart K

Joseph D. Nally
Nallianco LLC, New Vernon, New Jersey, U.S.A.

§211.204   RETURNED DRUG PRODUCTS

Returned drug products shall be identified as such and held. If the conditions under which returned drug products have been held, stored, or shipped before or during their return, or if the condition of the drug product, its container, carton, or labeling, as a result of storage or shipping, casts doubt on the safety, identity, strength, quality, or purity of the drug product, the returned drug product shall be destroyed unless examination, testing, or other investigations prove the drug product meets appropriate standards of safety, identity, strength, quality, or purity. A drug product may be reprocessed provided the subsequent drug product meets appropriate standards, specifications, and characteristics. Records of returned drug products shall be maintained and shall include the name and label potency of the drug product dosage form, lot number (or control number or batch number), reason for return, quantity returned, date of disposition, and ultimate disposition of the returned drug product. If the reason for a drug product being returned implicates associated batches, an appropriate investigation shall be conducted in accordance with the requirements of §211.192. Procedures for holding, testing, and reprocessing of returned drug products shall be in writing and shall be followed.

The intent of this section is:

1. To require an examination of the reasons for return in order to decide whether further action is required on the lot, on related lots, or to the storage and distribution chain.

   If the goods are known to have been handled within the normal range of conditions in the distribution chain, it may be adequate to redistribute after visual examination and confirmation of conditions. Consideration must be given to actual temperature, humidity, and storage conditions in relation to the labeled storage requirements. Product security and the potential for tampering must also be considered.
   Where the distribution or storage conditions are unknown or have been extreme, then return to stock, as is or after reprocessing, can only be considered after appropriate evaluation, which confirms no unacceptable deterioration. If the product testing is deemed necessary, a product stability test regimen is typically applied. However, testing alone does not provide absolute assurance every unit is within specifications.

2. To maintain full and comprehensive records to allow identification of returned goods distribution and accountability in the event of a recall. This places a considerable burden of work on those responsible for distribution, since return transactions can be separate from batch distribution records and returns can be of considerable volume (incorrect deliveries, ordered in excess, damaged in transit, nearing the end of shelf-life, etc.). The regulations require recording of the product details, disposition of returns, and also the reason for the return.

3. To remove from commerce portions of a lot which may have been adversely affected by atypical distribution conditions (see also §211.208).

Considering the liability of distributing substandard product, many firms have taken a cautious approach and do not allow redistribution of returns unless there is no doubt about the security, storage conditions, and product quality. If there is any doubt, the risk is not worth the economic return.

§211.208   DRUG PRODUCT SALVAGING

Drug products that have been subjected to improper storage conditions including extremes in temperature, humidity, smoke, fumes, pressure, age, radiation due to natural disasters, fires, accidents, or equipment failures shall not be salvaged and returned to the marketplace. Whenever there is a question whether drug products have been subjected to such conditions, salvaging operations may be conducted only if there is (a) evidence from laboratory tests and assays (including animal feeding studies where applicable) that the drug products meet all applicable standards of identity, strength, quality, and purity and (b) evidence from inspection of the premises that the drug products and their associated packaging were not subjected to improper storage conditions as a result of the disaster or accident. Organoleptic examinations shall be acceptable only as supplemental evidence that the drug products meet appropriate standards of identity, strength, quality, and purity. Records including name, lot number, and disposition shall be maintained for drug products subject to this section.

This section clearly states that products that have been exposed to improper storage conditions "shall not be salvaged and returned to the marketplace." As is the case of all agency rules and regulations, the word "shall" is mandatory while the word "may" is discretionary. However, this does not mean, for example, that in the event of a warehouse fire all goods stored there must be destroyed. If it is possible that some of the goods may not have been exposed to adverse conditions during the fire, then it is acceptable to evaluate these products, and if they do fully comply with the appropriate standards, they may be suitable for salvage.

EXAMPLES OF OBSERVATIONS FROM FDA 483 CITATIONS

1. Reason for product return not documented.
2. No evaluation of the cause of the returned "bad tablets."

Following are sections from FDA Compliance Policy Guides relating to product reconditioning:

SEC. 160.750 DRUG AND DEVICE PRODUCTS (INCLUDING BIOLOGICS AND ANIMAL DRUGS) FOUND IN VIOLATION OF GOOD MANUFACTURING PRACTICE REGULATIONS—RECONDITIONING (CPG 7153.14)

Background:

The question has arisen as to whether drug and device products that have been produced or held by methods or under conditions not in accordance with good manufacturing practice regulations (GMPRs), and consequently determined to be adulterated, may be reconditioned and returned to trade channels. Situations covered by this compliance policy guide (CPG) are those in which a "formal" judgment of adulteration has been rendered, e.g., drug and device products that have been seized and condemned pursuant to Section 304 of the Act due to good manufacturing practice deficiencies, drug and device products that have been recalled because they were found to be in violation of the current good manufacturing practice regulations (CGMPRs), etc. Although GMP deficiencies can be corrected in subsequent batches or lots of the involved product(s), it may be difficult or impossible to correct the effect of the deficiencies retrospectively in batches or lots already produced.

Policy:

The reconditioning of drug and device products found to be adulterated as a result of having been produced, processed, or held under conditions which are deficient with regard to GMPRs may be providing all of the following conditions are met as follows:

1. Any reconditioning proposal must be reviewed by all parties concerned (District, Center, OE, * OCC *) to determine whether the plan can reasonably be expected to bring the drug device product(s) into compliance.
2. In order to be acceptable, a proposed reconditioning plan must overcome any observed GMP deficiencies and correct any known product defects present.
3. If the lot to be reconditioned is held within the facility where the GMP violations occurred, the violative conditions must be corrected in advance of accepting a reconditioning proposal, or included as part of the reconditioning proposal.
4. If the lot is held in a facility separate from the one in which the GMP violations occurred and the separate facility is in compliance, a reconditioning proposal can be considered as provided for in paragraphs 1 and 2 earlier.

No product shall be released until all reconditioning commitments are fully met as verified by the FDA.

Material between asterisks is new or revised
Issued: 3/1/83
Revised: 3/95, 8/96

SEC. 448.100 RECONDITIONING OF NEW DRUGS, *WHICH DO NOT HAVE APPROVED*

New Drug Applications/Abbreviated New Drug Applications CPG (7132c.03)

Background:

Prior policy under the drug efficacy study implementation (DESI) program permitted the marketing of new drugs evaluated as effective upon the submission of a new drug application (NDA) or abbreviated new drug application (ANDA). This policy was challenged and overturned in a decision handed down on July 29, 1975, by the U.S. District Court for the District of Columbia (*Hoffman La Roche v. Caspar Weinberger, et al.*). The Agency implemented this order, Judge Green's decision, through Compliance Program 7332.26 covering products identical or related ("me too" drugs) to DESI drugs identified in a list published 1/76 [DHEW publication No. (FDA) 76–3009)].

Agency policy as set forth in the program required that such new drugs be discontinued from marketing and recalled if substantial stocks remain in trade channels. When responsible firms have failed to initiate the above actions after being warned by issuance of a *warning letter*, unapproved new drugs have been seized.

Although recall or seizure may have been necessary for uniform enforcement and protection of the public health, destruction of such recalled or seized material is not always required provided adequate safeguards are taken.

Policy:

In those instances in which an ANDA has been submitted and is currently pending, we will not insist upon destruction of recalled or seized material resulting from implementation of CPG 7132c.02 involving DESI effective drugs provided:

1. Recalled stocks are quarantined by the formulator and not held by consignees, i.e., substantial stocks in the hands of consignees must be disposed of either by return to the formulator or by destruction. Failure to do so will result in recommendation for regulatory action by the district, preferably seizure.

2. Recalled (quarantined lots at the formulator) or seized material may not be released until and unless all the following conditions are met:
   a. Approval of an NDA or ANDA is received.
   b. The firm can validate that the lots in question were manufactured in accordance with the specifications of the approved NDA/ANDA including the following:
      1. Compliance with current good manufacturing practice (CGMP).
      2. Affected lots meet all purity, potency, and labeling standards specified by the approved NDA/ANDA.
3. Where an unapproved new drug has been seized, under either Section 505 or 502, and as a result of subsequent ANDA approval, it is not in the public interest that it be condemned and destroyed under Section 304(d)(1), the Agency may consider entering into a stipulation of dismissal incorporating the following principles:
   (a) The claimant shall assure the Agency, by way of appropriate records, that the drug is in full compliance with the approved ANDA prior to dismissal of the complaint.
   (b) Where consistency with the approved ANDA requires labeling modifications, and compliance cannot be assured by a records review as in *"a."* above, the court may order the seized article be remanded to the custody of the claimant for the sole purpose of making the required modifications. If and when the agency is satisfied that the required modifications have been made, and the article is in all respects consistent with the approved ANDA, the complaint may be dismissed.
   (c) All activity undertaken to assure that the seized article is in compliance with the law shall be at the expense of the claimant, including investigatory and laboratory work performed by agency personnel. Current fee and mileage schedules shall apply, and payment shall be received prior to dismissal of the action and full release of the article.

Where no NDA/ANDA has been submitted, or if the district has information that quarantined or seized lots will not meet the above conditions or approval of a current pending application does not appear probable within a reasonable time frame (three months), then we will insist upon destruction of stocks. In the case of the latter, concurrence by Division of Drug Labeling Compliance (HFD-310) is required.

Except upon the specific conditions outlined above, nothing in these provisions shall be construed as altering agency policy that articles seized pursuant to Section 304 may not be reconditioned by agency consent without the entry of a decree condemning the articles and providing for reconditioning under agency supervision.

Material between asterisks is new or revised
Issued: 10/1/80
Revised: 3/95

13 | Repacking and Relabeling

Joseph D. Nally
Nallianco LLC, New Vernon, New Jersey, U.S.A.

The terms "repacking" and "relabeling" are commonly used to describe operations in which a drug product obtained from a manufacturer is packaged and labeled for distribution to a wholesaler or to a retail outlet. The repacking and relabeling operations are little different from the operations of the manufacturer who has drug product in bulk storage and packages and labels it in the final market container. The major difference is that the manufacturer has had control of the components and in-process material and, therefore, generally has more information about the quality and storage conditions of the drug product than the repacker. Additionally, the drug product has not been exposed to the hazards of transport by common carrier.

It seems important, therefore, that the repacker make special efforts to be assured of the quality of the product being repacked. There are three major areas of concern:

1. *Identity*: Is the product what it purports to be?
2. *Strength or potency*: What has happened to the drug product during transportation and bulk storage, both at the point of manufacture and at the repacker? Frequently, stability data for bulk storage are lacking. Expiration dates provided by the manufacturer for the drug product in the final market container are not applicable to material in fiber drums or different (composition) market containers. In the usual repacking operation, moreover, a drum may be opened several times before its contents are exhausted. Expiration dates apply only to unopened containers.
3. *Expiration date*: What is the proper method for setting an expiration date for a product that has been held for some time in one type of container, which may have been opened, and is now repacked in a different type of container?

The typical repacker who has little in-house laboratory capability has difficulty in addressing any of the three concerns.

The Food and Drug Administration (FDA) believes that a repacker should exercise control over incoming drug products for repacking and relabeling as is required for components in the current good manufacturing practice regulations (CGMPRs), essentially reading "drug product" for "component" throughout Subpart E, §211.86, §211.87, and §211.89. Since the regulations specifically apply to components and not to drug products [defined in §210.3(b)(4), "drug product" means a finished dosage form, for example, tablet, capsule, solution, and so forth, that contains an active drug ingredient generally, but not necessarily, in association with inactive ingredients] clearly distinguished from components [defined in §210.3(b)(3), "component" means any ingredient intended for use in the manufacture of a drug product], the FDA justifies its position by pointing out that repacking and relabeling operations are subsumed in the term "production" and also in the terms "packaging" and "labeling."

Whether the courts will uphold the FDA interpretation of the regulations is yet to be seen, and the situation may not be directly resolved. Under §210.1 the CGMPRs are "the minimum current good manufacturing practice" to assure drug quality. It is expected that the FDA will hold that even if the practices are not mandated by the regulations and even if the practices are not current with repackers, it is "good manufacturing practice" for repackers to treat drug products for repacking and relabeling as if they were components.

Whether a practice is "good" would seem to depend upon the exact circumstances. A cost–benefit decision must be made for each practice in its total setting as to whether the additional assurance of the drug quality afforded by the practice is worth the additional cost. Laboratory procedures that impose a trifling cost on manufacturers who already have laboratory facilities impose large direct and indirect costs on repackers who use consulting laboratories or impose capital costs for the equipping and staffing of an in-house laboratory. Since repackers may have new needs for disclosure on product labels under 21 CFR Part 201.1, the problem is met head-on in the identification of material to be repacked. In the past the FDA seems to have held that nonchemical identity testing is acceptable for domestically produced products; that is, visual comparison of the drug product with an authenticated (by the manufacturer) sample of the drug product is acceptable if there is little similarity between different drug products handled. It is unclear, whether "little similarity" applies to the repacker, the original manufacturer, or both. If the drug product is received from a foreign manufacturer, a chemical or physical identity test of each active ingredient in the drug product is required.

It is difficult to see why a dual standard is needed, since both domestic and foreign suppliers of drug products must comply with CGMPRs. Because of the physical similarity of many products, it seems reasonable to require a specific identity test for each active ingredient in a drug product to be repacked or relabeled. (The "little similarity" test can be checked by choosing a dosage unit present in a repacking plant and checking to see if it can be unequivocally identified by visual inspection alone.) Whether this seemingly reasonable requirement is also good is more difficult to decide. Obviously, a misidentified drug product will be mislabeled and will usually be a hazard to health. Since it is the clear intent of Congress that adulterated or mislabeled drug product not reach the marketplace, it is necessary to determine if visual examination for identity has ever led to the production (not distribution, since if a product exists, there is the probability that it will be distributed) of mislabeled drug product; that is, it should be determined if current practice is adequate to safeguard public health or whether good (better) practice should be required.

If the FDA should enforce a specific identity test for each active ingredient in a drug product to be repacked or relabeled, the intent to do so should be made explicit. The economic impact of such a requirement should be determined on repackers, since repackers generally do not have in-house laboratory facilities, repack a large number of different products, and compete primarily on the basis of low price. This economic study will be complicated by the existence of repackers who also do some manufacturing of drug products and, thus, already have minimal laboratory facilities and personnel. Because of the economic impact, it is obvious that enforcement of the requirement should be simultaneous for all repackers.

The problems of stability and expiration dating also must be faced by repackers who receive drug product in packages different from the final container-closure system or who repeatedly package from the same bulk container. If a manufacturer suggests a period of time in which the drug product packed in a particular container-closure system will meet all its quality specifications, what can be said about the product stored in bulk? How is an expected decline in product strength in bulk storage to be related to a specific expiration date on the final container? What happens to product quality if the bulk container is opened repeatedly for repackaging?

In the strictest sense, the questions can be answered only by assay of the drug product for active ingredients immediately before repacking and knowledge of the decomposition rate under ambient conditions in the final container-closure system. Since adsorption of moisture may affect stability of components in uncoated tablets, it is not safe to assume that a manufacturer's expiration date for bulk is satisfactory even for drug product packed in a more protective container-closure system, because moisture may be added to the tablet either due to repeated opening of the bulk container or by packaging at a higher relative humidity than that in which the tablets were manufactured. Again the lack of in-house laboratory facilities works against the repacker's having sufficient data and expertise to give an informed answer to the questions.

A temporary expedient might be to use on the repacked product an expiration date from the time of manufacture, based on stress tests in the final container-closure system (but not to exceed two years) and to follow the actual product decomposition at ambient conditions.

The dependency on moisture content should be determined and, with sufficient experience, some estimate of decomposition during bulk storage should be able to be made.

Another area of concern is drug product labeling. A repacker customarily prepares labels to match information on the label of the bulk container, a label supplied by the manufacturer, or a label supplied by the buyer. There is generally no independent label review to assure compliance with the regulations in 21 CFR Part 201—Labeling, §330.1(g) and Part 369—Interpretive Statements Re Warnings, and specific labeling requirements for classes of products for which monographs have been established (such as antacids, emetics, and daytime sedatives). If there is any error in the manufacturer's label, or more frequently, if there have been mandatory label changes (such as change in the official title of a component) between the time of bulk packaging and the time of relabeling, the final product will probably be mislabeled. This occurs because repackers generally do not make an independent check as to whether label copy is correct at the time of relabeling, and they do not have personnel who are aware of the need for such a check or the location of the pertinent information.

The repacker who is also a wholesaler has a problem with regard to distribution records. Distribution records are required by Section 211.196 to contain the name and strength of the product and a description of the dosage form, name and address of the consignee, date and quantity shipped, and lot or control number of the drug product. Wholesaler invoices, however, customarily do not have a description of the dosage form and do not have the lot or control number of the item shipped. Since the invoice usually has the item identification number from which the description of the dosage form can be obtained or the name of the product itself is a description of the dosage form, it can be reasonably argued that the distribution records do contain this bit of information. However, there is no arguing that a recorded lot number does not follow the drug product from wholesaler to retail distributor.

The reasons for this lack are easy to understand. Drug products shipped by wholesalers are frequently returned and reshipped. Keeping track of this would produce a cumbersome, expensive system of doubtful reliability. The problem is the number of accounts serviced by wholesalers as compared with the number of direct-to-retailer accounts serviced by manufacturers, and the small number of units of a given item in a single shipment to these accounts. A repacker–wholesaler generally will not have a record of the lot numbers of items distributed to individual retail accounts.

Clarification of the regulations is obviously necessary. An equitable interpretation might be that lot numbers are required to be recorded for products shipped by a manufacturer to repackers, wholesalers, or direct-to-retailer accounts and that repacker–wholesalers record lot numbers through the transfer of the repacked–relabeled drug product to the wholesale warehouse for retail distribution. This would give the same records for a drug product as are currently available when a manufacture ships to a wholesaler who later distributes to a retailer.

The lack of lot number on wholesaler invoices should not substantially affect ability to recall. Wholesalers customarily send recall notices to all accounts, or all accounts that purchased product after the date on which the specific lot was placed in retail distribution, or the manufacturer notifies all pharmacies in the area in which the lot was distributed.

Another area of concern relates to the usual lack of in-house laboratory facilities. Assuming that product specifications have been written and that the certificate of analysis provided by the manufacturer has been properly prepared, signed, and dated, validation of the certificate is required. This entails sending a sample to a consulting laboratory for analysis and comparison of the analysis by the consulting laboratory with the certificate of analysis from the manufacturer. Results will rarely be identical. Criteria for agreement between the two must be present in order to decide whether the certificate from the manufacturer is valid. If there is no in-house laboratory, there is probably no one sufficiently knowledgeable in analytical methodology to have an informed opinion of what degree of agreement constitutes validation. A knowledgeable consultant must then be employed.

If a consulting laboratory is used to provide analyses or stability studies not obtainable from the manufacturer, the problem of the validation of the results of the consulting laboratory arises and, thus, a second consulting laboratory must be used. If there is lack of agreement, a third or even fourth laboratory may be involved. It is suggested that initially all certificates have been shown to be valid, and thereafter not less than every fourth certificate should be validated.

Other areas of importance to repackers are standard operating procedures, documentation of formal training in current good manufacturing practice (CGMP) for all employees in the repacking–relabeling operations, and process validation (accuracy of label and labeling counters, fill accuracy, change in weight of tablets due to abrasion in packaging and its effect on compliance with weight variation requirements).

The standard operating procedures manual might contain the following sections:

1. Composition and scope of the quality control unit
2. Statement of laboratory facilities
3. How specifications are written, reviewed, and approved
   a. Components
   b. Drug products
   c. Containers
   d. Closures
   e. Packaging materials
   f. Labeling
   g. Job descriptions
4. Job descriptions, including consultants
5. Training in CGMP: employees, frequency
6. Protective clothing
7. Limited access areas: who is permitted to enter
8. Health check as per 21 CFR Part 211.28
9. Lighting description
10. Ventilating description
11. Proof of potable water (analysis of water supply)
12. Trash disposal (contract for removal)
13. Washing facilities as per 21 CFR Part 211.52 (add "toilet paper" and "container for used towels")
14. Sanitation program (contract with cleaners and exterminators)
15. Maintenance program (inside and out)
16. Equipment cleaning directions (including disassembly and reassembly)
17. Calibration of mechanical equipment
18. Controls on computer records
19. Receiving directions
20. Sampling directions
21. Testing specifications
22. Conditions for acceptance, re-examination, rejection
23. Assurance of first in/first out
24. Disposal of rejected materials
25. Validation of processes
26. Validation of supplier testing
27. Mechanisms for investigation and correction of deviations
28. Reprocessing procedures
29. Special procedures for gang printing of labels
30. Procedures for reconciliation of labeling
31. Procedures to prevent contamination of drug product
32. Procedures for packing and labeling
33. Setting of expiration dates: stability testing program
34. Directions for warehousing
35. Records review
36. Returned drug products
37. Complaints
38. Recall procedures

Although superficial consideration of repacking–relabeling might indicate that these operations have little effect on product quality, the more thorough examination listed above of what is involved in the operations and the probable impact on product safety and efficacy shows that repacking–relabeling is a critical operation that requires all of the safeguards of an adequate quality control unit.

EXAMPLES OF OBSERVATIONS FROM FDA 483 CITATIONS AND WARNING LETTERS (REPACKAGERS)

1. Employees have not been given training in the particular operations they perform.
2. Packaging and labeling controls for the usage and examination of materials are deficient.
3. Controls over labeling issuance are deficient.
4. Batch production and control records do not include complete information.
5. Expiration dates are not supported by sound scientific data (stability studies).

The following Draft Guidance Documents apply to Repackaged Drug Products:

REPACKAGING OF SOLID ORAL DOSAGE FORM DRUG PRODUCTS 2/1/92

The FDA was explicit on the requirements for applying the original manufacturer's expiration date. The Draft Guidance allows this practice but only under certain conditions:

- The original bulk container of the drug product was not opened previously and the entire contents are repackaged in one operation;
- Where the original bulk drug manufacturer's container was other than glass, the repackaging container was demonstrated to be equivalent to or exceed the original bulk manufacturer's container in terms of water vapor permeation and compatibility with the drug product, or where the original bulk manufacturer's container is polyethylene, the repackaging container meets current U.S. Pharmacopeia (USP) standards for high density polyethylene (HDPE) containers;
- The repackaging container meets or exceeds the original bulk manufacturer's container specifications for light transmission or meets the current USP standard for light transmission;
- The repackaging container meets or exceeds the special protective features of the original bulk manufacturer's container e.g., for preventing leaching of container material into the drug product or for maintaining low moisture; and
- The repackaging container-closure system meets the current USP standards for a "tight container" or a "well-closed container."

It is clear from the intent of the guidance that the repackaging firm must accept the responsibility for determining scientifically sound expiration dating.

EXPIRATION DATING OF UNIT-DOSE REPACKAGED DRUGS: COMPLIANCE POLICY GUIDE 5/31/2005 SEC. 480.200 (CPG 7132B.11)

Introduction:

Many companies repackage solid and liquid oral dosage form drug products into unit-dose containers. This guidance states the circumstances under which the FDA intends to exercise its enforcement discretion and does not intend to take enforcement action against such repackagers for failure to conduct stability studies to support expiration dates for these unit-dose repackaged products.

FDA's guidance documents, including this guidance, do not establish legally enforceable responsibilities. Instead, guidances describe the Agency's current thinking on a topic and should be viewed only as recommendations, unless specific regulatory or statutory requirements are cited. The use of the word "should" in Agency guidances means that something is suggested or recommended, but not required.

Background:

Unit-dose packaging systems are widespread in health care. Some unit-dose containers are available directly from manufacturers and repackagers; some drug products are repackaged into unit-dose containers by hospital or community pharmacies or shared service establishments. A shared service repackaging operation is one that exclusively serves one or more hospitals and/or related institutions, each having separate or no pharmacy services and each having responsibility for restricting distribution of the drug products received from the shared service establishment.

The nature of drug dispensing within hospitals in particular has made such unit-dose packaging useful and convenient in helping to ensure that medications are properly administered to patients. However, questions have arisen concerning the appropriate expiration dating for drug products repackaged into unit-dose containers.

FDA's CGMPRs for finished pharmaceuticals occur in 21 CFR Part 211, include §211.137 on "Expiration dating." Section 211.137(a) requires that each drug product bear an expiration date determined by appropriate stability testing, as described in §211.166. Under §211.137(b), a drug product's expiration date must be related to any storage conditions stated on the labeling, as determined by stability studies described in §211.166. Samples used for stability testing must be in the same container-closure system as that in which the drug product is marketed [§211.166(a)(4)]. This is to ensure the drugs' safety and efficacy over their intended shelf life.

The USP contains standards on expiration dating and beyond-use dating in its General Notices and Requirements section (http://www.fda.gov/cder/guidance/6169dft.htm#_ftn2#_ftn2). The USP directs dispensers of prescription drug products to place on the label of the prescription container a suitable beyond-use date to limit the patient's use of the product. The beyond-use date cannot be later than the expiration date on the manufacturer's container. The USP states:

For nonsterile solid and liquid dosage forms that are packaged in single-unit and unit-dose containers, the beyond-use date shall be one year from the date the drug is packaged into the single-unit or unit-dose container or the expiration date on the manufacturer's container, whichever is earlier, unless stability data or the manufacturer's labeling indicates otherwise (http://www.fda.gov/cder/guidance/6169dft.htm#_ftn3#_ftn3).

Discussion:

The FDA has considered the USP beyond-use standard and believes that similar conditions are appropriate for FDA's CPG 7132b.11 for expiration dating. The FDA believes that under certain specified conditions, it may be possible to assign appropriate expiration dating without conducting new stability studies on the nonsterile solid and liquid oral dosage forms repackaged into unit-dose containers. Therefore, the FDA does not intend to take action against any nonsterile unit-dose repackaging firm (including shared services repackaging operations) or drug product in a unit-dose container solely on the basis of the failure of the repackaging firm to have stability studies supporting the expiration dates used, provided that the repackager meets all other regulations applicable to repackaged drug products and:

1. The expiration date does not exceed (a) one year from the date of repackaging or (b) the expiration date on the container of the original manufacturer's product, whichever is earlier, unless stability data or the original manufacturer's product labeling indicates otherwise.
2. If the drug product repackaged is in solid oral dosage form, the formed unit-dose container complies with the Class-A standard described in the USP, General Chapter ⟨671⟩ Containers–Permeation, "Single-Unit Containers and Unit-Dose Containers for Capsules and Tablets."

3. The original bulk container of drug product has not been opened previously and the entire contents are repackaged in one operation.
4. The repackaging and storage of the drug product are accomplished in a controlled environment that is consistent with the conditions described in the labeling for the original drug product and the repackaged drug product. Where no temperature is specified in the labeling of the original drug product, a controlled room temperature (as defined in the General Notices and Requirements section of the USP) should be maintained during repackaging and storage of both solid and liquid oral dosage form drug products. Where no humidity is specified in the labeling of the original drug product, the relative humidity should not exceed 75% at 23°C for the repackaging and storage of solid oral dosage forms.

This CPG applies only to nonsterile solid and liquid oral dosage forms in unit-dose containers. Sterile products and other types of dosage forms and packages pose sterility and/or stability concerns that this CPG would not adequately address. Thus, this CPG does not apply to sterile products, other dosage forms, and other types of packages.

Liquid oral dosage forms should not be repackaged unless suitable materials are used and precautions are taken to prevent evaporation or solvent loss.

This CPG does not apply to nitroglycerin sublingual tablets or any other solid or liquid oral dosage form drug product known to have stability problems that preclude the product from being repackaged. This group of products generally would include any drug known to be oxygen sensitive or that exhibits extreme moisture or light sensitivity. In deciding whether a particular drug product is suitable for repackaging, the repackager should take into consideration any available information from the manufacturer, published literature, the USP, and the FDA.

The FDA's intent to exercise enforcement discretion concerning stability studies for repackaged products does not apply to any other requirements of Parts 210 and 211.

REPACKAGING CPG: SEC. 430.200 REPACKING OF DRUG PRODUCTS—TESTING/EXAMINATION UNDER CURRENT GOOD MANUFACTURING PRACTICES (CPG 7132.13)

Background:

Questions have periodically arisen regarding how various testing and/or examination requirements under the CGMPRs (21 CFR Parts 210 and 211) are to be applied to repackers of finished dosage form drugs. In particular, there have been questions regarding whether it is appropriate to apply various "component" requirements in the CGMPRs (such as those under Section 211.84 concerning identity testing and analysis or receipt of a report of analysis for purity, strength, and quality) to finished dosage form drugs that an establishment receives and repackages. It has also been questioned how the requirements under 211.165 are to be applied to repackers, insofar as the requirements for appropriate laboratory determination for identity and strength of each active ingredient prior to release are concerned.

We have carefully considered the suitability of applying the requirements concerning "components" in the CGMPRs to repackers of finished dosage form drugs. Due to the definitions of "component" under 210.3(b)(3) and "drug product" under 210.3(b)(4), we have concluded that the requirements for "components" under Part 211 cannot be suitably applied to finished dosage form drugs which are received by an establishment and repackaged without alteration to the "drug product" itself.

In the preamble to the final order for the CGMPRs, it is pointed out in regards to a manufacturer that there is no intent under 211.165(a), once the product is in its finished dosage form, to require potency testing of both the bulk and packaged drug product phases, and that manufacturers could choose to do potency assays at either phase (43 FR 45062, paragraph 389). We believe a similar principle is applicable to drug product repackers where the manufacturer of the finished dosage form in a bulk container is required to perform appropriate analytical testing for all appropriate specifications, including the identity and strength of each active ingredient; we do not consider it necessary for the repacker to repeat such testing upon such drug products he receives and repacks with label declarations consistent with those on the bulk container and without altering the properties of the finished dosage form product.

Policy:

Generally, we do not consider the CGMPRs (21 CFR Parts 210 and 211) to require repackers of finished dosage form drugs to perform analytical testing such as chemical identity tests or assays, or to require receipt of reports of analysis, on a batch-by-batch basis for drug products which are repacked under the following circumstances:

1. The incoming bulk containers of finished dosage form drug products are received in intact, undamaged containers which are completely and properly labeled as received, and there is no reason to suspect they have been subjected to improper storage or transit conditions prior to receipt;
2. The repacking operations are conducted under conditions which assure that the properties of the incoming drug product are not altered; and
3. The repackaged containers are labeled with the same substantive labeling declarations (e.g., identity, strength, and directions for use) concerning the properties and use of the drug product, which are consistent with the labeling on the incoming bulk containers.

Under such circumstances we consider that requirements for appropriate specifications and testing/examination procedures for repacked drug products will be met by an appropriate system involving examination of the labeling and sufficient organoleptic examination of the drug product to confirm its identity in accordance with corresponding specifications established by the repacker.

The policy in this CPG applies only to the question of adequate batch-to-batch testing/ examination criteria for routine acceptance and release of drug products, which are repacked. It does not alter any testing that repackers may be required to perform on drug products from other standpoints: stability test to establish appropriate expiration dates in the container-closure system used by the repacker, test to determine the suitability of the repacker's drug product containers and closures, test to establish appropriate time limits for the completion of each phase of production, or test on nonpenicillin drug products for the presence of penicillin.
Issued: 7/1/81

SEC. 430.100 UNIT-DOSE LABELING FOR SOLID AND LIQUID ORAL DOSAGE FORMS (CPG 7132B. 10)

Background:

In recent years, the pharmaceutical industry has responded to an increased demand for drug products, which are packaged for "unit dose" dispensing, i.e., the delivery of a single dose of a drug to the patient at the time of administration for institutional use, e.g., hospitals. The drug product is dispensed in a unit-dose container—a non-reusable container designed to hold a quantity of drug intended for administration (other than the parenteral route) as a single dose, directly from the container, employed generally in a hospital unit-dose system. The advantages of unit-dose dispensing are that the drug is fully identifiable and the integrity of the dosage form is protected until the actual moment of administration. If the drug is not used and the container is intact, the drug may be retrieved and redispensed without compromising its integrity.

In view of the intended use of unit-dose packaging, each unit-dose container is regarded as a drug in package form subject to all requirements of the Act and implementing regulations. However, the pertinent labeling regulations [21 CFR Parts 201.10(i) and 201.100] present problems in interpretation in that they are inconsistent with respect to exemptions for containers too small or otherwise unable to accommodate a label with sufficient space to bear all mandatory information. As a result of several recent regulatory actions emphasizing these inconsistencies, the regulations will be rewritten in the future to clarify the requirements.

Because of the general lack of uniformity in the labeling for unit-dose containers due to inconsistent interpretations of the regulations, or to a lack of knowledge of unit-dose labeling requirements, we are issuing this CPG.

This CPG does not encompass "Unit of Use" packaging, which is defined as a method of preparing a legend medication in an original container, sealed and labeled, prelabeled by the manufacturer, and containing sufficient medication for one normal course of therapy. (Reference: Proceedings Unit of Use Packaging Conference, January 24–26, 1979).

Policy:

Until the regulations are revised, the attached document describes the labeling requirements for oral solid and liquid dosage forms packaged in unit-dose containers. The requirements apply to all firms that package drugs into unit-dose containers.

Since unit dosage forms are primarily intended for institutional use rather than for sale to the general public, we will not require the warnings described in 21 CFR Part 369 or the statements described under item 6.b. (Section I and II) of Attachment A to be on the label; however, this information must appear elsewhere in the labeling.

Where unit-dose repacking is performed by a single facility for a closed membership or a group (e.g., "shared services"), a current package insert, bearing adequate directions for use, located on the premises of each member to whom the repacked goods are shipped is regarded as satisfying this requirement. The absence of such a current package insert on the premises of a member to which a drug product is shipped will cause that drug product to be misbranded.

Solid and liquid oral dosage forms in unit-dose containers shall be deemed misbranded under Section 502 of the Act if they deviate from the attached list of requirements.

Other unit-dose forms, e.g., topical ointments/creams, ophthalmic, etc. are not included in this document. They will be considered at a future date should circumstances warrant.

ATTACHMENT A

Unit Dose Labeling

I. Prescription Drugs (Solid and Liquid Oral Dosage Forms, e.g., Capsules, Tablets, Solutions, Elixirs, Suspensions, etc.)

The label of the actual unit-dose container must bear all of the following information (except item 9).

Note: A firm may not claim an exemption on the basis that the label is too small to accommodate all mandatory information if all available space is not utilized or the label size can readily be made larger, or if the type size on the label can readily be made smaller without affecting the legibility of the information.

1. The established name of the drug and the quantity of the active ingredient per dosage unit, if a single active ingredient product; if a combination drug, the established name and quantity of each active ingredient per dosage unit. In each case, the label must bear the established name and quantity or proportion of any ingredient named in Section 502(e) whether active or not. For solid dosage forms, a declaration of potency per tablet/capsule will suffice; for liquid dosage forms, the total volume shall be declared as well as the quantity or proportion of active ingredient contained therein, e.g., Cimetadine HCL Liquid 5 ml, 300 mg/5 ml or 300 mg/5 ml; or Septra/Bactrim Suspension 5 ml, contains Trimethoprim 40 mg and Sulfamethoxazole 200 mg/5 ml; or each 5 ml contains
2. The expiration date (see Attachment B). (Ref. 21 CFR Parts 201.17 and 211.137).
3. The lot or control number [Ref. 21 CFR Parts 201.100(b) and 211.130].
4. The name and place of business of the manufacturer, packer, or distributor as provided for in 21 CFR Part 201.1.

5. For a drug recognized in an official compendium, the subject of an approved new drug application (NDA/ANDA) or as provided by regulation:
   A. Required statements such as "Refrigerate", "Protect From Light", "Dilute Before Using", etc. [Ref. FD&C Act 502(f)(1)(D, 502(g), and 505].
   B. Any pertinent Statement bearing on the special characteristics of the dosage form, e.g., sustained release, enteric coated, chewable, suspension, etc. [Ref. FD&C Act 502(e), 502(a), 201(n)].
6. For any drug product not subject to 5:
   A. Any pertinent statement bearing on special characteristics of the dosage form, e.g., sustained release, enteric coated, sublingual, chewable, solution, elixir, suspension, etc. [Ref. FD&C Act 502(e), 502(a), 201(n)].
   B. While not required to be on the label per se, it is strongly recommended that:
      (1) Any pertinent statement bearing on the need for special storage conditions, e.g., "Refrigerate", "Do not Refrigerate", "Protect from Light", etc. [Ref. FD&C Act 502(f)(l)] appear on the label, and
      (2) any information needed to alert the health professional that a procedure(s) is necessary prior to patient administration to prepare the product as a finished dosage form, e.g., "Shake Before Using" [Ref: FD&C Act 502(f)(l)].
7. If more than one dosage unit is contained within the unit-dose container (solid dosage form), the number of dosage units per container and the strength per dosage unit should be specified (e.g., two capsules; each capsule contains 300 mg Rifampin).
8. The statement "Warning: May be habit forming" where applicable, the controlled drug substances symbol required by drug enforcement administration (DEA), and the name and quantity or proportion of any substance as required by Section 502(d).
9. The National Drug Code designation is recommended, although this is not mandatory.

In addition to all of the above (except item 9), the following information must appear on the outer package from which the unit-dose container is dispensed:

1. The number of unit-dose containers in the package, e.g., 100 unit doses. If more than one dosage unit is within each unit-dose container this should also be stated (e.g., "100 packets; each packet contains two tablets," or "100 packets of two tablets each").
2. Full disclosure information, as detailed in 21 CFR Part 201.100. Where unit-dose repacking is performed by a single facility for a closed membership or group (e.g., "shared services") a current package insert bearing adequate directions for use, located on the premises of each member to whom the repacked goods are shipped is sufficient to satisfy this requirement. The absence of such a current package insert on the premises of a member to which a drug is shipped will cause that drug to be misbranded.
3. The prescription legend.

II. Over the Counter Drugs (Solid and Liquid Oral Dosage Forms, e.g., Capsules, Tablets, Elixirs, Suspension, etc.)

The label of the actual unit-dose container must bear all of the following information (except item 9).

Note: A firm may not claim an exemption on the basis that the label is too small to accommodate all mandatory information if all available space is not utilized, the label size can be made larger, or if the type size on the label can readily be made smaller without affecting the legibility of the information.

1. The established name of the drug if it contains a single active ingredient; if a combination drug, the established name of each active ingredient. If a compendial drug, the label must express the quantity of each therapeutically active ingredient contained in each dosage unit, e.g., Aspirin Tablets, 325 mg, (USP-General Notices), and the quantity or proportion of any ingredient, whether active or not, as required by Section 502(e).
2. The expiration date (see attachment B).

3. The lot or control number.
4. The name and place of business of the manufacturer, packer, or distributor as provided for in 21 CFR Part 201.1.
5. For a drug recognized in an official compendium, the subject of an approved new drug application (NDA/ANDA), or as provided by regulation:
    A. Required statements such as "Refrigerate", "Protect from Light", "Dilute Before Using", etc. [Ref. FD&C Act 502(f)(1)(D, 502(g), and 505].
    B. Any pertinent statement bearing on special characteristics of the dosage form, e.g., sustained release, enteric coated, chewable, suspension, etc. [Ref. FD&C Act 502(e), 502(a), 201(n)].
6. For any drug product not subject to 5:
    (a) Any pertinent statement bearing on special characteristics of the dosage form, e.g., sustained release, enteric coated, sublingual, chewable, solution, elixir, suspension, etc. [Ref. FD&C Act 502(e), 502(a), 201(n)].
    (b) While not required to be on the label per se, it is strongly recommended that
        (1) Any pertinent statement bearing on the need for special storage conditions, e.g., "Refrigerate", "Do not Refrigerate", "Protect from Light", etc., [Ref. FD&C Act 502(f)(l)], appear on the label, and
        (2) Any information needed to alert the user that a procedure(s) is necessary prior to patient administration to prepare the product for use, e.g., "Shake Well", "Dilute Before Using" [Ref. FD&C Act 502(f)(l), 21 CFR Part 201.5].
7. If more than one dosage unit is contained within the unit-dose container, the number of dosage units per container should be specified (e.g., two tablets aspirin; each tablet contains 325 mg).
8. The statement "Warning: May be habit forming" where applicable, the controlled drug substances symbol required by DEA, and the name and quantity or proportion of any substance required by Section 502(d).
9. The National Drug Code designation is recommended, although this is not mandatory.

In addition to all of the above (except item 9), the following information must appear on the outer package from which the unit-dose container is dispensed:

1. The number of unit-dose containers in the package. If more than one dosage unit is within each unit-dose container this should also be stated (e.g., "100 packets; each packet contains two tablets," or "100 packets of two tablets each").
2. The labeling, i.e., the outer carton or a leaflet enclosed within the package must bear adequate directions for use as specified in 21 CFR Part 201.5 and should include:
    a. Statement of all conditions, purposes, or uses for which the drug product is intended.
    b. Quantity of dose, including usual quantities for each of the uses for which it is intended and usual quantities for persons of different ages and conditions.
    c. Frequency of administration.
    d. Duration of administration.
    e. Time of administration (in relation to time of meals, time of onset of symptoms, or other time factors).

ATTACHMENT B

Expiration Dating of Solid and Liquid Oral Dosage Forms in Unit-Dose Containers. (CPG 7132b. 11)

No action will be initiated against any unit-dose repackaging firm, including shared services, or drug product in unit-dose container meeting all other conditions of the FDA's repackaging requirements, solely on the basis of the failure of the repacking firm to have stability studies supporting the expiration dates used provided:

1. The unit-dose container complies with the Class A or Class B standard described in the Twentieth Edition of the United States Pharmacopeia, General Tests, Single-Unit Containers and Unit-Dose Containers for Capsules and Tablets (page 955);

2. the expiration date does not exceed six months; and
3. the six month expiration period does not exceed 25% of the remaining time between the date of repackaging and the expiration date shown on the original manufacturer's bulk container of the drug repackaged, and the bulk container has not been previously opened.

   This policy does not apply to antibiotics or to nitroglycerin sublingual tablets, which are known to have stability problems that preclude them from being repackaged.

Issued: 2/1/84

14 | Quality Systems and Risk Management Approaches

Joseph D. Nally and Laura L. Nally
Nallianco LLC, New Vernon, New Jersey, U.S.A.

This chapter will cover the two recent Food and Drug Administration (FDA) Guidance Documents on Quality Systems Approaches: *Quality Systems Approach to Pharmaceutical Current Good Manufacturing Practice Regulations, September 2006, Final Guidance* and *Compliance Program Guidance Manual for FDA Staff: Drug Manufacturing Inspections Program 7356.002.* These documents provide much insight into the FDA's current thinking and change in approach since the introduction of their Pharmaceutical Current Good Manufacturing Practices (CGMPs) for the 21st century initiative. To understand quality systems in practice also requires knowledge of fundamental work processes and process management. This chapter will explore the intent of the guidance documents and application to pharmaceutical quality systems and processes.

Quality Systems Approach to Pharmaceutical Current Good Manufacturing Practice Regulations, September 2006 was just finalized. The Introduction reads:

> "This draft guidance is intended to help manufacturers that are implementing modern quality systems and risk management approaches to meet the requirements of the Agency's current good manufacturing practice (CGMP) regulations (21 CFR parts 210 and 211). The guidance describes a *comprehensive quality systems (QS) model,* highlighting the model's consistency with the CGMP regulatory requirements for manufacturing human and veterinary drugs, including biological drug products. The guidance also explains how manufacturers implementing such quality systems can be in full compliance with parts 210 and 211. This guidance is not intended to place new expectations on manufacturers nor to replace the CGMP requirements. Readers are advised to always refer to parts 210 and 211 to ensure full compliance with the regulations."

The last two sentences are very important. The FDA is clearly saying that quality systems are not additional expectations or requirements and do not establish legally enforceable responsibilities. The quality system approach/model does not replace GMP regulation. However, the document does allow for more operational flexibility and use of modern quality concepts and business practices to meet GMP requirements. The Background section includes:

> "In August 2002, the FDA announced the Pharmaceutical CGMPs for the 21st Century Initiative. In that announcement, the FDA explained the Agency's intent to integrate *quality systems* and *risk management* approaches into existing programs with the goal of encouraging the adoption of modern and innovative manufacturing technologies. The CGMP initiative was spurred by the fact that since 1978, when the last major revision of the CGMP regulations was published, there have been many advances in manufacturing technologies and in our understanding of quality systems. Many pharmaceutical manufacturers are implementing comprehensive, modern quality systems and risk management approaches. The Agency also saw a need to address the harmonization of the CGMPs and other non-U.S. pharmaceutical regulatory systems as well as FDA's own medical device quality systems regulations."

Using modern quality management methods and harmonization principles is a welcome initiative and allows some departure from the test and control theme of the 1978 Drug GMPs. This guidance is a bridge between the 1978 regulations and current concept of quality systems. An important linkage between CGMP and robust modern quality systems is the quality by design (QBD) principle and the fact that testing alone cannot be relied upon to ensure product quality.

Harmonization with other widely used quality management systems and approaches (ISO 9000, Device Quality Systems Regulations, Drug Manufacturing Inspections Program, EPA Guidance for Quality Systems, etc.) brings into practice the science of process, systems, and quality management principles and allows for needed flexibility in applied GMP practices.

The goal of the guidance is to describe:

> "...a comprehensive quality systems model, which, if implemented, will allow manufacturers to operate robust, modern quality systems that are fully compliant with CGMP regulations. The guidance demonstrates how and where the requirements of the CGMP regulations fit within this comprehensive model. The inherent flexibility of the CGMP regulations should enable manufacturers to implement a quality system in a form that is appropriate for their specific operations."

Again the FDA is quite clear that this guidance is primarily based on sustainable GMP compliance and how that fits into modern quality systems approaches of running a business. A review of recent 483 citations confirms their approach. Each 483 item is under a quality system heading, but is written in traditional terms such as "Failure to have a quality control unit that shall have responsibility and authority ... " so on. They are not saying there is a management system failure and they are not performing a systems audit. However, they are saying there is a failure in fundamental GMP requirements. The root cause of that situation may indeed be a system and/or quality management failure but that is up to the firm to determine and remedy.

As with all guidance documents there are fundamental concepts and principles. There are seven in this document.

- *Quality*: achieving all product quality characteristics.
- *QBD and product development*: effective transfer of process knowledge from development to commercial manufacturing. This is in line with ICH Q8 Guidance.
- *Risk assessment and management*: especially for any product quality decisions. This is in line with ICH Q9 Guidance.
- *CAPA*: brings in ISO definitions of remedial corrections: CA—root cause analysis with corrective action to prevent recurrence, and PA—preventive action to prevent initial occurrence.
- *Change control*: managing change to prevent unintended consequences and creating a regulatory environment that encourages change towards continuous improvement.
- *The quality unit*: the collective name for quality control (QC), quality assurance (QA), and other quality organizations that have responsibility and authority for GMP activities and compliance.
- *Six system inspection approach* (Fig. 1).

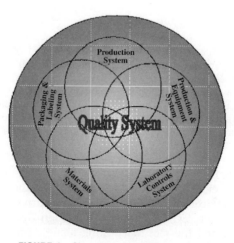

FIGURE 1   Six system inspection approach.

They further organize the six quality systems model into four sections:

- Management responsibilities
  - leadership
  - structure and organization
  - build/design quality systems to meet requirements
  - establish policies, objectives, and plans
  - review the system
- Resources
  - general arrangements (adequate resources)
  - develop personnel
  - facilities and equipment
  - control outsourcing operations
- Manufacturing operations
  - design and develop product and processes
  - monitor packaging and labeling processes
  - examine inputs
  - perform and monitor operations
  - address nonconformities
- Evaluation activities
  - analyze data for trends
  - conduct internal audits
  - risk assessment
  - corrective action
  - promote improvement

While still focused on product quality, the guidance brings in additional elements found in ISO and other quality standards (Chap. 24).

The model and the section requirements go beyond the basic GMP/CFR requirements in some areas:

- leadership
- general arrangements or providing adequate resources
- internal audits (other than data)
- preventive action
- promote improvement

Failures in these specific areas will not show up in 483 observations, but they are necessary parts of quality management and continuous improvement.

BASIC QUALITY SYSTEMS

At this point in the discussion, it is necessary to go back to basics to understand process, systems, and process management principles that are the foundation of a robust and sustainable system.

Any repeatable work activities can be characterized as a process (Fig. 2).

A process is defined as a repeatable sequence of activities with measurable inputs, value adding activities, and measurable outputs. Each process receives input from internal (previous

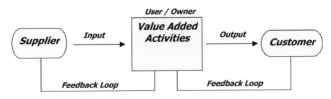

FIGURE 2 Process model.

operation) or external suppliers. Each process has a process owner who adds value to the input and is responsible for the output. The value added output goes to the next operation (internal or external) or customer (1). In a well-executed process, there are also information feedback loops to the supplier and from the customer on the quality and timeliness of the input or output.

Process management is defined as the ensemble of activities of planning and monitoring the performance of a business process.

A system is defined as a collection of components organized to accomplish a specific function or set of functions. A single process or collection or processes can make up a system.

A quality system is defined as a structured and documented management system describing the policies, objectives, principles, organizational authority, responsibilities, accountability, and implementation plan of an organization for ensuring quality in its work processes, products (items), and services. The quality system provides the framework for planning, implementing, and assessing the work performed by an organization and for carrying out required QA and QC activities (2).

The following table lists the six FDA quality systems and the typical GMP quality systems in a pharmaceutical manufacturing business.

FDA Quality System	30 Typical GMP Quality Systems

Facilities and equipment	Facilities and equipment management
	    Master planning
	    Commissioning, qualification, and validation
	    Drawings and document control

	Facilities cleaning Equipment maintenance Corrective and preventive maintenance systems for utilities and production equipment
	Equipment calibration—programs/systems for GMP and other equipment Facility environmental monitoring

Production	Manufacturing operations
	    Batch record execution and review
	    Document control
	    Product sampling
	    Equipment operation and clearance
	    Equipment cleaning

	Process validation Contract manufacturing (management) Technology transfer Reprocessing and rework

Packaging and labeling	Packaging operations
	    Batch record execution and review
	    Document control
	    Product sampling
	    Equipment operation and clearance

	Labeling control systems Receipt, inspection, release, issuance, control reconciliation, and storage
Materials	Raw material and components Receipt, sampling, test, release, and storage
	Warehousing and distribution Returns and salvage

Laboratory controls	Laboratory control systems
	    Sample management
	    Test methods and specifications
	    Method validation
	    Instrument qualification, calibration, and maintenance
	    Reference standards
	    Reagents, solutions
	    Data analysis and reporting

FDA Quality System	30 Typical GMP Quality Systems
	Failure investigations
	Glassware control
	Contract laboratories management
Quality	Policies and standards: creation and issuance
	Documentation control: Standard Operating Procedures (SOPs),
	protocols, records and reports, forms, and log books
	Regulatory reporting: new drug application (NDA), abreviated
	new drug application (ANDA)
	Training: GMP and job
	Change control: document, equipment, labeling, process, and computer systems
	Annual product review
	Audit program: internal, contractors, regulatory
	Complaint handling
	Failure investigations (other than laboratory)
	Batch record review and product release
	Management notification
	Product stability program management and reporting
	Computer system validation
	Recalls

The operational and quality systems in the above table are at an intermediate level. There are a number of subsystems possible in any of the 30 systems identified. For example, computer system validation has a number of subsystems that manage and control the computer system life cycle: validation master planning; DQ/IQ/OQ/PQ protocols, execution, reporting; periodic re-validation; change control; data center management and control; disaster recovery; and so on.

It is worthwhile to note that the vast majority of operational and quality systems are multi or cross-functional and involve more than one department for input, execution, and output. This is a reason why having SOPs only defined by department usually results in disconnects and incomplete system design and deviations/observation in performance. Policies, umbrella SOPs, multifunctional SOPs, or mapped and connected individual SOPs are needed to bridge those gaps and provide the communication links for a robust and sustainable system.

There are also system interdependencies that must be recognized. For example, the QA batch record reviews and product release process depends not only on a completed batch record (input) but also on batch-related information outputs from other control systems: laboratory out-of-specification investigations; process deviation or failure investigations, pending batch-related change controls, regulatory commitments, environmental monitoring and water testing results, product testing results, and so on. The quality and timeliness of that review and release process is dependent on the quality and timeliness of the input information.

To have effective, robust, and sustainable systems requires that the fundamental process elements are in place, are linked where needed, and sound process management and control is consistently being practiced.

KEY PROCESS/SYSTEMS ELEMENTS

- Process/system inputs are well defined, controlled, and monitored. In most pharmaceutical systems, the input is documented information; for example, a change control system input is a detailed change request. Information must be complete, accurate, and timely. The input quality can and should be measured where needed. In the change control example, a change request can be right at the first time or sent back for more information. That success rate can be measured and fed back to the suppliers.

- Process/system ownership, responsibility, and accountability is defined and accepted. This involves job role and responsibility definition in procedures, job descriptions, and role profiles. It also involves management leadership, planning, resource allocation levels, support organization levels, and process oversight and follow-up. The ownership, responsibility, and accountability must be consistently practiced.
- Process/system design is adequate for use. Simple designs by work processes are best but need to include/identify input information, activities, decision criteria, decision outputs, timeliness requirements, document requirements, and how to handle exceptions or deviations and fail-safe or stop criteria where needed.
- Level of process/system definition is adequate for use. Having the proper balance of enough information in SOPs, instructions, documents, and forms to achieve consistent execution by different people on different days in the goal. Refer to Chapter 7 for details on SOP content. SOPs should be concise, to the point, user friendly and be written for a trained operator. However, there must be enough "how to" detail to assure consistent execution. In the author's experience, operational SOPs are often good on what is supposed to be done but short on details of how it is done, which leads to varying approaches and unacceptable variation. Variation is the enemy of quality and consistency.
- Consistency in execution. If the previous elements are in place, consistent execution should follow. Audits and process metrics can be the measurement tools.
- Process performance and output should be monitored, measured, controlled, and reported where needed. Process performance and output can be measured by metrics. Metrics can be diagnostic or performance-related. In the change control system, for example, the performance metrics could be on time and right at the first time completion/approval of change control requests, change authorizations, and change close out. The ultimate performance metric is no adverse impact in product quality or compliance as a result of the change. Diagnostic measures of change control process performance may be types of changes submitted, departmental breakdowns, overall cycle times and so on.

COMPLIANCE PROGRAM GUIDANCE MANUAL FOR FDA FOOD AND DRUG ADMINISTRATION STAFF: DRUG MANUFACTURING INSPECTIONS PROGRAM 7356.002

This guidance closely follows the approaches of risk management and quality systems laid out in the Quality Systems Approach guide and the 21st century FDA initiatives.

- Background section states that the guidance is structured to provide for efficient use of resources devoted to routine surveillance coverage, recognizing that in-depth coverage is not feasible for all firms on a biennial basis.
- The inspection is defined as audit coverage of two or more systems, with mandatory coverage of the quality system. Coverage of a system should be sufficiently detailed so that the system inspection outcome reflects the state of control.
- There are two inspections options: The first is a Full Inspection, which is a broad and deep evaluation of CGMP when little or no information exists on the firm or there is reason to doubt the compliance efforts (history of objectionable conditions, recidivism, warning letters, etc.). The other option is an Abbreviated Inspection, which is designed to provide an efficient update evaluation of CGMP. Decisions for the type of inspection are made by the District Office.
- The guidance further lists subsystems and compliance requirements/expectations in each of the six quality systems (quality, laboratory controls) production, packaging, and labeling, materials, facilities, and equipment.

As indicated previously, recent 483 observations support the new approach. They are listed in quality systems buckets but are written up in traditional GMP context.

To have an effective internal audit program to evaluate conditions and level of risk and most importantly to gain prompt corrective action, the authors would recommend taking a different approach to internal audits than is typically practiced. The traditional approach is to mimic a regulatory inspection and cite GMP deficiencies usually by department. Reports go to the functional department head for correction. This does result in fixing the observation or symptom but often does not address the system deficiencies, root causes, or lasting improvement. One only needs to read current warning letters available on the FDA website to see the pattern of comments from the FDA continually citing firms for inadequate response to 483 observations because they are applying patches to procedures to fix 483 observations and not considering the root cause system fix.

Performing process audits and techniques can provide for better identification of system deficiencies, root causes, and a more effective level of corrective action (3). However, this approach requires a different knowledge/skills base for a typical compliance auditor.

An example is presented to illustrate the difference in approach. An analytical laboratory was audited and after the first day there was an observation that the secondary reference standard incubator had expired reference standard vials co-mingled with in-date vials. The auditor was ready to write the observation as is. The likely action taken to that observation would have been to go through the incubator and remove the expired standards. This would have done little to fix the problem from recurring. The lead auditor took charge the next day and performed a system review of the total reference standard program because reference standards are used in a direct quantitative comparison with product and would be a high-risk activity in any simple risk analysis. The following system elements were challenged:

- Quality of reference standard system inputs (compendial and other standards);
- Primary and secondary reference standard program ownership, responsibility, and accountability;
- Overall system design;
- Level of definition and detail in the SOPs;
- Consistency in practice;
- Management oversight of the process.

As a result of this evaluation, the auditors found that the system failure was not an isolated incident. It had been happening for some time. The failure was due primarily to the fact that the SOP required one person to be in charge of primary standards and another person to be in charge of secondary standards and in practice no backups were designated or assigned. The person in charge of secondary standards was out on extended leave and no one was picking up the responsibility. The overall causes involved a questionable system design and poor system oversight and resource allocation.

The resulting observation cited the deficiencies in design, oversight, and resource allocation. The CAPA now had to address those system deficiencies instead of fixing only the symptom (outdated standards).

The recommended systems audit approach is to challenge high-risk or value-adding systems with the fundamental system elements that should be in place rather than just looking for nonconformances (see Key Process/Systems Elements). However, be advised that this will require a different skill level for the auditor, different sets of questions being asked, and, most importantly, management support of the concept.

QUALITY RISK MANAGEMENT

Risk identification and management methodologies have been used for a number of years and applied in many different areas including investment, finance, safety, and medicine.

Quality risk management in the pharmaceutical industry is a relatively new concept and has had limited application to date. The FDA recognized in 1999 that the time was right for a new risk management framework when it came to the safety of medical products. They published a report in May 1999 on *Managing the Risks from Medical Product Use* (available at: http://www.fda.gov/oc/tfrm/riskmanagement.pdf). In this report, the FDA expressed the risk of medical products in terms of roles. The FDA is responsible to evaluate the benefit/risk for the population; the providers (including manufacturers) are responsible to evaluate the benefit/risk to the patient; and the patient evaluates the benefit/risk in terms of personal value. The risk management concept was utilized within the FDA and in August 2002, they announced their new major initiative for drug quality regulations on *Pharmaceutical CGMPs for the 21st Century: A Risk Based Approach.* This is available at: http://www.fda.govoc/guidance / gmp.html.

The FDA has since used the risk management approach and methodology in the prioritization of CGMP inspections of pharmaceutical manufacturing sites. In the initial concept paper issued on the initiative, the FDA identified "a risk-based orientation" as one of the guiding principles that would drive the initiative. The concept paper stated that "resource limitations prevent uniformly intensive coverage of all pharmaceutical products and production" and that "to provide the most effective public health protection, the FDA must match its level of effort against the magnitude of the risk" (4). Based on the basis of this analysis, the FDA determined the top three priorities for their inspection program: firms that produce sterile products, firms that produce prescription drugs, and firms that have not been inspected previously.

Applying risk management approaches to pharmaceutical manufacturing operations and decisions makes good business sense and should benefit the company and the patient. The importance of quality systems has now been recognized in the pharmaceutical industry and quality risk management is a valuable component of an effective quality system. The ICH Q9 Draft Consensus Guideline on Quality Risk Management (available at http://www.nihs. go.jp/dig/ich/quality/q9/q9-050322_e.pdf) describes the general quality risk management process, tools, and application in pharmaceutical operations.

Before embarking on quality risk management approaches, there are two very important cautions:

■ The time, effort, formality, and documentation of the quality risk management process should be commensurate with the level of risk. One can spend more time on the process than mitigating the risk. Although a systematic approach and use of tools are preferred, informal processes can be acceptable, especially for more obvious risks.
■ The quality risk management process should not be used as an excuse to delay or avoid compliance gaps/issues.

Figure 3 is an overview of the risk management process (from ICH Q9).

The primary principle of risk management is that the evaluation of risk to quality is based on the risk to the patient. From a manufacturing perspective, anything that has a high impact or is very close to the product will be high risk. For example, weighing of active ingredients in pharmaceutical production operations is a high-risk process worthy of compliance monitoring.

In the world of GMP compliance, there are at least three types of risk to consider. Patient and product-related are obviously the highest risk and must always be considered. Collective risk is another type for consideration. One can have a series of risks or failures identified that individually may not appear serious or have direct product impact but collectively could have direct product impact. An example would be a weak or incomplete change evaluation process, coupled with an inconsistent periodic re-validation process and incomplete historical product records and data. In combination, these deficiencies could lead to product failure. During the risk assessment phase, in addition to ranking individual risks, it is sometimes important to look at the collective risk, especially when systems and interdependencies are involved. Compliance failures are another GMP risk, especially in companies that have received warning letters or who are in consent decrees. Patterns of failure in GMP compliance, regardless of individual severity, may have an adverse impact on the business if the FDA perceives that the systems are still not in control.

Initiating a quality risk management process usually involves establishing a multidisciplinary team dedicated to the task. Key leaders and decision makers need to assure risk management has cross-functional participation. The process begins by defining a problem or risk

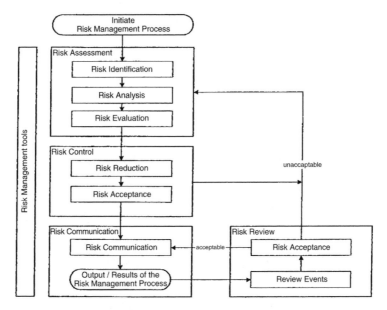

FIGURE 3 Risk management process.

and assumptions. Background information and data are then collected. A team leader needs to be identified and project timelines and deliverables established.

The first phase of the project is *Risk Assessment*, which includes risk identification, analysis, and evaluation. It is very important that the process starts with a well-defined problem description or risk question. This will help facilitate the gathering of information and data and to choose the correct tools for analysis.

Risk identification typically involves asking three questions: What might or could go wrong? What is the probability or likelihood it will go wrong? and What is the severity or consequence?

Risk analysis involves focusing on the last two questions and estimating the associated risk and ability to detect.

Risk evaluation can involve a qualitative (high to low) or qualitative (numerical probability) approach. The identified and analyzed risk is evaluated against the defined criteria. The output of the risk assessment phase is an estimate of risk for a quantitative approach or a range or risk for a qualitative approach.

The second phase is *Risk Control* where the goal is to eliminate or reduce the risk to an acceptable level. Risk control focus on four questions: Is the risk above an acceptable level? What can be done to reduce and control or eliminate the risk? What is the correct balance between risk, benefits, and resources? and Are new risks introduced as a result of these efforts? Risk control involves risk reduction (actions taken to mitigate or avoid the risk) and the risk acceptance decision. In some cases, it may not be possible to eliminate the risk altogether but short-term remedial actions may reduce it to an acceptable level or make sure it is detected.

Risk Communication is the third phase. If a team has been working together on the problem there should have already been communication between the decision makers and stakeholders. However, there may be a need for a more formal process of notification for other parties involved in or impacted by the decisions and changes.

Risk Review is the final phase. The output of the risk management process should be documented, especially when a formal process is used. The output and results should be reviewed for new knowledge and lessons learned. The changes and results should be monitored and if needed the risk management process can be re-engaged to handle planned or unplanned events. Risk management should be an ongoing quality management process.

Similar to failure investigations (Chap. 11) and process improvement projects, a number of useful tools and techniques can be used.

- Flowcharts, process mapping, check sheets, and cause and effect diagrams can help organize information and facilitate decision making.
- Failure mode effects (and criticality) analyses (FMEA and FMECA) evaluate potential failures and likely effect. Can be used for equipment, facilities, manufacturing, and system analysis.
- Fault tree analysis identifies root causes of an assumed failure. Can be used in failure and complaint investigations or deviations.
- Hazard analysis and critical control points (HACCP) was developed in the food industry and is a seven step systematic and preventive methodology that is used primarily for chemical, biological, and physical hazards.
- Hazard operability analysis (HAZOP) is used in cases of suspected deviation from design or operating intentions. It has been used for safety concerns regarding facilities, equipment, and manufacturing processes.
- Preliminary hazard analysis (PHA) uses past knowledge to help identify future failures. Can be used for product, process, or facility design, especially when information is scarce.
- Risk ranking and filtering breaks down the basic risk question into its components.

Throughout the process, statistical tools can be used to gather and analyze data; for example, control charts and process capability (Cp, Cpk) analysis.

Quality Risk Assessment is being increasingly adopted by the FDA and the pharmaceutical industry. The FDA has actively used it in prioritizing CGMP inspections as a result of the increasing demand for inspections and the finite level of staff to cover them. Time will tell if that approach is effective.

An industry example of the use of the process and tools was a recent assessment done on environmental monitoring. Using three different tools/techniques (HACCP, FMEA, Conceptualization Table) the group determined environmental criticality factors and monitoring frequencies for microbiological monitoring in a manufacturing facility. The risk analysis approaches were not only concerned with selecting environmental locations but also involved a complete review of the facilities, operations, and practices. The approaches recognize a risk, rate the level of the risk, and can help provide a plan to minimize, monitor, and control the risk (5). Other areas of more popular use of risk assessment tools are with failure investigation analysis and quality systems remediation and improvement.

REFERENCES

1. Nally J, Kieffer R. GMP Compliance, Productivity and Quality. Chapter Ch. 13. Interpharm Press, 1998, 443.
2. http://www.epa.gov/swerffrr/documents/data_quality/ufp_sep00_appx_e.htm, accessed 7/2/06.
3. Nally J, Kieffer R, Stoker J. From audits to process assessment—the more effective approach., Pharm Technol, 1995; 19 (9), 128.
4. Dills DR. Risk-based method for prioritizing CGMP inspections of pharmaceutical manufacturing sites—a pilot risk ranking model. J GXP Compliance 2006; 10 (2), 75.
5. Sandle T. Environmental monitoring risk assessment. J GXP Compliance 2006; 10 (2), 54–73.

SUGGESTED READINGS

1. Quality Systems Approach to Pharmaceutical Current Good Manufacturing Practice Regulations, September 2006 (Final Guidance).
2. Compliance Program Guidance Manual for FDA Staff: Drug Manufacturing Inspections Program 7356.002.
3. The ICH Q9 Draft Consensus Guideline on Quality Risk Management.
4. Bhatt V. GMP Compliance, Productivity and Quality. Interpharm, 1998.
5. Field P. Modern Risk Management A History. Risk Books, 2003.

6. Vesper JL. Risk Assessment and Risk Management in the Pharmaceutical Industry: Clear and Simple, July 2006.
7. Bhote KR. The Power of Ultimate Six Sigma, Amacom, 2003.
8. Russell JP. The Process Auditing Techniques Guide, ASQ, 2006.
9. Kausek J. The Management System Auditor's Handbook. ASQ, 2006.
10. Cobb CG. Enterprise Process Mapping, ASQ, 2005.
11. Imler K. Get It Right. ASQ, 2006.
12. Nally J, Kieffer R, Stoker J. From audits to process assessment—the more effective approach, Pharm Techno 1995; 19 (9), 128.

15 | Clinical Trial Supplies and Current Good Manufacturing

Graham Bunn
GB Consulting LLC, Berwyn, Pennsylvania, U.S.A.

The pharmaceutical industry is focused and striving to bring new products to market to treat diseases but it is also a business, which needs to make a profit to prosper. Without significant revenue coming in to continue producing products to sell, there is no chance of developing new ones to replace the marketed ones in the future. Clinical trial supplies (CTS) are tomorrow's revenue generators and are as essential as the CGMPs that support them.

The Declaration of Helsinki and good clinical practices (21CFR50 and 56) protect patient interests and safety in clinical studies. The pharmaceutical CGMPs are defined in the Code of Federal Regulation 21CFR211 for the manufacturing, testing, packaging, labeling, and distribution of drug products. These regulations are there to protect patients and ensure that only quality products are introduced into interstate commerce. The revised regulations introduced into law in 1978 stemmed from the repeated problems experienced in the industry and provided the basis on which today's regulations, regulatory expectations, and industry standards are implemented. The fundamental difference between CTS and the commercial prescription drug products is the commercial product has more physical and pharmacological data collected and has been used in many more patients. There is no difference in the quality expectations by patients whether taking a CTS versus a prescription product.

Step back from the differences and examine the similarities from a business perspective. There is no doubt in anyone's mind in the pharmaceutical industry that the output of CTS manufacturing is a product fit for human consumption. Without this objective there are no CTS, as someone has chosen to differentiate between them and commercial drug products based solely on the number of patients using the drug product.

Examination of the entire product development process for CTS enables the identification of those areas of the CGMPs for commercial products where there is no difference, and instances when a degree of interpretation of the regulatory requirements is needed. There is no question that any differences must have no impact on the safety of the product being administered.

From a business perspective, it makes unquestionable sense to collect data and information during the development process in an accurate and complete manner. Meaningful decisions cannot be made in formulation selection, mixing parameters, and other criteria if the data are not accurately recorded or are open to a wide degree of interpretation. A wrong decision at the least could ultimately delay approval to market and lose revenue income and not forgetting delays to treating patients.

BACKGROUND TO PRODUCT DEVELOPMENT

Identification of potential therapeutic agents has changed dramatically with the development of high volume screening techniques and the ability to predict with a degree of certainty the therapeutic nature of the compound. Serendipity still plays a role in the discovery of new products and clinical indications. Preclinical testing (nonhuman) under good laboratory practices (21CFR58) establishes the safety data and other information. These data/information are essential in establishing that the administration of the compound does not expose the human trial subjects to unnecessary risks. According to Federal law, an approved marketing application must be approved for a drug to be entered into interstate commerce and transported across state lines. As the components go into the new dosage form and ultimately the final drug product will be transported across state lines in order for clinical trials to be performed, an exemption from the legal requirement is needed. The exemption process is defined in 21CFR312: Investigational New Drug (IND) Application, which is subject to section 505 of the Federal Food, Drug, and Cosmetic Act or to the licensing provisions of the Public Health Service Act [58 Stat. 632, as amended (42 U.S.C. 201 et seq.)]. Part 312 includes applicability, labeling, principles of the submission, content and format, withdrawal of the application, safety, responsibilities of investigators, record keeping, and several other areas. The reader is encouraged to visit the Food and Drug Administration (FDA) web site (www.FDA.gov) to obtain further information and understanding of the IND. There are three types of INDs (investigator, emergency use, and treatment), which are divided into two categories (commercial and research). There are three main sections of the IND application [animal pharmacology/toxicology data, clinical protocols (description of the methodology of conducting the human study), and details of producing the product].

After the sponsor (person/company responsible for the study) submits the IND to the FDA, they must wait 30 calendar days before initiating any clinical trials.

The 1978 CGMP regulations in 21 CFR Parts 210 and 211 are the interpretation of the statutory requirement for drugs to be produced in compliance with GMPs, found in the *Federal Food, Drug, and Cosmetic Act* (1). No distinction is made in the act between commercial and investigational products. The definition in 21CFR210.3(b)(4) implies that the regulations applied to CTS as placebos are only used in clinical trials: "Drug product means a finished dosage form, for example, tablet, capsule, solution, etc., that contains an active drug ingredient generally, but not necessarily, in association with inactive ingredients. The term also includes a finished dosage form that does not contain an active ingredient but is intended to be used as a placebo." The FDA did state at the time that the regulations applied to all types of pharmaceutical production (2) and that additional regulations relating to investigational drugs used in clinical studies were being considered.

In March 1991, the FDA issued *Guideline on the Preparation of Investigational New Drug Products (Human and Animal)*. However, the guideline omitted discussing all manufacturing situations and expectations that overall manufacturing controls would change as further data and information became available through the drug development stages toward the submission of the New Drug Application (NDA). Although formulation would be expected to change little, the greatest change would be scale of production and finalization of the packaging configurations (more so for oral products).

There is also reference to clinical production of new drugs in the FDA's pre-approval inspection compliance program (3):

Inspections/Audits
1. Manufacturing Process
    (a) Drug Product (Dosage Form)
         In many cases, clinical production or trial runs of a new drug are produced in facilities other than the ones used for full scale production. The facilities and controls used for the manufacture of the batch(es)* (2) must be audited. For a generic drug product, the biobatch(es) are required to be manufactured in production facilities, using production equipment, by production personnel, and the facility is to be in conformance with CGMPs. Accurate documentation is essential so that the production process can be defined and related to the batch(es) used for the early clinical, bioavailability, or bioequivalence studies of new drug or generic drug products.
              FDA has also published guidance for compliance with GMPs for the production of investigational drugs. Refer to this document for added assistance.

---

*Generic product biobatches are ANDA batches that are compared to the originator/reference product to establish their equivalence. NDA biobatches are NDA batches comparing the product planned for marketing with that studied during clinical trials to establish their equivalence.

The batch records submitted in the application must be audited as part of the inspection to assure that the proposed production process is the process that was used for the manufacture of the bio/stability batches. Some manufacturers have historically made small batches that were used for biostudies and stability studies and misrepresented them as larger batches in submissions. Documentation sometimes has included R&D notebooks and/or batch records. Inventory records and/or receiving records of drug substances have been found to be of value in documenting the accountability of drug substances used in the early batches.

On January 17, 2006, the FDA issued the draft document *Guidance for Industry INDs— Approaches to Complying with CGMP During Phase 1* for comment. This guidance outlined key manufacturing control points to be covered during development for the different types of products (pharmaceutical, biological). The FDA linked the draft guidance to the CGMP for the 21st century initiative (4) which covered among other topics risk-based approaches, quality management systems, quality systems approach, international collaboration, and electronic records. Contents of the draft guidance included personnel, quality control (QC) function, facilities/equipment, control of components, production and documentation, laboratory controls, containers closures and labeling, distribution, record keeping, and special production situations. In addition to the guidance document, the FDA proposed amending its CGMP regulations for human drugs, including biological products, to exempt most investigational Phase 1 drugs from complying with the requirements in FDA's regulations.

During the comment period, the FDA received "significant adverse comments" and on May 2, 2006 withdrew the rule to amend its current good manufacturing practice regulations for human drugs, including biological products, to exempt most investigational Phase 1 drugs from complying with the requirements in the FDA's regulations.

The comments received by the FDA varied, and although most of the industry welcomed the document and found it useful, concerns/comment expressed disagreement with fundamental responsibilities of the QC unit and that Phase 2 and sometimes Phase 3 products are made in the same facilities under the same quality systems. There was concern that there was a difference in the approach when compared with the recent EMEA Clinical Trial Directive requiring that manufacturing of products for clinical trials be in essence the same as commercial products. Patient safety and related topics mentioned in several comment letters are the focus of the industry and remains independent of the phase of development. Without patient safety, there is no industry.

There are four phases of product development under INDs:

- *Phase 1 studies*: typically 20 to about 80 healthy volunteers in a controlled medical environment to determine dosing, drug metabolism/excretion, and identify any acute side effects.
- *Phase 2 studies*: typically 100 to 300 participants who are treated for the condition to gain further safety data, efficacy, and benefit:risk ratio. Placebo controlled studies are also performed.
- *Phase 3 studies*: typically several thousand people with the condition being treated to determine effectiveness, identification of further side effects from longer-term treatment, and perform testing comparator/placebo to prove efficacy.
- *Phase 4 studies*: may be performed after drug approval and marketing to determine long-term risks, optimize use, and determine applicability for specific populations.

There are multiple sections in an NDA including the Chemistry, Manufacturing, and Controls Documentation where the processes, specifications, analytical methods, stability, etc. are defined, justified, and supported with data. The submission of the NDA is the formal request to the FDA to review the information and data and to consider the product for marketing approval.

The FDA has 60 days from receipt of the NDA to decide if it can be filed so it can be reviewed. In addition to the documentation review, the FDA can inspect the facilities where the drug product will be manufactured, analyzed, and packaged as part of the approval process. Successful review of the documentation and resolution of any questions combined

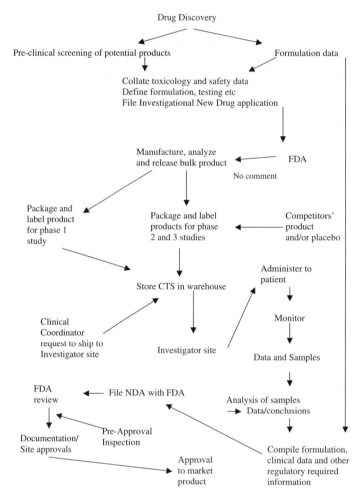

FIGURE 1   Overall drug development process to NDA approval.

with an acceptable inspection of the drug product facilities can ultimately result in approval of the application. The approval enables the sponsor of the NDA to ship the commercial product into interstate commerce. The approved IND enables the sponsor to ship drug product across state lines without the approved NDA. The overall drug development process to NDA approval is shown in Figure 1.

SUBPART A GENERAL PROVISIONS

§211.1   SCOPE

As mentioned previously, CTS are currently included within the regulations. The FDA withdrew the proposed revision to the requirements for Phase 1 products in May 2006 following "significant adverse comments."

§211.3 Definitions, see Chapter 2.

SUBPART B ORGANIZATION AND PERSONNEL

§211.22   RESPONSIBILITIES OF QUALITY CONTROL UNIT

(a)   There shall be a QC unit that shall have the responsibility and authority to approve or reject all components, drug product containers, closures, in-process materials, packaging material, labeling, and drug products, and the authority to review production records to assure that no errors have occurred or, if errors have occurred, that they have been fully investigated. The QC unit shall be responsible for approving or rejecting drug products manufactured, processed, packed, or held under contract by another company.

(b) Adequate laboratory facilities for the testing and approval (or rejection) of components, drug product containers, closures, packaging materials, in-process materials, and drug products shall be available to the QC unit.

(c) The QC unit shall have the responsibility for approving or rejecting all procedures or specifications impacting on the identity, strength, quality, and purity of the drug product.

(d) The responsibilities and procedures applicable to the QC unit shall be in writing; such written procedures shall be followed.

The testing functions of QC are defined in part (b) above and parts (a) and (c) are covered by quality assurance (QA).

The specific title of this section relates to the QC department and QA is not mentioned in the regulations. However, current industry standard has been to establish two separate functions. The QC unit has primary responsibility for the testing functions and QA for the overall independent responsibilities for quality.

It may still be possible in smaller CTS organizations to assign these responsibilities separately. It is an expectation that irrespective of the size of the organization the Quality Unit will have an independent reporting structure of that of production so as to provide an impartial and unbiased decision on the acceptance or rejection of components, materials, and product. Some companies further separate the reporting structure by having QA report directly to a corporate function and on an administrative basis to local management.

A challenge to the QA function is the outsourcing of testing to a third party contractor. The same standards and requirements are required for in-house and contracted processing. This is further discussed in chapter 19. There is no question that the final responsibility for release resides clearly with the client's quality function as confirmed in the preamble to the CGMPs in comment #97 by the Commissioner (5). §211.25 Personnel qualifications, see Chapter 3; §211.28 Personnel responsibilities, see Chapter 3 and §211.34 Consultants, see Chapter 3.

SUBPART C BUILDINGS AND FACILITIES

§211.42   DESIGN AND CONSTRUCTION FEATURES

There are some specific expectations relating to CTS because of the amount of data available. Consideration and careful review of handling of the active ingredient, intermediates, and final products needs to be performed. Manufacturers and contractors will have both systems and procedures for the evaluation of the new chemical entity before it is approved for entry into the facilities. For the contractor, this is critical, as this could be the first time that they have handled this type of material. A material safety data sheet listing precautions for handling, toxicology data, cleanup procedures, and hazard levels is available and personnel need to be trained in the requirements. The contractor will have multiple clients' products and it is essential that there is no potential for cross-contamination. Although this is also required for commercial products, they have more experience in handling the product.

Facilities are also used for multiple CTS in various stages of production in relatively short runs compared to commercial operations, which can also be performed in dedicated facilities.

Additionally, there needs to be adequate segregation for CTS and commercial operations. Normally, these are in different buildings with their own set of standard operating procedures (SOPs) covering management of operations. §211.44 Lighting, see Chapter 4; §211.46 Ventilation, air filtration, air heating, and cooling, see Chapter 4; §211.48 Plumbing, see Chapter 4; §211.50 Sewage and refuse, see Chapter 4; §211.52 Washing and toilet facilities, see Chapter 4; §211.56 Sanitation, see Chapter 4; §211.58 Maintenance, see Chapter 4.

SUBPART D EQUIPMENT

§211.63 EQUIPMENT DESIGN, SIZE, AND LOCATION

It is critical for product development that variables are controlled so as to provide a degree of consistency. For example, an appropriate size mixer reflective of the relatively smaller size batches with CTS is needed. Testing data (e.g., blend uniformity) are used to determine mixing speed and time. Failure to use the appropriate size may cause data to be misleading and formulation problems experienced at scale up. The same problems are not experienced in commercial production with a validated process.

§211.65 EQUIPMENT CONSTRUCTION

There are no specific equipment construction requirements other than ensuring it is of a suitable material so as not to react with the product or any of its components.

§211.67 EQUIPMENT CLEANING AND MAINTENANCE

Cleaning verification is a critical parameter that must be supported by adequate documentation and data. An SOP must define the requirements for generation of the protocol and be separately approved by the client if this process is performed by a contractor. Although the contractor is familiar with all the characteristics of all its clients' products relating to cleaning, the individual client is primarily concerned with ensuring that the product made in the equipment before theirs has been sufficiently removed. The client can review the procedural requirements/approach to cleaning and request confirmation that the results met acceptance criteria. Additionally, the hold time between end of manufacturing and cleaning needs to be established from data generated in early batches to support subsequent manufacturing runs. Equipment and particularly batch size in equipment will change as batch size increases during development. Cleaning needs to be assessed for adequacy with each change with cleaning validation being performed in the final stages.

§211.68 AUTOMATIC, MECHANICAL, AND ELECTRONIC EQUIPMENT

Calibration of equipment used for CTS needs to be performed, otherwise there is no assurance that the analytical results of the product are valid. Readings of temperature, pressure, and speed are captured into computer systems as evidence and may also control parameters of the process, e.g., tablet weight adjustments. If the measuring devices are not calibrated, then there is no assurance that the product can consistently be produced. Process validation is performed towards the end of product development but an incorrect measurement/data during development could result in the project leader making an incorrect decision based on the invalid data. Subsequent batches would be rejected, time lost, and process development work repeated. Sterile processes must have supporting validation data and supporting documentation as part of a submission. §211.72 Filters.

SUBPART E FILTERS, CONTROL OF COMPONENTS AND DRUG PRODUCT CONTAINERS AND CLOSURES

See Chapter 6:
§211.80 General Requirements
§211.82 Receipt and Storage of Untested Components, Drug Product Containers and Closures
§211.84 Testing and Approval or Rejection of Components, Drug Product Containers and Closures.
§211.86 Use of Approved Components, Drug Product Containers and Closures
§211.87 Retesting of Approved Components, Drug Product Containers and Closures
§211.89 Rejected Components, Drug Product Containers and Closures
§211.94 Drug Product Containers and Closures

SUBPART F PRODUCTION AND PROCESS CONTROLS

See Chapter 7:
§211.100 Written Procedures; Deviations
§211.101 Charge-in of Components
§211.103 Calculation of Yield
§211.105 Equipment Identification
§211.110 Sampling and Testing of In-process Materials and Drug Products
§211.111 Time Limitations on Production
§211.113 Control of Microbial Contamination
§211.115 Reprocessing

SUBPART G PACKAGING AND LABELING CONTROL

§211.122 MATERIALS EXAMINATION AND USAGE CRITERIA

Labels and labeling (printed materials) must be controlled to prevent mix-ups and incorrect content. Initiation and changes need to be managed and approved through a change control system to ensure that the correct labels are used. All required labels/labeling are defined in the clinical protocol. A CTS project manager normally works closely with clinical coordinators during the drafting of the protocol to ensure that the label/labeling requirements are clearly understood and practical. A draft of each label/labeling and any additional patient instructions are generated from the protocol and may be sent to a contractor for entry into their label system before returning for proof approval. The final proof copies must be physically compared to those in the clinical protocol to ensure that they are an exact copy. Once acceptable, the CTS project manager approves the copies for printing. All labels/labeling printed at approved contractors must be quarantined at receipt and follow SOPs for inspection to ensure they meet the predefined specifications. If acceptable, they are released, and if unacceptable, they are destroyed. The same procedure for quarantining, inspection, and release/reject is followed for labels generated within the company.

With CTS labels, there is an added complexity as there are two main types: open labels (CTS drug product name and strength included) and blinded labels (no CTS drug product name and strength). Some trials are blinded to either the patient or both the patient and clinical investigator. In either case, the drug name and strength does not appear on the label. A unique patient number and often a period number are included and linked to a randomization code where the identification of the drug product is maintained. Some labels have a removable covering that reveals the identity of the drug product in emergencies. The cover cannot be replaced and hence the investigator could easily tell if the blinding had been broken. Alternatively the unique code on the label is linked to the specific code breaker which identifies the drug product (drug, strength or placebo).

Computer validation of the systems used to generate blinded labels is required to ensure that the correct sealed identification corresponds with the patient number. Line clearance must be performed and documented to ensure that no extraneous materials are present in the printing area. Documentation of each print run and reconciliation of labels printed, inspected, rejected, and issued is required.

§211.125 LABELING ISSUANCE

Issuance of labels is one of the most closely controlled operations. As discussed in other sections, the labels for CTS are closely controlled from design to attachment on the container and initially appear to be no stricter controls than those for commercial labels. However, the CTS labels could be blinded and hence have no product identification that introduces additional controls/challenges. Although labels are required to be reconciled within limits for commercial, there is no margin for error with CTS labels. Every label is unique in blinded and most open studies and must be accounted for in the reconciliation. All discrepancies must be investigated. Labels are normally printed for a specific "arm" of the study and therefore any excess destroyed at the end of the packaging operation (6). If unique labels are damaged during operations, they must be replaced under the same strict controls that the originals were issued. It is imperative that sufficient checks are implemented to ensure that the replacement label is exactly the same as the original unless an error was noted. The request to the printing department (or contractor) must include all the required information, the label checked and released as per SOP. It is normal practice to attach the original defective label to the packaging record as evidence that it was not used.

CTS labels contain the product name and strength for open studies but not for the blinded study label as shown below and includes the caution statement as required by 21CFR312.6(a):

Period 1   **30 Capsules**
pt no. 12

Take one capsule each morning with breakfast
Caution: New Drug-limited by Federal law to investigational use
Store at room temperature (59–89°C)

Pkg ID 199
Protocol 264
Gpharm, 356 Dibone Rd, Wynber PA 19543

Note that the labels in a blinded study (as illustrated above) will appear the same with the exception of the patient number and period number.

211.130   PACKAGING AND LABELING OPERATIONS

A line clearance/inspection is required for all packaging/labeling operations to ensure that there is no drug product remaining from the previous batch and that packaging and labeling materials not suitable for subsequent operations have been removed. Results of the clearance/inspection are documented in the production batch record. The master production records provide a description of the drug product including size, shape, distinguishing markings, and color. Blinded CTS do not contain markings and placebos are made to resemble the drug products. There must be no doubt at any time as to the identity of the CTS, either active or placebo.

Production controls for the generation of active drug products for CTS incorporate many of the same requirements used in the packaging of commercial products. CTS packaging operations are performed in separate operations and then combined. There needs to be checks when packaging/blistering operations are performed to ensure the correct identification of the drug products. For the study below, the drug products could be packaged into blister strips of seven dosages in separate blistering operations. The sealed blisters are inspected for physical defects and leak tested to ensure adequate sealing before bulk packaging and sealed in labeled cartons. Reconciliation of the bulk and calculation of yield are required to ensure that limits are met. Some companies perform identification of samples from each of the blistering operations. In the example below, the sealed cartons of the two blistered products (CTS: active product and placebo) would be brought into the carding room and checked for correct identification using the carton label and quantity confirmed according to SOPs. These checks would be documented in the production record. The two different blister strip cartons are segregated physically on either side of a carding table. Three CTS blister strips would be placed face down in the card in the specified locations. Either a second check is performed at this time or a template is used so as to prevent the operator placing a blister strip in dose 4 location. The carding table rotates and the second operator places one placebo blister strip in dose 4 location. The table rotates again and is checked to ensure all locations are filled before placing a backing card over the blisters and sealing in a heated press. Carded blisters are inspected for physical defects before being placed in a labeled carton. Following completion of carding, samples may be taken for identification of the correct location of the CTS drug product and corresponding placebo. The confirmation of product identity relies on strict documented controls for the location of the strips.

Dose	1	2	3	4
Day				
1	CTS	CTS	CTS	CTS placebo
2	CTS	CTS	CTS	CTS placebo
3	CTS	CTS	CTS	CTS placebo
4	CTS	CTS	CTS	CTS placebo
5	CTS	CTS	CTS	CTS placebo
6	CTS	CTS	CTS	CTS placebo
7	CTS	CTS	CTS	CTS placebo

Automated machines fill specific blister wells in blister with different products, and vision systems confirm that correct locations are filled. The equipment must be qualified to confirm that it consistently performs as required. §211.132 Tamper-Resistant Packaging Requirements for Over the Counter (OTC) Human Drug Products: Not applicable to CTS; §211.134 Drug Product Inspection, see Chapter 8.

§211.137 EXPIRATION DATING

Some companies refer to the date as "recertification" rather than expiration, which is normally associated with a finite date. Drug products used in phase 1 and early phase 2 will have limited recertification dates and ongoing stability will be used to continually evaluate shelf life and extend it when justified. A link between the approval of recertification date extensions and the shelf life maintained in the warehouse stock control system needs to be established. Unlike commercial products, which have established expiry dates, CTS dates may be extended, or sometimes shortened and withdrawn should analysis confirm a shorter date. Labels on CTS containers do not contain a date beyond which the product cannot be used because it may be extended and therefore not require additional relabeling. CTS are normally shipped with a defined minimum time before the recertification date to enable shipment, receipt, and use by patients. A procedure needs to be established to notify clinical coordinators of the finite date beyond which CTS must not be used and either sent for destruction or returned to the distributor for counting and arrangements for destruction.

SUBPART H HOLDING AND DISTRIBUTION

§211.142 Warehousing Procedures, see Chapter 9.

§211.150 DISTRIBUTION PROCEDURES

The process to distribute the correct product to the specific clinical investigator requires careful monitoring and control. As described in §211.130 (packaging and labeling issuance), CTS are packaged with either open or blinded labels with unique patient numbers. Clinical coordinators determine which investigator receives each patient numbered product and send written confirmation/authorization to ship the products. Regulations in 21CR50 cover good clinical practices, including the regulations relating to CTS which link to those of good manufacturing practice.

SUBPART I LABORATORY CONTROLS

§211.160 GENERAL REQUIREMENTS

Any in-process or final testing is required to have defined and approved methods and specifications. Development work performed prior to making the CTS batches provides the initial data to support the limits defined at the time of testing. As further development and adjustments are made in the formulation/manufacturing process, the methods and specifications are revised accordingly. Although the specifications may be initially relatively "broad," they still must be justifiable and supported with adequate data.

In 21CFR312.23 (a)(7)(iv)(b), the requirements for the Chemistry, Manufacturing, and Control Information section of an IND Application for a Phase 1 Drug Product are defined. It includes the acceptable limits and analytical methods used to assure the identity, strength, quality, and purity of the drug product. A brief description of the proposed acceptable limits and the test methods used should be submitted. It is acknowledged that the tests will vary according to the dosage form. The FDA has issued additional guidance for IND applications (7). §211.165 Testing and Release for Distribution, see Chapter 10.

§211.166   STABILITY TESTING

SOPs define the requirements of the stability program to ensure that there is adequate data to confirm that CTS maintain a defined shelf life at the required storage conditions. Limited stability data are known when new drugs enter the development program. Phase 1 studies are done in relatively small numbers of patients in a single location over a short time period. There needs to be sufficient stability of the dosage form to enable the clinical study to be performed. A stability protocol is generated defining product name, packaging details, number of samples for each storage condition, testing methodology (indicating stability and other parameters: sterility), frequency of testing and number of required sample container, and acceptance criteria. CTS are placed in accelerated conditions to gain further data as real-time data are also collected. As data are collected at each stability testing, point analysis is performed to confirm the assigned recertification date. This date can be extended as further data are analyzed and continue to meet predefined acceptance criteria. §211.167 Special Testing Requirements, see Chapter 10; §211.170 Reserve Samples, see Chapter 10; §211.173 Laboratory Animals, see Chapter 10; §211,176 Penicillin Contamination.

SUBPART J RECORDS AND REPORTS

§211.180 General Requirements, see Chapter 11.
§211.182 Equipment Cleaning and Use Log, see Chapter 11.
§211.184 Component, Drug Product Container, Closure and Labeling Records, see Chapter 11.

§211.186   MASTER PRODUCTION AND CONTROL RECORDS

The regulations require a master record for each drug product, including each batch size, and there are clear requirements for the minimum contents. There are times when the formulation or manufacturing process may only be used once during the drug development process. There is a temptation to use this as an excuse for only generating a single copy and minimizing the effort/resource required to generate the record. However, it is not only a requirement but also makes good business sense to maintain a master copy. A well-controlled process for the generation of master records is critical for maintaining a systematic documentation of the product's development history. Change control must be utilized to manage all changes to the master record with the appropriate approvals according to the SOP.

Content of a CTS manufacturing record may be similar to that of a commercial product except for the scale. Records used for some of the first manufacturing batches will contain sufficient instructions for the trained operators to understand how to make the product. Often the experience of the development personnel is utilized to train the operators in the manufacturing process prior to executing the record. There are also development personnel who have also been trained in GMP manufacturing requirements and can provide support to CTS manufacturing personnel. Early batch records contain apparently broad parameters, e.g., mixing speeds, but these will be reduced in future batch records as more data are collected.

§211.188 BATCH PRODUCTION AND CONTROL RECORDS

The most distinguishable difference between CTS and commercial is found in packaging records. Clinical trials require an approved protocol, which describes the clinical aspects, clinicians, and other patient-specific requirements and testing/monitoring to be performed. Also included in the protocol is the design of the study, with a detailed description of the products, their packaging configurations, and dosing frequency of the subjects. The content and layout of the each label applied to the primary container, package, and shipper is defined. In phase II and III trials, the CTS are compared to competitors' products and placebos to demonstrate advantages/efficacy. A placebo tablet is made to look like the comparator product but contains no active ingredient. When the competitors and/or placebo is used in a clinical study with the CTS, they are often blinded so that neither the patient nor the clinical investigator know the identity of the product being administered.

Illustration of a "Cross Over" study involving CTS and competitor products with corresponding placebos

Period 1 pt no.	Period 2 pt no.	Treatment arm			
		CTS	Competitor	CTS placebo	Competitor placebo
1	2	XXX		X	
2	10	XX		XX	
3	8			XXXX	
4	5		XX		XX
5	6		X		XXX
6	1				XXXX
7	12		X		XXX
8	7				XXXX
9	4			XXXX	
10	11		XX		XX
11	9	XXX		X	
12	3	XX		XX	

There is often a high degree of manual operations in the packaging of CTS products into final containers. Although automated/semiautomatic operations are utilized where possible, the number of containers to be packaged does not always warrant setting up the equipment, e.g, bottle and blister packaging. It is essential that the records contain all the adequate directions for the operations and are clear/unambiguous to the operators.

To add to the complexity of packaging the clinical studies, there are often different treatment "arms" with administration of different products and doses, as shown in the previous illustration.

Each X above indicates a dose administration. When the study is blinded, the dosage forms appear the same, and in the above illustration, the CTS, competitor, and placebo could be blinded by encapsulation. Sometimes, there is a need to have two different dosage forms, which either cannot be encapsulated or the sponsor of the study wants the two different forms, i.e., tablet and capsule. In this case, it would necessitate the need for two placebos that resemble the active dosages. The study above involves administration of the two dosing levels of the active drugs and placebos. After a defined period (e.g., one month), there would be a "wash-out" period when no drug product is administered. Then the patients are assigned a second treatment period as defined by a randomization code. The code is known to a limited number of people and is only unblinded after the study results database has been "locked." There is access in the event of an adverse event relating to patient safety. Blinding of the actual drug product is designed to remove bias from the patient and investigator. Hence, it is critical that there is no apparent difference between the drug product and the corresponding placebo and also the individual packaging/labeling.

After the dosage forms have been identified, the process of packaging them with the required labels can be developed. Records must include sufficient instructions to perform the operations and capture adequate data/information to confirm what was done. Complexity of clinical studies requires additional checks and often multiple records for a single study. §211.192 Production Record Review, see Chapter 11; §211.194 Laboratory Records, see Chapter 11.

§211.196   DISTRIBUTION RECORDS

There must be adequate records maintained to enable the complete distribution of the product to be tracked. This is critical in the case of a "stock recovery," which is the same as recall of a commercial product. §211.198 Complaint Files, see Chapter 11.

SUBPART K RETURN AND SALVAGED DRUG PRODUCTS

§211,204 Returned Drug Products, see Chapter 12.
§211.208 Drug Product Salvaging, see Chapter 12.

REFERENCES

1. Federal Food, Drug and Cosmetic Act, as amended, 21 USC 301 et seq., Sections 201(g) [21 USC 321 (g)] and 501 (a)(2)(B) [21 USC 351(a)(2)(B)].
2. Preamble to the CGMP 1978, Comment #49. "The Commissioner finds that, as stated in 211.1, these CGMP regulations apply to the preparation of any drug product for administration to humans or animals, including those still in investigational stages. It is appropriate that the process by which a drug product is manufactured in the development phase be well documented and controlled in order to assure the reproducibility of the product for further testing and for ultimate commercial production. The Commissioner is considering proposing additional CGMP regulations specifically designed to cover drugs in research stages."
3. FDA. Compliance Program 7346.832. Preapproval Inspections/Investigations, April 5, 2005.
4. http://www.fda.gov/cder/gmp/21stcenturysummary.htm.
5. Preamble to the CGMP 1978, Comment #97: Comments asked for modification in the last sentence of 211.22(a) relating to production of drugs under contract by another company. They suggested that the contractor could release drugs for distribution on his own responsibility if a certificate of analysis showed compliance with appropriate specifications. Others suggested that the responsibility for approving or rejecting drug products produced by contractors appeared in this paragraph to rest with the contractor.
6. Bunn G. Controlling clinical trial labels. Am Pharmaceutical Rev 1999; 2(4).
7. FDA. Content and Format of Investigational New Drug Applications (INDs) for Phase 1 Studies of Drugs, Including Well-Characterized, Therapeutic, Biotechnology-Derived Products. 1995.

16 | Contracting and Outsourcing

Graham Bunn
GB Consulting LLC, Berwyn, Pennsylvania, U.S.A.

One thing that is common to the small virtual company and the multinational company marketing commercial product is that they both use contractors and outsource one or more operations. The virtual company with its first product in development is solely dependent on contractors and, because of this, needs to minimize any potential regulatory-related risks that may jeopardize product development or regulatory submission review. The multinational company can have product manufactured and/or packaged and analyzed by a contractor; here too there is a regulatory risk associated with the operations if they are not in compliance with the regulatory filing or current good manufacturing practice (CGMP) requirements and regulatory expectations. While the contractor would be required to respond to any observations observed during a regulatory inspection, there is also an impact for the client, virtual or commercial, having their product manufactured at the facility. Outsourcing a process to a contractor does not devoid the client (sponsor) of responsibilities for compliance.

During the last 10 years, there has been a significant change in the way pharmaceuticals are developed and brought to market for the consumer/patient. One of the biggest changes is the expanded use of contractors for one or more of the developmental steps (premarket) and, increasingly, for the entire process. The outsourcing industry used to support virtual companies with no/limited capabilities of their own, and some products from large companies were contracted out because of logistics and limited in-house capabilities. Today, there is increasing demand for commercial products to be produced by contractors and also by business alliances.

The last several years have also seen a number of major acquisitions and mergers involving larger pharmaceutical companies, which has resulted in demands for cost constraints. An outfall of this is amalgamation of capabilities (production, analysis, packaging/labeling, and distribution) and selling selected facilities. Additionally, executive management uses the opportunity to streamline the drug development processes, resulting in increased demand for contractor capabilities.

Contractors have gained knowledge and experience over the last 10 years enabling them to manage the requirements/requests from multiple clients simultaneously. New operations attract experienced management personnel to successfully exceed client expectations. Outsourcing operations offer distinct advantages that have to be utilized to their maximum benefit.

While there has always been a need for clinical trial supplies contractors, the increase in business, regulatory, and public expectations for rapid time to market has had a significant impact on this section of the industry. Additionally, regulatory authorities are requiring increased data and information from clinical trials to support product indications. Clinical studies have increased in complexity and resulted in additional control requirements during packaging. Increased safety focus added additional requirements for further and more complex studies.

Key support companies identified the need for specialized services, for example, clinical supplies labels with all the complexity and controls for blinded labels. This service not only covers initial issuance of thousands of labels, but may also include different languages and different arms of a study, that is, development drug, comparator drug, and sometimes two placebos. Additionally, analytical testing and manufacturing capabilities have expanded from "conventional" pharmaceuticals to more complex biological systems used for the generation of therapeutic treatments. Biopharmaceutical contractors have gained valuable experience and expertise and are now able to offer a wide range of services covering today's product requirements.

OUTSOURCING

The following chapter discusses different topics and can be adapted for individual processes that are outsourced to a contractor. The initial response of "This cannot be performed at this facility within those timelines" may actually have many other underlying reasons to outsource the process.

Key reasons to consider outsourcing include the following:

- Capabilities outside company facilities/operations or portfolio: for example, a pharmaceutical company with a biotechnology product needs to develop an injectable formulation.
- Specialized facilities/operations needed: for example, filling/lyophilizing, containment requirements.
- Accelerate drug development.
- Reduce time to market.
- Production capacity exceeded or fluctuates with seasonal demand.
- Capital investment limited or not prepared to be committed at this stage of development. Purchasing or building facilities specifically for a product that later fails to meet acceptance criteria can ultimately result in an empty building and loss of capital investment, which could have been utilized in developing other products.
- Greater return on investment.
- Maximize patent life by decreasing time to initiate clinical studies and accelerate decision points.
- Flexibility and ability to respond to changing developmental requirements based on marketing and clinical responses.
- Finances redeployed to other strategic areas, including product development.
- Complement or provide redundancy in the supply chain.
- Support uncertain product approvals without tying up internal resources.
- Smaller volume products.

Due to the degree of outsourcing, complexity of the process, and degree of risk, some companies are dedicating resources specifically to manage contractors. While this is a distinct advantage, many companies cannot afford the expense, especially small virtual companies. For small virtual companies, there are major obstacles to managing contractors with personnel in different departments with sometimes an ineffective process. One of their largest challenges is communication and contractor management.

Companies are using contractors' capabilities to free internal resources for other drug product development projects. Another advantage that contractors offer is the ability to provide specialized services (e.g., containment, biological handling, and lyophilization), which would take capital investments, other resources, and mainly time. A contractor has the ability to initiate services in a fraction of the time taken at a company facility and can do so with highly experienced staff. This has a distinct time-saving advantage for the company. Additionally, if development of the product is stopped (either for clinical or business development reasons), the company can, with the appropriately worded contract, stop development and associated expenditure promptly.

Some contractors also dedicate a contact person (customer project manager) within their organization who is responsible for interfacing with clients. Although this does not prevent direct contact with key personnel, for example, QC laboratory manager, it may appear that the contractor is preventing access to specific personnel. Without the customer project manager, there is the potential for breakdowns in communications and multiple requests/questions being delivered to the wrong personnel. The customer project manager has a critical role in ensuring that customer needs are met, questions resolved, and meetings held with appropriate key personnel. An organized and client-focused contractor employs dedicated customer project managers.

TYPES OF OUTSOURCING

Testing, Manufacturing, Packaging, and Distribution

Virtual companies (those with only "office-based" personnel) having no facilities of their own for developing, manufacturing, packaging/labeling, and testing and distribution of products rely solely on contractors. One or more of these operations are contracted to one or more contractors. It is not uncommon for commercial operations involving multiple steps (e.g., granulation, pelletization, tableting, liquid formulation, and vial filling) to involve more than one company at locations that can be in different countries. The complexity of coordinating with one contractor is multiplied several times when another is required for a specific processing step.

A larger pharmaceutical company may choose to manufacture an oral solid product at a contractor and ship to one of its own facilities for final manufacturing steps or packaging/labeling.

The decision to develop an injectable formulation for an existing drug is a challenge. If formulation development is successful, there is still the logistics of finding a suitable sterile clinical trials contractor. To build a sterile manufacturing facility requires capital and enormous running cost and commitment. The last thing a company needs when a drug development program is cancelled is an empty facility drawing cash from the budget. A specialized sterile clinical trial contractor can offer the ability to deliver sterile dosage forms. However, there are a limited number of these facilities available because of the overhead costs, specifically trained personnel, and other factors. Selecting a contractor, especially for long-term projects, must be performed with a systematic evaluation.

Capabilities, Selection, and Evaluation

Key processes, which are outsourced, include the following.

Entire Process
This is where the owner of the regulatory filing (Investigational New Drug or New Drug Application) performs no operations within their facilities. The owner could be a virtual company, or one that does not have the specific capabilities. The process may be performed at one or more contractors.

Manufacture of the Active Substance and Unique Excipients
Both these type of components are often sourced from contractors. There has been a significant change in the outsourcing of active substances to contractors, especially the Far East and India. Many large and smaller pharmaceutical companies no longer choose to maintain facilities and operations for the production of active substances. Additional challenges are generated when sourcing from any company outside the United States. It is the client's responsibility to ensure that they have sufficient confidence through physical verification and discussions with the contractor that their specific requirements are completely understood.

Manufacture of Intermediates and Final Dosage Forms
Intermediates may be manufactured at a contractor and either sent to another contractor for process completion or sent to the client's facility.

Manufacture of the dosage form can be initiated at any stage following discovery, production of the initial dosage form for phase I studies, and through all subsequent phases of development to commercialization. Ideally, all this type of work should be performed at a single contractor, but it is not uncommon for more than one contractor to be involved. Formulation work, including analytical testing, may be performed by one contractor and transferred to another because of logistics of scale or other factors. Some specialized testing may be performed at another contractor. Stability storage and testing can be performed at a different contractor. Clinical trial supplies are often packaged by more than one contractor because of timelines, logistics, specialized equipment, and capabilities. Add to this the complexity that any one of these functions could be performed in a different country and local language on

FIGURE 1   Contractor capabilities.

labels is required for clinical studies. The management of the contractors just increased ten-fold.

A product manager would have key responsibilities for working with contractors and the company regulatory affairs to ensure timely manufacturing, testing, packaging/labeling, and delivery of clinical supplies to investigator sites for clinical studies.

REGULATORY SUPPORT

The entire process from compilation of pre-human testing, formulation/development, stability studies, clinical study results through to the inclusion of all the required information/ data as defined by regulatory requirements and reviewers are shown in Figure 1: Contractor capabilities.

CONTRACTOR SELECTION

The decision to outsource one or more processing capabilities originates from strategic business development or commercial logistics. A company either cannot or does not want to provide the resources (facilities, equipment, personnel, etc.) necessary to progress development of the product to the next phase or support the supply of a commercial product. A commercial product may not be financially viable although there is a patient need. These types of products are designated "orphan drugs" by the Food and Drug Administration (FDA). A company will supply the product because of multiple reasons, including compassionate use and good public relations. In this case, the company may elect to use a contractor for the manufacturing and packaging of the product. This releases company scheduling, space, capital investment, and other resources.

A standard operating procedure (SOP) must define the process, requirements and responsibilities for the selection, and evaluation and approval of contractors. A company is transferring an operation or the entire process to another company with whom they have an approved agreement, including business and regulatory requirements. Legal and financial aspects are often a separate agreement. There is a risk (legal, financial, regulatory, etc.) associated with the outsourcing, and this needs to be evaluated and managed so as to minimize this for the client. The later the product is in development, that is, phase III or it is a commercial product, the greater the risk to the client, as delays and problems have greater and more visible impact to regulatory authorities, public, and share holders.

Considerations for selecting a contractor include the following:

- Reputation.
- Regulatory inspection history, both domestic and international.
- Financial stability.
- Broad client-base, supported by several years of experience.
- Can provide innovative suggestions to product development and input to resolve technical problems.

- Ability to provide a wide selection of technologies for specific nonsterile and sterile product requirements.
- Personnel with proven experience in technology (e.g., publications and presentations) and management/coordination.
- Potential to provide international shipments through partnerships or other facilities of the company.
- Reliability.
- Ability to meet and manage timelines.
- Strong communicator.
- Accountable for commitments.
- Complete understanding and commitment to quality. This is a critical component because of regulatory implications.
- Sufficient staffing and equipment to perform the specific work.
- Openness.
- Preparedness to build and maintain a strong and lasting business relationship.

Request for Proposal

Once the decision to outsource a capability has been made, due diligence is performed to determine potential outsourcing companies. Before this can be done, a business project plan needs to be defined with input from multiple sources (e.g., technical support, quality, compliance, and regulatory). The plan defines the requirements of the outsourcing company in relation to the originating company standards and expectations. After the capabilities of suitable companies have been identified as potentially fitting the requirements, preliminary discussions take place, which may necessitate the joint signing of confidentiality agreements (CDAs) between the companies. Some discussions may take place without a CDA in order to decrease the resource requirements utilized to identify the top two or three potential contractors. The technical capabilities of the companies determine potential candidates. Due diligence would involve determining:

- Business history, including financial
- Regulatory history
- Capabilities for current project requirements and potential to accommodate expansion of capacity/volume
- Ability of contractor to listen to client's needs
- Ability of contractor to develop a long-term business relationship/partnership
- Any liabilities or pending litigation

Selection and maintenance of a contractor is shown in Figure 2.

There is both a business and compliance risk if regulatory action is taken against the contractor. This in turn could be linked to the client. It is therefore essential that a GMP quality assessment be performed to establish the regulatory compliance status of the selected company/companies.

Once the two or three potential contractors have been identified, the compliance function of the client arranges and performs a compliance quality assessment against CGMP requirements and company standards. It is essential that the results of the assessment be analyzed for potential regulatory risks, and recommendation for use or not is issued. There may be limited choices of companies when specialized capabilities are required, but compliance gaps/observations are made. As with all business partners, it is essential that the contractors work with them to ensure that there is complete understanding of expectations/requirements. This can take on several forms, from independent function of the contractor to a significant Quality Assurance (QA) oversight by the sponsor. The "quality" function makes the final recommendation for use of the contractor.

There will be times when the client will have no opportunity to select the contractor because there is only one. Contractors who provide specialized services clearly appreciate they have a niche in the market with "captured" clients, but do not use this to their advantage.

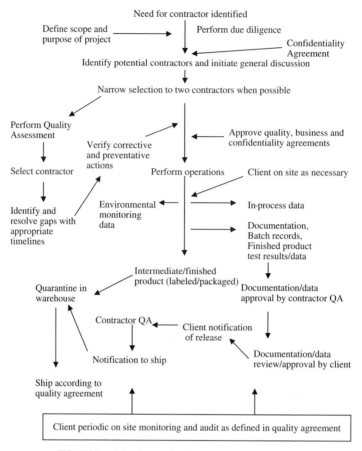

FIGURE 2 Selection and maintenance of a contractor.

Agreements

To successfully outsource a process it requires careful planning, selection, communication, maintenance, and above all management of the contractor.

In addition to business and legal agreements, there must be a clear understanding of responsibilities relating to CGMP defined in a quality agreement. Agreements define essential requirements of the project and the process for managing when things do not go according to plan. The content of the agreement must be clearly understood and signed/approved by appropriate personnel (senior management: production and QA) of both parties. These requirements are dependent to a certain extent on the type of outsourcing. Key sections of an agreement include but are not limited to the responsibilities, timelines, and documentation for the following.

- Regulatory
    - Authority interaction and notification to client: need to define process for notification of the client if a regulatory agency indicates that they are inquiring or on-site for the client's product.
    - Statement of the regulatory requirements to which the process defined in the agreement is to be performed.
- Clear definition of the processes, tasks, and especially any limitations/exclusions and scope.
- Facility addresses.
- Contact personnel of both parties with titles, e-mail, and telephone numbers (with designees).
- Identification of the party responsible for performing each process/task or other key steps.

- Change control requirement notifications to the client: Provide examples of changes that need to be promptly notified to the client in writing. Emergency changes will be notified retrospectively. A client may request that their next batch be delayed until they have reviewed and agreed with the changes. The changes need to be assessed in the client's change control system with the appropriate internal approvals.
- Recall responsibilities and the process steps with timelines.
- Notification of client for significant events (e.g., failures of sterility, environmental, out-of-specification results, and manufacturing deviations) with timelines. Provide examples and ensure the contractor is clear of expectations. The process/timelines for client review and comment need to be defined, and also how the deviation/investigation is closed.
- Annual product review: define in sufficient detail the data/information that the contractor is required to supply and the due date.
- Compliant management: normally, the client handles the initial contact and contacts the contractor if the problem could be related to the operations performed by the contractor. Equally, if the contractor receives a complaint, they are required to notify the client immediately with follow-up documentation.
- New Drug Application field alert responsibilities.
- Material/product release responsibilities.
- Storage and shipment requirements.
- Timelines for key milestones in the process or for achieving each sub-step in the process. This would also include delivery of intermediates, product, and other materials. Include key performance indicators by which the contractor will be measured. These may include defined timelines and number of batch documentation errors requiring correction.
- Definition of the documentation and materials/components to be supplied by the client and those to be supplied by the contractor. This ensures that there are no gaps in the exchange of information/data between the two parties.
- Agreement for access of the client to perform quality assessments. These are normally performed at least every two years and "for cause" as necessary.
- Integrity of electronic information/data exchange between the two parties.
- Use of subcontractors by the contractor: this is normally prohibited without prior approval from the client.
- Confidentiality and intellectual property ownership are often included in the business agreement, which also contains details of payments, penalties, and other business-related items.

Agreements are controlled documents and as such must be version-controlled, signed by appropriate management of both parties, and contain an effective date. These must be revised as needed to reflect changes in the agreement and at least every three years. Ensure that the contractor has trained appropriate personnel on the requirements of the agreements, and the client has an equal responsibility to ensure their personnel are trained. It is not only embarrassing to either party if the other identifies that an agreement requirement has not been met due to lack of awareness, but also detrimental to the business relationship.

Dispute resolution should be managed at early stages to prevent escalation and "finger pointing" to assign blame. If the quality agreement does not address the issue and it cannot be resolved through discussion with regular contracts, then it needs to be managed through dispute resolution as defined in the legal agreement.

MANAGEMENT OF CONTRACTORS

Never underestimate the resources needed to adequately monitor and manage contractors. Failure to provide the required resources will be detrimental to the client. The management varies according to the type of outsourcing and the individual contractors. Frequent and in-depth management of operations at the contractor at the initiation of the project should be able to be decreased in frequency, as confidence and business relationship are built. Regular meetings and monitoring visits ensure operations run according to plan. There should be infrequent surprises for either party with open communication. Contractors welcome open and honest clients to build long-lasting business relationships. Management of the contractor is critical in maintaining compliance with regulatory commitments and the regulatory filing of the product. It is critical that the client does not squeeze the budget putting the contractor in a position of cutting cost that impacts on the quality of work and staff morale.

Technical transfer is the process of transferring knowledge from the client to the contractor efficiently, completely, and with clarity—lack of, without ambiguity is essential for everyone's success. Contractors are conscious of client's success and want to share in it. A contractor's goals include those clients with products that were brought from development to commercial approval in their facilities. The contractor's reputation in the industry is enhanced each time one of their clients obtains product regulatory approval. Regular on-site meetings with the appropriate key client personnel are necessary to establish business relationships, respect, and to clarify details of logically working together. Regulatory compliance with the filing and CGMPs is part of the overall package between the two companies. Without understanding each other's requirements, the product lifecycle will surely fail to meet criteria and expectations. While the client needs the assurance that development/production is meeting timelines without major compliance problems, the contractor needs to be informed of proposed changes and above all if they are not meeting client requirements. These points can only be met by open and honest communication and by putting everything in writing.

At the center of a long-term business relationship between the client and the contractor is a degree of trust. This does not mean that the need to confirm requirements is not in writing, because communication and agreement are as essential as trust. The client must build an open and strong rapport with the contractor and they will respond accordingly, as it is in their business and financial interest to keep clients and build long-term partnerships. The client is ultimately responsible for compliance relating to their products. Data-related questions concerning compliance to CGMP requirements could seriously impact a drug application under review by the FDA. Equally, if the FDA issues FDA483s or even more importantly a warning letter to the contractor where a commercial product is being processed or analyzed, there is a risk to the client. This risk must be assessed promptly to determine if there is any potential impact to the client's product. Observations in FDA483s and warning letters are related to system failures, and although not always directly mentioned against the client's product, they can still have a significant impact because system problems normally have no restrictions. Additionally, further regulatory actions could jeopardize the capabilities of the contractor.

Monitoring client operations when in progress is part of the ongoing relationship with the contractor. All contractors welcome clients to their facilities because without this welcome there is no business relationship. While the client does not want to disrupt operations, there is the understanding that the client has a need and to an extent an invitation to monitor their product being processed. The visit is normally pre-planned and may be for specific steps in the process. Some clients have arrangements to have personnel on-site continually monitoring the process. Key to monitoring is confidentiality and communication. Confidentiality is needed because it is almost impossible to be in a facility and not detect/identify some form of information from another client. It is not difficult because that is the business the contractor is in: managing multiple clients simultaneously. Respect confidentiality and ensure that the contractor is promptly aware of any concerns that may arise during on-site monitoring.

The client must not be under any illusion of the magnitude of resources needed to maintain a business relationship with a contactor. Initial audit and any remediation, agreements, defining and organizing materials/components, confirming timelines, approving final labels, and regular meetings with defined agendas/actions are initial examples. At some point in the growth of contracting, the sponsor organization may want to consider dedicated staff for contract administration. This threshold can be reached unexpectedly with the surprise of a failed project/batch or major delay.

While cost is also a determinant for selecting a service provider, it typically ranks at or near the bottom of survey responses for outsourcing selection criteria.

OTHER TYPES OF CONTRACTORS

Other areas where contractors are used are as consultants (outside the scope of this chapter) and in performing functions relating to engineering or technical support. The latter areas cover instrument calibration, weight certification, high efficiency particulate air (HEPA) filters, equipment maintenance, and calibration. The process for the selection of services from a contractor in this category has the same approach as those for the manufacturing.

- Define project requirements and scope.
- Perform due diligence to identify potential contractors who can provide the required services.
- Perform selected audits of contractors with a team comprising of quality CGMP, technical and engineering experts. This may be performed initially by a questionnaire, depending on the SOP for contractor approval, and based on a risk assessment. Additional documentation from the contractor can also be requested in support of the selection.
- Write the audit report and make recommendations.
- Select contractor and obtain written agreement for scope of service and requirements.
- Manage the contractor according to SOP requirements.
  - *On-site visit by contractor*: It is essential that the contractor is adequately supervised while on site. This includes following site procedures for security, safety, and access to specific areas. Contractors should have some degree of training (documented) on the requirements of performing work on site. Escorting and closely supervising contractors is essential in maintaining a controlled environment/conditions. Often, contractors will send the same person for the periodic work, which enables the host to have consistency, as the contractor is aware of the client's requirements.
  - *Separate operating facilities.* These include those providing services (e.g., calibration and weight verification), manufacturing operations where the product is shipped back to the client for completion of operations (packaging/labeling) or further processing. Equipment can be shipped back to a vendor/manufacturer for maintenance and calibration.

DOCUMENTATION/DATA REVIEW AND ACCEPTANCE

It is the ultimate responsibility of the client to ensure that all the documentation and data received from the contractor is acceptable. The quality agreement must define the documents, contents, approvals, and timelines for delivery. Copies of manufacturing/packaging records may be reviewed on site, perhaps initial batches and then periodically as agreed in writing with the contractor. The objective is to build confidence with the documentation produced to support the client's product. Minimum documentation requirements to release the batch must be clearly defined.

Specific testing results are added to the core template contents shown in Figure 3. Certificates of Compliance contain the following:

- Product description/strength and batch number.
- The regulatory requirements under which the product was made.
- List the number of each deviation, investigation, and out-of-specification investigation related to this product and the status (open or closed). The client needs to confirm all of these by obtaining copies for review.
- Name, job title, and signature of the author/date.
- Name, job title, and signature of QA representative/date.

The documentation and data must meet at least the requirements of the client's SOP for documentation principles. The contract will define the requirement to provide the raw data, including copies of charts, printouts, and laboratory records. The contractor's QA unit must review and approve all documentation, including release of the batch to the client according to the contractor's SOP.

If it can be shown that the contractor's "quality" unit released product for distribution, even though it is aware of CGMP deviations, (e.g., it is aware that the test methods used by the contractor have not been validated); it then would be appropriate to cite the manufacturer for the CGMP deviation of releasing the inadequately tested drug.

Product Description	Film coated 10 mm diameter white, with 'BC5'		
NDC #	26728-0000		
Container	PVC 100 ml with PVC screw cap, cardboard liner Ref # 5667 and 890-08873		
Pack size	100		
Batch #	07-2987		
Finished Product Label	Attach sample here		
Carton label	Attach sample to reverse		
Shipper label	Attach to this document		
Package insert	Attach to this document		
Finished product, Batch #:	Test method	Specification	Result
Diameter			
Disintegration			
Dissolution			
Friability			
Weight uniformity			
Hardness			

FIGURE 3 Certificate of analysis for bulk "Bungatic" 5 mg tablets.

In both cases, the contracting firm also could be held responsible for shipping adulterated drugs in interstate commerce.

CHANGE CONTROL

Compliance with regulatory requirements includes managing changes. This applies equally to the contractor with respect to operations/products and to the client for informing the contractor. The contractor will have a change control SOP, which must be reviewed, and any concerns responded to before operations commence. It must be remembered that the contractor needs to be able to work daily with the SOP. The contractor's SOP will then interface with the specific change requirements, especially the scope of notification to individual clients. These requirements are defined in the quality agreement, and the client needs to confirm with the contractor how their staff will consistently implement the requirements. Clients must ensure specific requests are managed by the contractor and fully incorporated into their systems. Ask, clarify, and confirm before work begins. Monitor progress and resolve any concerns with the contractor.

CONTRACTOR CERTIFICATION

Although there is no specific GMP regulatory requirement for a contractor certification program, there is an expectation that clients will evaluate contractors to ensure that they can meet requirements. Evidence that the contractor has been evaluated may be requested during a regulatory. It is essential that the program is defined in a SOP, its scope clearly defined, and implemented with supporting documentation. A program for certification should include the process for the initial evaluation of the contractor and recommendations. The recommendation may be: Approved, Improvement Needed, and Not Approved. Where improvements are required, these should be discussed with the contractor to ensure that expectations are clear. Some improvements may need to be compromised in content and timelines, but this does not relate to unacceptable risk. It is critical to assess the potential risk factor, as this can result in holding operations until the required changes have been made and confirmed. Figure 4 shows the overall process for a finished product, but can equally be applied to another process with different categories.

Ongoing monitoring of contractors for compliance with CGMPs and agreement requirements is performed and documented by the client. Deviations and concerns need to be promptly discussed and resolved. Failure to resolve these problems can result in changes to a contractor's status. Appropriate action, including informing senior management, must be

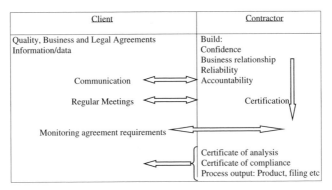

FIGURE 4 Overall outsourcing process for a finished product.

defined in a SOP. The client and contractor should work together to resolve differences. Sometimes, the client may make the decision to transfer operations to a new contractor.

CONCLUSIONS

Clients need contractors to meet their company goals for increased productivity and, above all, for making a profit. Consumers/patients in turn rely on these clients to make products available at competitive prices at the right time. Both parties rely on service and quality products.

Clients are responsible for ensuring that the contractor has a clear understanding of the scope and content of the project. Contractors are responsible for working with their clients to understand their specific requirements. It should be remembered that contractors have multiple clients "pulling" them in many different directions at the same time. Communication is at the center of all the projects and the responsibility of both parties. If there is any doubt, ask the other party to confirm in writing. A lost batch because one person failed to inform another about a required change is a lost opportunity to continue building a strong business relationship. Be open with the contractor and they will in turn want to keep an open rapport with clients. It is in everyone's interest to build and not erode relationships.

17 | Active Pharmaceutical Ingredients

P. Denis Celentano
Percels Consulting Inc., Orlando, Florida, U.S.A.

This chapter is intended to provide guidance on current good manufacturing practices (CGMPs) for the manufacturing of active pharmaceutical ingredients (APIs) and the current thinking by the Food and Drug Administration (FDA).

FOOD AND DRUG ADMINISTRATION REQUIREMENTS

FDA requirements cover APIs that are manufactured by chemical synthesis, extraction, cell culture/fermentation, and recovery from natural sources.

Historically, it has been recognized that there is a significant difference between the processes used for API manufacturing and those used for the manufacture of dosage forms. API processes are frequently diverse chemical operations. These operations often range from final purification/isolations steps to complex processes that employ multiple organic synthesis steps prior to the purification, isolation, and packaging of the final API. These chemical entities are usually better defined thus making the task of testing and quality evaluation easier.

The quality of an API must be assured by building quality into the manufacturing process. Long gone are the days when the industry believed quality could be achieved through testing of APIs. Manufacturing processes must include systems that assure a state of control is achieved and maintained. This state of control is built into the operations by designing the CGMPs into the manufacturing processes.

For years the CGMP regulations (21 CFR 210 and 211) acted as a general guide for API manufactures since the definition of "drug" in the Food, Drug and Cosmetic Act includes both dosage forms and APIs, and Section 510 (a)(2)(B) requires that all drugs are manufactured, processed, packed, and held in accordance with the GMPs. In 1991, in an effort to established consistency among inspectors, the FDA issued a revised version of the Guide to the Inspection of Bulk Pharmaceutical Chemicals. This was not a regulation, but this document provided guidance and support for FDA field investigators. Compliance with this guide effectively became a requirement for approval in FDA Pre-Approval inspections. With the issuance of the Guidance for Industry Q7A Good Manufacturing Practice Guidance for Active Pharmaceutical Ingredients (available at: http://www.nihs.go.jp/dig/ich/quality/q7a/q7astep4e.pdf) in 2001 the CGMP expectations of API manufactures became better defined.

In the introduction Guidance for Industry Q7A Good Manufacturing Practice Guidance for Active Pharmaceutical Ingredients, the FDA states that this document is intended to provide guidance regarding good manufacturing practice (GMP) for the manufacturing of APIs under an appropriate system for managing quality. It is also intended to help ensure that APIs meet the quality and purity characteristics that they purport, or are represented, to possess. Manufacturing is defined to include operations of receipt of materials, production, packaging, repackaging, labeling, relabeling, quality control, release, storage, and distribution of APIs and associated related controls. All APIs manufactured or used in a drug product should be produced following CGMP stated in this guidance.

API starting materials are defined as raw materials, intermediates, or APIs that are used in the manufacturing process of an API. These materials typically become a significant structural fragment of the final API structure. Normally API starting materials have well-defined structures and chemical properties.

Defining when the API process begins and associated critical process steps is crucial in assuring compliance with regulatory expectations. Companies must document their

2222

57923456789

23456789234567892345678923456789

5678923456789234567892345678923456789

rationale on designating the point at which API manufacturing begins. In order to minimize the potential for disagreements with regulatory agencies, companies should provide a strong scientific rationale in defining the start of an API process. Critical process steps that have been determined to impact the quality of the API should be validated. At times, the company may choose to validate other steps to assure a state of control throughout the process or facility. However, it should be noted that choosing to validate a process step does not indicate that step is critical. This CGMP guidance does not apply to steps prior to the introduction of an API starting material. The following table indicates the application of the CGMP guidance to API manufactures.

Type of manufacturing	Application of this guidance to steps (shown in italics) used in this type of manufacturing				
Chemical manufacturing	Production of the API starting material	*Introduction of the API starting material into the process*	*Production of intermediates*	*Isolation and purification*	*Physical processing, and packaging*
API derived from animal sources or extracted from plant sources	Collection processes	Cutting, mixing, and/or initial processing	*Introduction of the API starting material into process*	*Isolation and purification*	*Physical processing, and packaging*
Biotechnology: fermentation/ cell culture	Establishment of master cell bank and working cell bank	*Maintenance of working cell bank*	Cell culture and/or fermentation	*Isolation and purification*	*Physical processing, and packaging*
"Classical" fermentation to produce an API	Establishment of cell bank	Maintenance of the cell bank	*Introduction of the cells into fermentation*	*Isolation and purification*	*Physical processing, and packaging*

Abbreviation: API, active pharmaceutical ingredients.

INCREASING GMPs

The details provided in Guidance for Industry Q7A Good Manufacturing Practice Guidance for Active Pharmaceutical Ingredients are intended to provide guidance regarding GMP for the manufacturing of APIs under an appropriate quality management system.

Responsibilities of the Quality Unit

As with drug products the FDA expects that the quality unit of an organization will be involved in all quality-related matters. The quality unit is expected to be an independent organization and their main duties are not to be delegated. Key duties would include but are not limited to the following:

1. Releasing or rejecting raw material, purchased intermediates, packaging components, labels, and all APIs.
2. Approving master batch records, standard operating procedures, and all specifications.
3. Reviewing and approving batch records and laboratory records.
4. Reviewing and approving validation protocols and summary reports.
5. Reviewing and approving all investigations of all critical deviations and ensuring appropriate corrective actions are implemented.
6. Review of results of the stability-monitoring program.
7. A review of all quality-related returns, complaints and recalls.

Buildings and Facilities

Buildings and facilities used to manufacture intermediates and APIs should be designed and constructed to facilitate manufacturing, cleaning, and maintenance. Facilities should be designed and operated to minimize the potential for contamination. There should be adequate space for the placement and storage of equipment and materials. Defined areas to ensure control of activities such as receipt of materials, storage, sampling, cleaning, productions, and packaging should be established to prevent mix-ups or contamination.

Laboratory areas are typically separated from manufacturing areas. Laboratory areas used for in-process controls can be located in the manufacturing area provided they do not adversely impact the accuracy of the laboratory measurements.

Critical utilities that can impact product quality (e.g., steam, gas, compressed air, heating, ventilation, and air conditioning) should be defined by the company. These utilities should be qualified, monitored, and appropriate action taken if defined limits are exceeded. HVAC systems and exhaust systems should be designed and constructed to minimize risks of con-tamination and cross-contamination and should include temperature and humidity controls as appropriate for the stage of manufacturing being performed.

Water used in API manufacturing should be demonstrated to be suitable for its intended use.

Unless more stringent requirements are needed to ensure the quality of the API, potable water may be used provided it meets the regulatory requirements for drinking water. Where purified water is used, the process must be validated and include microbial controls. Water used in the final isolation and purification steps should be monitored and controlled for total microbial counts, objectionable organisms, and endotoxins.

When highly sensitizing materials, such as penicillins or cephalosporins, are being pro-duced, dedicated processing areas, including air-handling equipment, should be used.

Process Equipment

The FDA places a high value on the cleanability of equipment to prevent the possibility for con-tamination. Written procedures should be developed and put in place for the cleaning process of equipment and utensils. Cleaning procedures must include sufficient details to enable oper-ators to clean each type of equipment in a reproducible and effective manner. The procedures should define who performs the cleaning, cleaning frequency, cleaning agents to be used, and recording of any critical cleaning parameters. Acceptance criteria for residues and cleaning agents should be defined and justified. The cleaning activities should be documented either in the batch record, a separate worksheet, or a logbook. Dedicated equipment should be used for production when equipment is not readily cleanable. Using appropriate signage, equipment should be identified as to its contents and cleanliness status.

As with drug products, written maintenance and calibration procedures should be estab-lished. Key equipment and critical instruments should be defined by the company. Control, weighing, measuring, monitoring, and testing equipment critical for ensuring the quality of intermediates or APIs should be done following an established schedule. Deviations from approved standards of calibration on critical instruments should be investigated to determine if any impact on quality occurred.

Computerized Systems

A focus on ensuring computerized systems employed in both manufacturing and laboratory operations are validated for their intended use has increased over the past decade. GMP-related computerized systems must be appropriately validated. The depth and scope of validation will depend on the complexity and criticality of the computerized application. Written procedures for the operation and maintenance of computerized systems must be developed and implemented. Changes to the automation must be made in adherence to established change control procedures. Records should be maintained of all changes that demonstrate the system is maintained in a validated state.

Documentation and Records

Over the past decade, it has become the expectation that companies employ a systems approach to documentation. During regulatory inspections, the failure of companies to produce proper documentation has led to regulatory citations and action. A robust and effective documentation system is critical for the overall compliance posture of the organization. Written procedures must be established for preparing, reviewing, approving, and distributing documents associated with the manufacture of intermediates or APIs.

The documentation system should establish a retention schedule for all CGMP documents (e.g., batch records, laboratory records, validation protocols, raw material testing, calibration records for critical instruments, etc.). The documentation system should establish norms of recording information. Training on proper documentation should be delivered to both operating and technical personnel to ensure good documentation practices are routinely followed. Entries of information or data should be made with indelible ink in the space provided at the time the activity is performed, and should identify the person making the entry. Corrections to entries should be dated and signed and leave the original entry still legible.

The company should develop and implement written procedures for the review and approval of batch production and laboratory control records, including packaging and labeling, to ensure the compliance of the intermediate or API with established specifications. The quality unit must review and approve all completed batch production and laboratory control records of critical process steps prior to release of product.

Materials Management

In API manufacturing deliveries of materials routinely consist of large numbers of containers. A system to identify the status of the material should be established. This system would typically be used as an acceptable alternative to a designated quarantine area. Container of materials or group of containers should be properly identified and labeled with a distinctive code (e.g., batch number, lot number, receipt number, etc.).

It is acknowledged that the specific batch identity will be lost with the storage of some materials and solvents in bulk. However, prior to mixing bulk deliveries with existing stocks (e.g., solvents or stocks in silos), they should be tested and released. Procedures should be available to prevent the discharging of incoming materials wrongly into the existing stock. Inventory levels and date and time of discharging of the incoming materials should be recorded to provide traceability of the material. External storage is also acceptable provided labeling remains legible and that containers are cleaned prior to opening.

Testing to verify the identity of each lot of material should normally be conducted. A supplier's certificate of analysis can be used in place of performing additional tests. However, if a supplier's certificate of analysis is used, the company must have a system in place to evaluate its suppliers.

Release of hazardous materials with reliance on the manufacturer's certificate of analysis is also acceptable. To establish the identity of the hazardous materials a visual inspection of storage containers, labels, and recording of lot numbers is routinely included as part of the system. The company should develop and document appropriate justification for not testing these materials.

Storage of raw materials, intermediates, and APIs should be done in a manner to prevent degradation and contamination. Storage conditions should be defined and documented by the company for a period that has no adverse effect on their quality. The company should have an inventory system employed that assures the oldest stock of material is routinely used first. Appropriate space needs to be provided for the storage to allow for cleaning and inspection of storage facilities. To reduce the potential of damage and contamination, materials stored in fiber drums, bags, or boxes should be stored off the floor. The company should establish expiry periods for materials and written re-evaluation procedures to determine their suitability for use.

Production and In-Process Controls

As with drug products, the FDA expects companies to assure that materials are weighed, measured, or subdivided in a manner to assure the quality of the material and to prevent contamination. After dispensing and weighing, materials should be properly labeled (e.g., material name, item code, receiving number, weight, re-evaluation date, etc.). Critical weighing and dispensing activities require a verification step. The operation can be witnessed by a second person or should be subject to equivalent control, such as electronic confirmation of weights and activities performed.

In the API process expected yields should be defined and appropriate yield ranges should be established for designated steps based on previous laboratory, pilot scale, or manufacturing data. Actual yields should be compared with expected yields. Deviations in yield should be investigated to determine their impact quality, root cause, and corrective actions, as appropriate.

It is not unusual to recover additional material from of mother liquors in API processes. Often the recovery is designated as a second crop isolation and is reused in subsequent production batches. Additionally, solvents are routinely recovered and reused. These materials are frequently mixed prior to further processing. Appropriate documentation and/or testing should be established to track these materials and ensure their use does not adversely impact the quality of the API produced. In-process mixing of fractions from single batches (e.g., centrifuge loads from a single crystallization batch) or combining fractions from several batches for further processing is considered to be part of the normal production process and is not considered to be blending.

Blending is the process of combining materials within the same specification to produce a homogeneous intermediate or product. As with dosage forms, blending of out-of-specification or nonconforming material with conforming material is not acceptable.

In API processes, it is acknowledged that some carryover from one batch to another may occur due to the physical inability to completely empty a process tank, centrifuge, or other processing vessel. This carry over should not include degradants or microbial contamination that could impact the quality of the API or its API impurity profile.

In order to ensure a process is maintained in a state of control, critical process steps and parameters should be defined based on the information gained during development, scale-up, or from historical data and documented. In-process controls and their acceptance criteria should be put in place to monitor and control the performance of critical processing steps. Deviations from established ranges should be investigated to ensure no impact to the quality of the product. Since many API processes are multistep, less stringent in-process controls may be appropriate during the early processing steps.

As with drug products, in-process testing by manufacturing personnel is acceptable provided the quality unit has the final decision making responsibility for release or rejection of product.

Packaging and Labeling
Labeling requirements are similar to those for drug products with respect to storage of labeling, issuance, and reconciliation. Written procedures that describe the receipt, identification, quarantine, sampling, examination and/or testing, release, handling of packaging, and labeling materials should be developed and implemented. Labels should have a defined storage area with access limited to authorized personnel. Label procedures will include reconciliation of label quantities issued, used, and returned. All discrepancies found between the numbers of containers labeled, and the number of labels issued must be investigated with final approval by the quality unit.

Laboratory Controls

Laboratory controls for API manufacture have evolved to be more consistent with the expectations of drug products. Due to the diverse nature of API processes, specifications and testing requirements for different raw materials can vary significantly depending on the nature of the raw material and its criticality in the synthetic process. API specifications are expected to be more comprehensive and should include limits for solvents. Specifications should be scientifically sound and appropriate to ensure quality and purity of the material produced. Specifications and test procedures should be consistent with those presented in regulatory filings.

When an API has a specification for microbiological purity, action limits for total microbial counts and objectionable organisms should be established and met. When an API has a specification for endotoxins, appropriate action limits should be established and met. Out-of-specification result obtained for any test should be investigated and documented according to written procedures and final approval obtained from the quality unit.

Analytical methods are expected to be validated in accordance with a written protocol that has been approved by the quality unit.

Stability Testing

An on-going stability-testing program should be established to monitor the stability characteristics of APIs. Since APIs are frequently stored in commercial containers that are not practical for stability (e.g., 50 Kg fiber drums, carboys, silos, etc.), it is an accepted practice to store stability samples in containers that "simulate or approximate the market container." The container/material contact surfaces should be the same and where possible offer an equivalent amount of secondary protection. For example, polybags used in small cardboard containers may be considered equivalent to a polyliner in a 50-kilo fiberboard drum. The results should be used to confirm appropriate storage conditions and retest or expiry dates.

Reserve Samples

Reserve samples for each API batch are to be retained for three years after distribution is complete or for one year after expiry date of the batch, whichever is longer. Additionally, for APIs with retest dates, reserve samples should be retained for three years after the batch has been completely distributed.

VALIDATION

API manufactures must establish a validation policy that defines their method and approach to validation. This policy should be documented and approved by the quality unit. The policy should include but is not limited to validation of manufacturing processes, cleaning procedures, analytical methods, test procedures, computerized systems, persons responsible for design, implementation, approval, and documentation of each phase of validation.

The critical parameters and quality attributes are typically identified or defined during process development, scale-up, or from historical data. During the technology transfer or process demonstrations, expected ranges for each critical process must be identified and subsequently used by manufacturing routine and process control to ensure the quality of the API produced. Operations determined to be critical to the quality and purity of the API are normally validated.

CHANGE CONTROL

A written change control system that evaluates all changes that could impact the production and control of the intermediate or API must be developed and put in place to ensure a state of control is maintained. In addition to evaluating the potential impact of a proposed change (specification, test procedures, production processes, and equipment) with respect to validation and regulatory impact, the need to notify the customer is included.

After the change is approved, there should be careful evaluation of the initial batches manufactured using the modified process. This evaluation should be built into the change control process.

EXAMPLES OF OBSERVATIONS FROM FDA 483 CITATIONS

1. The company violated CGMP requirements by failing to use scientifically sound standards and test procedures to ensure the purity of the product.

2. The firm was cited for CGMP violations in the firm's sterilization process during production of product A.
3. There is a failure to ensure that automatic equipment will perform the function satisfactory since the equipment did not have a complete installation qualification.
4. There is a failure to follow procedures. Inspection and testing of equipment do not conform to the operating procedures.
5. The firm lacked a full justification or validation of the synthetic process.
6. The manufacturing facility has no formal written standard operating procedures (SOPs) relating to supplier qualification.
7. There is a failure to assure that the air filtration system, including prefilters, is working correctly when used in the finished API area.
8. Written procedures for cleaning and maintaining equipment used in the manufacturing process are inadequate.
9. The master validation plan for PRODUCT A does not address the use of cleaning solution B and the Validation Summary report does not address the removal of ITEM C from the equipment. You did not simulate actual bulk storage conditions for stability studies.
10. There is a failure to date and sign the Master Production and Control record.
11. There is a failure of the production batch record to include a statement of theoretical yield and a specimen of a copy of the label.
12. A failure to have written procedures for complaints.

18 | Bulk Pharmaceutical Excipient GMPs

Irwin Silverstein
IBS Consulting in Quality LLC, Piscataway, New Jersey, U.S.A.

BACKGROUND

As noted in Chapter 17: Active Pharmaceutical Ingredients, there appears to be universal agreement that there are significant differences between the processes used for the manufacture of dosage forms and those used for the manufacture of the API. There are even greater differences between excipient processes and those of either the dosage form or API. These differences include:

- Diversity of bulk pharmaceutical excipient (BPE) manufacturing operations, which range from minerals that are mined and ground, to modified natural substances and synthetic molecules produced via multi-step processes.
- The BPE process, like that of the API, is often designed to improve purification, whereas this is not possible in dosage-form manufacture.
- Unlike the API which is usually a single chemical entity, the excipient is often a complex mixture and thus more difficult to evaluate for quality.
- Virtually all BPEs have other nonpharmaceutical applications often representing higher volumes and sales dollars, whereas APIs seldom have other significant nonpharmaceutical uses.

Often the BPE is nonhomogeneous within the lot for one or more properties, such as moisture or particle size, unlike the API where homogeneity is easier to achieve and confirm since the API is usually a single entity.

The current good manufacturing practices (CGMP) regulations (21 CFR 210 and 211) apply only to dosage forms and are to be used only as a general guide for excipient manufacture. However, the definition of "drug" in the Food, Drug and Cosmetic Act includes both dosage forms and their components, and Section 510 (a)(2)(B) requires that all drugs be manufactured, processed, packed, and held in accordance with the GMPs. This creates a problem since implementation of the Act requires compliance but there are no GMPs specific to excipient ingredients. Unlike APIs, where the FDA has recently adopted Q7A as guidance for their manufacture, there is no similar guidance accepted by the FDA for the quality system used in the production of excipients.

As noted in the chapter on APIs, it is clearly stated that 21 CFR 210 and 211 should be applied to APIs at a point that "it is reasonable to expect GMP concepts to start to become applicable ... where a starting material enters a biological or chemical synthesis or series of processing steps, where it is known that the end product will be a (API)." As noted earlier, unlike the API, excipients are usually also sold into other industries. Excipient manufacture may involve using the same equipment and processing at least in early stages that lead to other saleable nonpharmaceutical products. Therefore, the application of excipient GMP requirements may begin later than in the beginning of the chemical synthesis or series of processing steps. Consideration for the starting point of the application of full GMP requirements is dependent upon whether or not there are other markets for the excipient or its starting materials. Therefore,

1. Where there is no nondrug commercial use for the excipient or its precursors:
   a. That step in a chemical synthesis where a significant structural fragment is formed or
   b. When a chemical reaches a point in its isolation and purification where it is intended that it will be used in a drug product.

2. Where there are other nondrug commercial sales of the chemical or its precursors:
   a. That step in a chemical synthesis where the final molecule is formed or
   b. When the chemical undergoes its final purification.

It is expected that the excipient manufacturer will document their justification for the selection of the point in the manufacturing process where full GMP requirements are to be applied.

Excipients are typically produced by companies whose primary business is not pharmaceutical ingredient manufacture let alone pharmaceutical dosage-form manufacture. Seldom does the excipient manufacturer even produce APIs. The excipient manufacturer is usually a chemical manufacturer or a supplier to the food industry. These manufacturers are often refiners of inorganic chemicals, large-scale manufacturers of commodity chemicals, specialty chemical companies, food-additive producers, or food processors. As such, it should be recognized that their understanding of the quality system needed to produce pharmaceutical ingredients is often minimal. Likewise, it is important to note that pharmaceutical companies often do not disclose to their excipient suppliers the dosage forms in which the excipient is used. Also excipient producers typically lack a clear understanding of the functionality performed by the excipient in the pharmaceutical formulation.

It is important for the excipient manufacturer to understand that the prime concern of their pharmaceutical customers relative to the quality of the excipient is consistency. For the pharmaceutical manufacturer, the quality characteristics of the excipient, the compositional analysis as described by its purity and impurities should remain unchanged from lot to lot. By contrast, it is also important for the pharmaceutical manufacturer to appreciate that the excipient manufacturer is driven by competitive pressures to reduce their cost of manufacture and perhaps also improve the quality of the nonpharmaceutical grades of the product. These conflicting forces can be difficult for the excipient manufacturer to reconcile where both pharmaceutical and nonpharmaceutical grades are produced using the same equipment.

INTERNATIONAL PHARMACEUTICAL EXCIPIENTS COUNCIL

To help address the issues, the International Pharmaceutical Excipients Council (IPEC), a trade association of excipient producers and pharmaceutical manufacturers, was formed in 1991. IPEC quickly became a global organization with associations in the U.S.A. (IPEC-Americas), Europe (IPEC-Europe), and Japan JPEC. There has recently been talk of organizing similar associations in India and China.

An early objective was to address the lack of guidance concerning the quality system to be used in the manufacture of excipient ingredients. The IPEC-Americas GMP Committee immediately was organized and began work on development of excipient GMP requirements. The GMP Committee, comprised of representatives from excipient producers as well as pharmaceutical manufacturers, issued *The IPEC Good Manufacturing Practices Guide for Bulk Pharmaceutical Requirements* in 1995. This guideline was organized according to the International Organization for Standardization (ISO) 9002:1994 guidelines since many excipient suppliers were either ISO 9000 certified or were in the process of seeking certification. Organizing the GMP requirements in this manner simplified the understanding of conformance requirements for the excipient supplier. The guideline was adopted by all three IPEC affiliates and was distributed under IPEC, the umbrella organization.

The focus of excipient GMP requirements was first to ensure that the quality of excipient ingredients was suitable to assure their safety and efficacy in the dosage form. Secondarily, the GMP requirements established the documentation and records necessary for confirmation that these ingredients were properly produced under an appropriate quality system. Thus the focus of excipient GMPs is somewhat different from the dosage form where the safety and efficacy of the drug formulation will have already been established through clinical trials. The drug applications then lay the foundation for the manufacture of the drug product. The drug GMPs, as specified in 21CFR211, have as their primary focus the documentation and record-keeping requirements for verification that the drug was produced in conformance with the drug application.

When the ISO revised the ISO 9000 standard in 2000, IPEC reissued the excipient GMP guideline in alignment with this change in 2001. The latest revision of this guideline was issued in 2006 when the excipient quality system requirements were harmonized with those of the Pharmaceutical Quality Group, Institute of Quality Assurance, a British trade association. The guideline is now called the IPEC-PQG Good Manufacturing Practices Guide for Pharmaceutical Excipients.

In 2000, the United States Pharmacopeia (USP) issued excipient GMP requirements in General Chapter 1078. This chapter falls within the section of voluntary requirements. However, since the USP requires that appropriate GMPs be applied to the manufacture of ingredients listed in the USP/NF, it is implied that for excipient ingredients, the requirements of 1078 would be followed. The USP has been working with IPEC to update the excipient GMP requirements as appropriate.

Unlike the situation for manufacturers of dosage forms and even the API, there is a dearth of regulatory and compliance information for the excipient producer. Therefore, IPEC has been working to provide guidance documents for the industry. This initiative received additional impetus with the glycerin poisoning in Haiti[a] where over 80 children were poisoned with acetaminophen syrup contaminated with diethylene glycol. The source of this poison was traced to the glycerin excipient, which was shown to contain approximately 50% diethylene glycol. As a consequence, the U.S. FDA asked IPEC to help assure the safety of excipient ingredients.

Subsequently, IPEC issued the following guidance documents:

- The IPEC *GMP Audit Guideline for Distributors of Bulk Pharmaceutical Excipients*, 2000. This guideline provides details concerning the application of excipient GMP requirements to distributors of excipients as well as guidance concerning auditing their operation. This guideline has subsequently been replaced by the *IPEC Good Distribution Practices Guide for Bulk Pharmaceutical Excipients*, 2006. This updated guidance provides details on conformance to the World Health Organization (WHO) *Good Trade and Distribution Practices for Pharmaceutical Starting materials*, 2003.
- The *IPEC-Americas Certificate of Analysis Guide for Bulk Pharmaceutical Excipients*, 2000. This guideline establishes the requirements for the content of an excipient Certificate of Analysis (COA). Among other items, it requires that the COA provide for traceability of the excipient back to the excipient manufacturer.
- The *IPEC-Americas Significant Change Guide for Bulk Pharmaceutical Excipients*, 2004 and subsequently revised for 2006. This document provides a framework for the evaluation of changes in the manufacture of excipient ingredients and establishes requirements for notification of such change to the customer and regulatory authorities.

GOOD MANUFACTURING PRACTICES GUIDE FOR PHARMACEUTICAL EXCIPIENTS

The remainder of this chapter will follow the organization of the 2006 excipient GMP guide with the emphasis on the unique issues for excipient manufacture.

General Guidance

The excipient manufacturer should establish the point in the manufacturing process where full conformance to the excipient GMP requirements begins. The rationale for this decision should be justified and documented.

---

[a]http://www.drugnet.com.hk/tox/tox_gly.htm (accessed April 2006) and Malebrance R et al., Fatalities associated with ingestion of diethylene glycol–contaminated glycerin used to manufacture acetaminophen syrup-Haiti, Nov 1995-Jun 1996, Morbidity and Mortality Weekly Report, Centers for Disease Control and Prevention, Aug 2, 1996; 45(30).

Management of Excipient Quality Systems

A system should be in place for the control of documents such as procedures, work instructions, process documentation, test methods, and specifications. The system should protect the documents from deterioration, facilitate their retrieval, and assure that only current copies are available for use. Revision to these documents should be made only after review and approval by qualified personnel. An electronic documentation system is acceptable if it assures a comparable level of assurance that only properly authorized documents are available for reference.

The same document control system should be suitable for the control, retrieval, and use of records. This includes completed checklists, forms, test data including instrument output, and so on which must be retained as part of the production record for each lot of excipient produced. Such records should be retained for at least one year beyond the retest or expiry date of the excipient lot.

Finally, there should be a change control process such that all changes that can impact the quality of the excipient are evaluated. Changes, such as to manufacturing instructions and equipment, specifications, sampling plans, instruments, and test methods, should be approved prior to use with final approval given by an independent organization such as the Quality or Regulatory function. All changes should be evaluated for their potential impact to the excipient in accordance with the IPEC-Americas *Significant Change Guide for Bulk Pharmaceutical Excipients* and should be communicated to customers and regulatory authorities as recommended in the guide.

Management Responsibility

There should be a stated management commitment to produce excipient ingredients in conformance with appropriate GMP requirements. This should be demonstrated through the implementation of an excipient quality system. Materials should not be labeled as compendial grade merely by confirming the lot when tested meets the requirements of a monograph. It is also important that the lot be produced in accordance with the requirements of the excipient quality system.

Top management is expected to supply adequate resources for the manufacture of excipient ingredients. This includes an independent quality unit with defined responsibilities for overseeing the proper manufacture of the excipient. Top management should conduct periodic reviews of the performance of the excipient quality system where review inputs include such items as internal and external audit findings, customer feedback, production process capability, laboratory testing, and change control. Adjustments to the quality system should be made to improve conformance as indicated.

Resource Management

Management should provide the resources necessary to produce excipient ingredients in conformance to excipient GMP requirements. To this end, there should be employee training in both their job responsibilities as well as to provide an appropriate understanding of excipient GMPs. Records should demonstrate that employees were trained to perform their tasks and were also trained in relevant excipient GMP requirements. It is important to provide employees with GMP training on a continuing basis. For this industry, ongoing training requirements can be met by providing a refresher course in appropriate excipient GMP requirements every two years.

During manufacture, the excipient is often only exposed to the environment at packaging. As a consequence, unless there is exposure of the excipient to the employee during processing no special attire need be specified. At packaging and other areas where there is exposure to the excipient, the employee should wear appropriate clothing including a clean outer garment, head covering, and gloves to protect the excipient.

A major emphasis in the construction and maintenance of excipient facilities is the potential for contamination from other chemical processes or from airborne contaminants. Since much processing occurs in closed systems dedicated to the production of the excipient and lesser grades of the chemical, the risk of such contamination occurring is usually limited to raw material charging, purification and drying steps, and packaging. Unlike APIs, air-handling systems are seldom a concern since they are usually dedicated to the manufacture of the BPE. However, justification or appropriate data should be available to demonstrate that operational practices do not constitute a potential for cross-contamination.

Oftentimes, chemical plants that produce excipient ingredients also produce a range of other products including those that are poisonous, including pesticides. The manufacturer should take appropriate precautions to ensure that other products produced on-site, particularly if harmful, cannot contaminate the excipient. Such precautions might involve enclosing the excipient manufacturing area, redirecting air intakes, and so on.

Excipients can be derivatives of animal, vegetable, or mineral raw materials. Although it is recognized that these starting materials may be contaminated with objectionable matter such as insects or filth, appropriate efforts should be made to minimize their presence in the finished excipient. These efforts usually involve purification of the excipient.

While it is advisable to use potable water in the preparation of BPEs, it is recognized that it is not always feasible since excipient production may represent a rather small portion of the total output of the facility. Also potable water may not be available at the site. Finally, many chemical processes use water where the time and temperature of processing to which the water is exposed would exercise control over microorganisms introduced from the water. Where such microbial control of nonpotable water is exercised during processing, a study should be conducted to demonstrate that the use of water not shown to meet the potable drinking water standard is suitable. The study might demonstrate a reduction in microbes present to almost sterile conditions or continuous microbial monitoring of the excipient shows no elevated levels of microorganism.

Where processing conditions do not exercise control over microbes in nonpotable water and the site wants to confirm the water meets portable standards, the manufacturer should test the process water to confirm it meets the drinking water standard. Such testing should include all EPA requirements for potable water and should be conducted at least once each season.

Where potable water is supplied from a municipality, the FDA has allowed reliance on data provided by municipal water authorities to demonstrate the water is potable. However, recent experience with FDA visits indicate it is appropriate, from time to time, to recheck the water quality at the point of use in the plant since this may be a long distance away from the municipal testing point. Where purified water is used, the process must be validated to demonstrate continuing conformance to the specification for the purified water as well as for adequate control of microbes.

Maintenance is important not only to the quality of the excipient but also to the safe operation of the facility. It should be recognized that where processing occurs within the equipment and the material is not exposed to the environment, the exterior equipment surface condition should be treated merely of cosmetic importance. Even so, it should be remembered that such appearances do reflect upon the maintenance activities at the site and create a first impression to the auditor.

Oftentimes, excipient production occurs outdoors, particularly for continuous processing, which is acceptable provided that the processing occurs in closed equipment. Where the excipient is exposed to the environment, such as during charging of ingredients and packaging, the area should be under environmental control to prevent airborne contamination. Environmental control involves protecting the excipient from precipitation, airborne contamination, and overhead accumulations of dust and dirt. Generally, operations where the excipient is exposed should be conducted in a room with washable walls, floor, and ceiling. Also the area should be maintained under positive pressure with air filtered through a furnace-type filter, which is generally sufficient for excipient production. The environmental control should be demonstrated to be adequate to protect the excipient.

Computer systems used for GMP purposes should be qualified. A documented study should be conducted that demonstrates the computer system performs as designed. There should be procedures governing the backup and archive of copies of both the program files and data. Also unauthorized use of the computer system should be prevented through secure user identification and passwords. Finally, changes to the computer system, including both hardware and software, should be governed by change control with details of the change to be made reviewed and approved and the impact of the change assessed to confirm it meets expectations.

Where computer systems are used to issue documents through electronic signature, the requirements of 21CFR Part 11 should be met. As noted above, the computer should be qualified and access should be secure. Also the computer system should time-out and be password protected for relogin. The computer system should be validated, have an audit trail, and record the identification of anyone making changes to the data. Finally, the computer system should link approval of the record, such as the lab test results or Certificate of Analysis, to the user ID.

Housekeeping can be important to excipient quality wherever the excipient is exposed to the environment. For such areas, it is suggested that there be procedures in place for area cleaning and sanitation including a description of schedules, methods, equipment, and materials. Waste should either be promptly disposed or stored in appropriate containers.

From the processing step where full GMP requirements are applied, wherever the excipient is exposed to the environment, or where the packaged excipient is stored, there should be a pest-control program in place. Oftentimes, the principle area of excipient exposure is at packaging. Where excipient is shipped to the customer in bulk such as tankers, no insect and rodent control measures are expected aside from routine precautions.

Typically, the pest control program involves having an exterminator inspect the areas where excipient is exposed or stored in packages and apply appropriate pest control measures using only FDA-approved materials. The exterminator should leave a report of their findings concerning insect and rodent activity.

Product Realization

Conformance to excipient GMP requirements includes good communication with the customer. There should be mutual agreement to excipient specifications including any additional requirements the customer may have requested. Of particular importance is an agreement to notify customers of significant changes as discussed earlier.

The excipient manufacturer should assure that appropriate documents are controlled. Such documents include customer specifications and technical reports. Also, there should be a system for responding to customer inquiries and for handling complaints from the customer. Feedback of the conclusions from complaint investigation should be promptly provided to the customer.

Whereas GMP requirements clearly begin for the pharmaceutical manufacturer, at the very latest, at the receiving area of pharmaceutical grade ingredients, the same expectation cannot be applied to excipient operations. As discussed earlier, the excipient manufacturer should determine the production step where full excipient GMP requirements must be applied. However, it should be recognized that certain good manufacturing principles should be applied beginning with the receipt of raw materials. Generally, these earlier operations should be conducted in conformance with the requirements of an ISO 9001 quality system. As such, there should be a quality assessment of all key raw material suppliers to confirm that their quality system is sufficient to assure delivery of excipient raw materials in conformance with specifications agreed to by both the supplier and excipient producer.

Wherever possible, raw materials should be accepted after Quality Control has at least confirmed their identity and verified from the COA data that the lot meets specification. Consideration should be given to conducting a site audit when it is not feasible to sample incoming raw material important to the quality of the excipient, such as for pipeline or bulk delivery. This audit should confirm the supplier's quality system provides adequate assurance of the quality of material shipped from the plant. Where the raw material is deemed too hazardous to sample, there should be confirmation that the paperwork from the supplier, such as the Bill of Lading, matches the COA, that is, raw material name, grade, and lot.

As for APIs, a physical quarantine of unapproved raw materials is often not feasible. Typically, the excipient manufacturer will assure that unapproved raw material is not released for use through the status of the raw material lot in the computerized inventory system.

This is an acceptable practice only where the accuracy of the inventory locator system is high. Where the locator system is unreliable, consequentially for convenience the warehouse worker is often tempted to pick a raw material lot that was not designated by the computer system and which thus may not have been released by Quality Control.

Oftentimes, raw materials are delivered in bulk and stored in tanks until consumed. It is recognized that there is a loss of lot identity associated with this practice, which is acceptable as long as records are sufficient to provide an indication, at a point in time, of the lot that was available for consumption in the production of the excipient. It should be recognized that the impact of not being able to trace raw materials by lot to the excipient so produced, in the event of a problem associated with bulk raw material, is that a greater quantity of excipient will have to be evaluated.

Storage of raw materials, especially in bulk, often occurs outdoors. These raw material storage tanks should be protected from entry by precipitation, insects, or birds. Preferred venting of the tank is to use a conservation vent where a suitable gas creates a slight positive pressure at the vent. Where a gooseneck vent is used for a tank under ambient conditions, the opening should be screened to prevent drawing in birds or insects as material from the tank is discharged.

Outdoor storage of drummed raw materials is acceptable provided the container labels remain legible. However, it should be recognized that allowing water to collect atop a drum containing liquid raw material presents a risk that a sharp drop in the temperature of the contents can result in water being sucked in through the bung opening.

Excipient that is sold in a solid state, such as powders, granules, gelatin, and so on, usually is packaged in contact packaging within an outer container. Oftentimes, the excipient is filled into polyethylene liners inside a drum, bag, or box. Since the excipient only comes into contact with this polyethylene packaging, it is especially important that the manufacturer assures the contact packaging conforms to specification. A sample of each lot of contact packaging should be tested by Quality Control to confirm it meets specification. Where such testing is not feasible, a site audit of the contact packaging manufacturer should be conducted to verify their quality system is adequate to provide materials in conformance with the specification.

Where the excipient is a liquid, it may be packaged directly into the container. Since it is not feasible for Quality Control to inspect each container to confirm it meets specification, it is important to evaluate the container manufacturer's quality system through on-site audit. In addition, packaging personnel should be instructed by quality unit personnel to look for obvious signs the container is not suitable.

Excipients are produced either via batch or continuous processing or a combination of the two. In any event, there should be instructions provided to the operators along with records that are to be kept to demonstrate these instructions were followed. Thus, the documentation should be sufficiently complete to identify all materials used and operations performed to manufacture the excipient lot. These records should be reviewed as part of lot release conducted by the quality organization or their designate. Where continuous processing is conducted, this presents a problem since these records are usually either a continuously maintained log with no direct relationship to the finished lot or an electronic file containing operating settings and parameters. Even though it is difficult and inconvenient to provide the information to the quality unit for review, means should be taken to meet the intent of the requirement, such as periodic audit of those records.

The trend in monitoring process performance is towards conducting in-process testing at the production unit. Where such testing is performed with on-line instruments, the instruments should be properly calibrated and maintained and should have been shown to provide suitable measurements. Where the testing is performed at the production line by the personnel, there should be an assurance that production personnel have the ability to provide measurements comparable to that produced in the Quality Control laboratory.

Unlike API manufacture where the emphasis of equipment construction is on cleanability and the potential for cross-contamination, excipient producers should focus on demonstrating sufficient equipment cleaning and maintenance. Equipment used for excipient manufacture, even when dedicated, must be periodically cleaned if only after maintenance activities or prior to start-up after periodic maintenance shutdown. Requirements for equipment cleaning should be outlined with sufficient specific detail to assure it is done repetitively. Excipients are often produced in equipment whose size and shape precludes easy cleaning. Excipient equipment is almost always cleaned in place, often with restricted access to the interior surfaces not only for the purposes of cleaning but also subsequent inspection to verify efficacy. However, the excipient manufacturer should demonstrate the efficacy of their cleaning procedures through suitable means.

Packaging and labeling operations present a risk of mix-ups between different lots and different materials. Therefore, before such operations begin, there should be an inspection of the area for the presence of materials used for prior packaging activities and if any are found, they should be removed from the area.

Records of equipment use should be maintained. However, unlike the pharmaceutical dosage-form manufacturer where logs are kept for each major piece of equipment, this is seldom the case for the excipient manufacturer. The maintenance department will often keep records of equipment service and repair, whereas production maintains records of equipment cleaning and production use. Even so, the excipient manufacture should be able to reconstruct from applicable records regardless of where they are kept, the sequence of activities with the subject equipment.

Validation is an area just now beginning to receive the attention of excipient manufacturers. The qualification of equipment is a problem for the industry since much of the equipment in use was installed before equipment qualification became an expectation. Even so, it expected that the manufacturer will present evidence to support the proper installation and operation of equipment that can affect excipient quality from the point at which full GMP requirements apply. Such assurance can be provided through a documented review of maintenance and production records.

Validation activities by excipient manufacturers range from the continuous monitoring of process capability to conducting validations using approved protocols and issuing reports. Generally, it is expected that the excipient manufacturer has demonstrated the process is capable of producing excipient ingredient lots in conformance to quality requirements from the point at which full GMP requirements have been determined to begin.

Traceability requirements are similar to those for pharmaceutical dosage forms and API. However, it should be recognized that raw materials are often stored in bulk tanks, which makes unique traceability to a raw material lot unfeasible. Also, excipients are often produced using continuous processing, which again makes traceability to a lot of raw materials or intermediates impractical. In either event, the excipient manufacturer is expected to be able to place boundaries around a finite amount of material that may have been consumed into the excipient lot being traced.

Finished excipient is sometimes stored in bulk, typically stored outdoors. This is acceptable where this practice has been demonstrated not to have an adverse effect on excipient quality. Where excipient storage, whether discrete or bulk, requires controlled conditions such as temperature and inert atmosphere, records should demonstrate those specified conditions are being achieved.

Excipient packaging systems should be demonstrated to protect the excipient during transport to the customer and under the recommended storage. Studies should demonstrate that the packaging system continues to assure the excipient meets quality requirements during its shelf-life. Also, there should be tamper-evident seals to facilitate the identification of packages that have been opened or otherwise breached.

Measurement, Analysis, and Improvement

There should be an internal audit program that includes scheduled audits of the excipient quality system. Such audits should be conducted annually and corrective and preventive measures should be identified. Such measures should be reviewed to assure that the planned improvement activities were undertaken and have been shown to be effective.

The excipient manufacturing process should be monitored through a combination of in-process operating controls, on-line measurements, and at-line testing. These data should confirm that the process remains in a state of control.

Finished excipient quality should also be monitored through a combination of finished product testing and confirmation that all manufacturing-related operations were performed in accordance with the requirements of the excipient quality system. The review of laboratory operations should include documentation confirming all testing were conducted, all raw data including instrument output were retained, and all calculations were verified. Out-of-specification test results should be investigated in accordance with a site procedure that assures a thorough and objective review that supports the final disposition of the discrepant test result.

There should be a review of all records associated with the production, packaging, labeling, and testing of the excipient prior to offering the lot for sale. The review should confirm that there are records showing that the manufacturing and packaging instructions were followed, that there is a retained copy of the package label, and finally that all required lab testing was performed. Also, there should be confirmation that procedures are in place to assure that a retain sample of the excipient lot is properly stored for at least one year beyond the expiry date of the excipient lot.

The stability of the excipient should be demonstrated through a scientific study. This usually involves setting aside the excipient in the market container under the recommended storage conditions to show that the excipient continues to meet the intended specification through its retest interval. This study is often conducted as a kinetics experiment, in which the initial samples are taken frequently, with the interval between sampling increasing as the duration of the study increases.

Although it is expected that stability testing of dosage forms will involve the use of stability-indicating test methods, such testing is not possible for many excipients. It is not unusual for an excipient to be very stable especially under normal warehousing. Excipients such as inorganic salts do not degrade due to temperature or humidity conditions from normal warehousing. Also, it can be technically impossible to measure decomposition products without affecting the development of those products. This currently is the case for Crospovidone whose measurement of the degree of cross-linking would involve conditions that cause the breaking of these bonds.

The excipient manufacturer should conduct a study of the impurities typically present in the excipient. The collection of these impurities is referred to as the impurity profile. It is important that the quantity of impurities present in the excipient remain constant since, oftentimes, an impurity may be important to the functionality of the excipient, at least in some customer formulations.

Nonconforming excipient lots should be isolated and the cause for the lot's failure to meet specification should be investigated. Even though the excipient manufacturer will often regrade nonconforming material for use in other industries or reprocess the excipient, an investigation into the reason for failing to meet specification should still be conducted. Likewise, returned excipient is often downgraded and sold to other industries. However, if the material is to be offered for sale as excipient grade, there should be a documented review to show that the excipient packaging has not been compromised and that the excipient remains suitable for sale.

EUROPEAN REQUIREMENTS

As noted earlier, the IPEC-PQG excipient GMP guide is intended to be applicable globally. The guide has been approved by all three affiliated organizations, IPEC-Americas, IPEC-Europe, and JPEC.

As in the U.S.A., the European Commission has adopted ICH Q7a[b] as the standard under which the API is to be manufactured. Also like in the U.S.A, there is currently no European regulatory requirement that excipients for use in medicines must comply with excipient GMP requirements. The section titled "Basic Requirements for Medicinal Products"[c]

---

[b]http://pharmacos.eudra.org/F2/eudralex/vol–4/pdfs–en/2005_10_03_gmp–partII–activesubstance.pdf (accessed Apr 2006).
[c]http://pharmacos.eudra.org/F2/eudralex/vol–4/home.htm (accessed Apr 2006).

establishes various general requirements for the use of starting materials in drug products, including their purchase from approved suppliers, sampling, and retention, but the document does not stipulate that excipient ingredients be produced in accordance with Good Manufacturing Practices.

However, Directive 2004/27/EC (Article 46)[d] requires development of a list of excipients that must be produced in conformance with GMP. Meetings between industry and regulators were conducted at which it was recommended that this list be developed based upon an assessment of the probability and severity arising from the potential risks posed by the listed excipients. To date, it has been proposed that the following be included on this list[e]:

1. Excipients derived from humans or animals with a potential risk of viral contamination.
2. Excipients derived from animals with a potential risk of transmissible spongiform encephalopathy.
3. Excipients claimed to be sterile and used without further sterilization.
4. Excipients claimed to be pyrogen or endotoxin controlled.
5. Propylene glycol.
6. Glycerin.

The last two excipients were included in the list due to past incidents of their contamination.

It is important to note that whereas the directive requires the development of a list where the excipient is to be produced in conformance to GMP requirements, the document is silent as to what those requirements will be. It is assumed that these excipients will be required to conform to the IPEC-PQG excipient GMP quality system requirements.

REFERENCES

1. The Joint IPEC-PQG *Good Manufacturing Practices Guide for Pharmaceutical Excipients*, 2006.
2. The IPEC Good Distribution Practices Guide for Bulk Pharmaceutical Excipients, 2006.
3. The IPEC-Americas Certificate of Analysis Guide for Bulk Pharmaceutical Excipients, 2000.
4. The IPEC-Americas Significant Change Guide for Bulk Pharmaceutical Excipients, 2006.

---

[d] http://pharmacos.eudra.org/F2/review/doc/final_publ/Dir_2004_27_20040430_EN.pdf (accessed Apr 2006).

[e] Atzor S. GMP for Excipients in the European Union Legal Framework, IPEC Conference, Cannes, France, 2006.

19 | Recalls, Warning Letters, Seizures and Injunctions: CGMP Enforcement Alternatives in the United States

Joseph D. Nally
Nallianco LLC, New Vernon, New Jersey, U.S.A.

In Chapter 1, we noted numerous remedies available to the Food and Drug Administration (FDA) to enforce CGMPs and the Federal Food, Drug and Cosmetic (FFDC) Act generally. However, the FDA has no authority, under the Federal Food Drug and Cosmetic Act, to order a firm to recall a violative product without the aid of a court. Although the FDA will press for voluntary recall action by a firm, they also have the option of charging the proprietors of a firm with a crime under the statute, lack of intent notwithstanding, or using the special remedial statutory actions as for adulterated products. They could make multiple seizures of the product in virtually as many geographic areas as were feasible for it to prosecute and infeasible for the manufacturer to defend. Since every adulterated product is also misbranded, the opportunities for fines could be complicated and significant. The FDA could also seek injunction to bring the product distribution to a halt.

In fact, new legislation in 1984 and 1987 amended the U.S. Code to greatly increase penalties for all federal offenses. The maximum fine for individuals is now US $100,000 for each offense and US $250,000 if the violation is a felony or causes death. For corporations the amounts are doubled. Under the recent legislation, fines in successful government prosecution on behalf of the FDA have totaled in the millions and tens of millions considering consent decree fines.

Many of the criminal offenses are prosecuted in the same philosophical vein of Dotterweich and Park, that is, holding corporate officers personally culpable for failure to properly supervise and protect the enterprise from the carelessness, the criminal activity, or the corruption of persons in their realm of supervision. Under the Prescription Drug Marketing Act, the larcenous behavior of the manufacturer's employees in stealing and trafficking in prescription drugs, that insecure conditions of manufacture and distribution make possible faces serious criminal prosecution and tremendous fines.

Looking at the trends and results of regulatory actions provide some insight into how the FDA is operating and how the industry is reacting. Tables 1 to 3 illustrate recent data released by the FDA for warning letters, recalls, seizures and injunctions.

Examining the data for trends shows an obvious decrease for warning letters and seizures and a flat or "saw-toothing" trend for recalls. Is industry getting better or are FDA enforcement activities decreasing? The truth probably lies somewhere in-between. In general, the industry has improved their GMP practices and FDA enforcement activities have changed focus to a risk-based enforcement approach. Risk-based enforcement is prudent use of enforcement resources that will provide consumer protection and sustainable compliance. The FDA's inspection workload has increased substantially over the last 10 years with the largest increase being in foreign inspections. Because of budget constraints, staffing could not increase at the same level to accommodate the growth; hence a smarter approach was needed.

The FDA is now relying on existing enforcement tools and more collaboration and cooperation with foreign governments, other federal agencies, and state and local governments to assure consumer protection. They are also relying more on voluntary action (such as comprehensive responses to 483 observations) in light of a reasonable compliance history. However, once identified they will vigorously pursue any indication of fraud, gross negligence, or intentional violative actions. Another way of looking at the risk-based enforcement is that there may be less chance of getting caught, but the penalties will be greater if you do.

TABLE 1   Warning Letters

Year	All divisions	CDER
2001	1032	70
2002	755	58
2003	545	29
2004	737	30
2005	535	18

Source: From Ref. 1.

TABLE 2   Recalls

Year	All divisions	CDER
2001	4588	319
2002	5105	437
2003	4628	342
2004	4670	288
2005	5338	502

Source: From Ref. 2.

TABLE 3   Seizures and Injunctions

Year	All divisions (seizures)	CDER seizures/injunctions
1992	183	
1993	117	
1994	98	
1995	73	
1996	36	
1997	46	
1998	35	
1999	25	
2000	36	
2001	27	15/4
2002	13	3/5
2003	25	3/6
2004	10	0/4
2005	20	6/7

Source: From Ref. 3.

WARNING LETTERS

This is the FDA's primary tool to achieve timely, voluntary compliance. Warning letters require a response with corrective action "within 15 working days" and typically state that "You (addressed to President/CEO of firm) should take prompt action to correct deficiencies at your facility. Failure to do so may result in further regulatory action without notice. These actions may include seizure of your products or injunction." Although the corrective action is still voluntary, the possibilities of seizure, injunction, consent decree, or any business disruption should generate serious concern with senior management.

The four-fold decrease in drug firm warning letters since 2001 is an indication of some mix of improved compliance activities by firms and use of the FDA's risk-based approach in inspections and enforcement.

Of compliance interest is the breakdown of warning letter citations by CFR subpart and quality system for drug firms in 2005:

No.	Part 211 sections	Quality system
31	Records and reports	Quality (40)
9	Organization and personnel	
23	Laboratory controls	Laboratory controls (23)
12	Equipment	Facilities and equipment (23)
11	Buildings and facilities	
22	Production and process controls	Production (22)
7	Containers and closures	Materials (7)
5	Packaging and labeling controls	Packaging and Labeling (5)

Quality, laboratory, facilities, equipment and production systems account for 90% of the citations. These are both the area of focus in FDA inspections and where most of the compliance issues reside.

SEIZURES AND INJUNCTIONS

While there have been few seizures and injunctions in recent years, the impact on the involved firms is substantial.

One very effective tool in the FDA's enforcement actions is the consent decree of permanent injunction. A consent decree is: "Status imposed by the FDA in serious violation of federal regulations and related safety and quality standards. A company must agree to a series of measures aimed at bringing its manufacturing standard into compliance with federal regulations. Until agreed-upon conditions are met, a company may be forbidden to distribute its products in interstate commerce, except for those products deemed essential for the public health" (4).

Consent decrees are very costly and in addition to the fines often result in the discontinuation of products and substantial loss in sales. As an example, the Schering Plough Corporation paid a record US $500 million to the U.S. Treasury after the consent decree (May, 2002) had been entered by the court. The company also agreed to suspend manufacture of 73 products and pay up to an additional US $175 million if it failed to adhere to the established corrective action timelines established in the decree (5).

Needless to say, achieving and maintaining a state of sustainable compliance and avoiding a consent decree is the preferred course of action. Typically, there are a number of warning signals before the FDA takes this course of action. The most common symptom is a lack of effectively correcting deficiencies previously cited. This is evidenced by repeat 483 observations and/or warning letters on the same subject or condition. Management in the firm promise corrective action and on the next inspection the same or similar problems are found again. If it is found a second and especially a third time, warning letters and consent decrees can likely follow. Repeat events are not corrected because the root cause of the deficiency has not been identified and/or corrected. Observations are almost always symptoms of underlying problems and fixing or patching the symptom (answering only the observation) will not correct the problem.

RECALLS

Recalls have an interesting background in both legislative and administrative history and have become a major contributor to the FDA's remedial success; the process of development is worth a review.

In 1978, the FDA published a "final rule," under Title 21, Part 7, Enforcement Policy. It adds somewhat to the confusion between substantive and interpretive regulations. Indeed, since it deals with recalls, a remedy not provided for under the agency statute, it may be further in a hybrid area of definition than any other regulations. Here, by excerpts from Federal Register, Vol. 43, No. 117, Friday, June 16, 1978, are some basic considerations as the FDA interprets and expresses them. At the outset, you should be aware that these are not universally shared outside the agency.

> Title 21—Food and Drugs
> CHAPTER I—Food and Drug Administration, Department of Health, Education, and Welfare
>
> SUBCHAPTER A—General PART 7—ENFORCEMENT POLICY
> Recalls (Including Product Corrections)—Guidelines and policy, procedures, and industry responsibilities
> AGENCY: Food and Drug Administration
> ACTION: Final rule.
> SUMMARY: This document establishes regulations intended as guidelines that set forth the agency's policy and procedures for product recalls and that provide guidance to manufacturers and distributors of products regulated by the FDA, so that they may more effectively discharge their responsibilities. Recall is the most expeditious and effective method of removing or correcting defective FDA-regulated products that have been distributed commercially, particularly when those products present a danger to health. The guidelines apply to all FDA-regulated products (i.e., food, including animal feed; drugs, including in vitro diagnostic products; cosmetics; and biological products intended for human use) except electronic products subject to the Radiation Control for Health and Safety Act.
> EFFECTIVE DATE: July 7, 1978.

The position of the FDA can best be understood by separate discussion of the three principal areas of recall-related authority: (i) to order recalls; (ii) to prescribe procedures and requirements concerning the conduct of recalls; and (iii) to require the making of reports to the FDA concerning recalls.

As mentioned in the beginning of this chapter, the FDA has no authority under the FFDC Act to order a firm to recall a violative product without the aid of a court. Thus, where the agency requests a recall under these regulations, it has no authority to impose or seek sanctions for a firm's refusal to carry out the recall. (the FDA may, of course, take legal action respecting the underlying violation that led to the agency's recall request; for example, it may seize an adulterated drug and/or prosecute those responsible for distributing the drug.)

Secondly, the FDA does have authority under both the FFDC Act and the Public Health Service Act to prescribe mandatory procedures and requirements that, among other things, facilitate the conduct of recalls. The agency is not fully exercising this authority in this document in which the provisions set forth are merely guidelines rather than mandatory requirements. The agency has authority to prescribe mandatory procedures and requirements concerning the conduct of recalls because such procedures and requirements prevent the introduction into commercial channels, or facilitate the removal from commercial channels, of adulterated, misbranded, or otherwise violative food, drugs, devices, and cosmetics.

In the preamble to the proposal, the Commissioner cited *National Confectioners Association v. Mathews*, CCH Food, Drug, and Cosm. L. Rep. ¶38.062 (D.D.C. April 14, 1976) in support of these regulations. In the National Confectioners case, the court upheld mandatory FDA regulations concerning good manufacturing practices for cocoa products and confectionery that included a number of recall-related requirements. The regulations, among other things, defined production lot, required coding on shipping containers or finished product packages identifying at least the plant where packed and the product lot or packaging lot, and required maintenance of distribution records. The court held that "[t]he statutory scheme as a whole and §§402(a)(4) and 701(a) in particular clearly provide an adequate statutory basis for the promulgation of the regulations."

The district court decision in National Confectioners was recently upheld by the United States Court of Appeals for the District of Columbia Circuit (*National Confectioners Association v. Califano*, No. 76-1617, January 20, 1978). The court of appeals decision also strongly supports the position that the Federal Food, Drug, and Cosmetic Act provides the FDA with the authority to impose requirements that facilitate recalls. The court held that,

[t]he voluntary nature of recalls does not foreclose their regulation. When accommodation between the FDA and private industry has produced an efficient procedure for enforcing the Act, and when the procedures emphasize voluntary cooperation in lieu of a more disruptive and cumbersome remedy specifically authorized by the Act, the FDA may regulate the procedure of voluntary cooperation.

Moreover, it is proper for the FDA to conclude that it cannot rely exclusively on voluntary compliance to protect the public interest. Regulations that require source codes and distribution records may be based legitimately on the need to expedite seizure when voluntary recalls are refused.

National Confectioners is one of many cases holding that FDA regulations under section 701(a) of the FFDC Act are enforceable. See, e.g., *Weinberger v. Hynson, Westcott, and Dunning, Inc.*, 412 U.S. 609 (1973).

The guidelines published in this document have purposes similar to the regulations upheld in the National Confectioners case, i.e., "to prevent the introduction of adulterated [articles] into commercial channels," "to facilitate the withdrawal by the manufacturer of contaminated or suspect [articles] from the market and to enable FDA to monitor such withdrawal," to "facilitate public warning where necessary," and to "increase the capability of both the FDA and the manufacturer of locating the lots which may be adulterated." The Commissioner believes that many of the provisions in this document for the conduct of recalls (e.g., having a strategy for each recall, notifying customers of the recall, and having a current written plan for affecting recalls) could be promulgated as mandatory requirements if experience under the guidelines proves that mandatory requirements are necessary.

Several comments thought that the Commissioner was relying on the National Confectioners case in support of an argument that the FDA could promulgate regulations under the FFDC Act, enabling the agency to order a recall. As explained above, the FDA does not believe that this act provides authority to promulgate regulations enabling it to order a recall although it can, as indicated in National Confectioners, promulgate mandatory requirements to improve the efficiency of recalls once begun.

Thirdly, the FDA has specific authority to require the making of reports to the FDA concerning recalls, such as notification that a recall is occurring, for some of the product it regulates, but not for all. The Commissioner maintains that the FDA has clear authority to require such reports where there is specific statutory provision authorizing such a reporting requirement, e.g., section 505(i) and (j) [21 U.S.C. 355(i) and (j)] with respect to new drugs and sections 519 and 520(g) [21 U.S.C. 360(i) and 360(g)] with respect to devices; where there is specific statutory authority authorizing records inspection that includes inspection of recall-related records, e.g., section 704 (21 U.S.C. 374) with respect to prescription drugs and restricted devices; or where the product is subject to a licensing or permit requirement and reporting of recalls is a condition to the license or permit, e.g., under section 351 and 361 of the Public Health Service Act (42 U.S.C. 262 and 264) with respect to biologies. The Commissioner points out that the reporting provisions in this document are guidelines for all FDA-regulated products, despite the agency's specific authority to promulgate mandatory reporting requirements for certain products. Other FDA regulations may contain mandatory requirements concerning recall reporting.

In its preamble to the regulations the FDA quoted *U.S. v. Park*, the modern restatement of the *U.S. v. Dotterweich* propositions of earlier FDA years. This was probably inappropriate. Thus, the FDA goes on to say:

3. A comment stated that the sanctions available to FDA are already spelled out in the FFDC Act and nowhere is there any hint that recall is one of the, or any basis to conclude that Congress intended to include recalls as a means of assuring compliance. The comment further noted that the Supreme Court's decision in *United States v. Park*, 421 U.S. 658 (1975), makes no reference to product recall and citing this decision in the preamble to the proposed regulations cannot in any logical manner bolster the claim that recalls can be imposed upon offenders of the act. The comment asserts, therefore, that if FDA should at some future date, consider mandatory requirements necessary, the agency should seek legislation to provide the authority, which it does not possess now.

   The Commissioner emphasizes that nowhere in the proposal nor in this final rule it is implied that recall is a sanction that FDA can order administratively under the FFDC Act and seek criminal prosecution of persons who do not comply with the order (excepting the repair, replacement, and refund authority for medical devices discussed in paragraph 20 below in this preamble). It is true that the preamble to the proposal cited dicta in *United States v. Park* to support the agency's position that manufacturers and distributors have the responsibility "to seek out and remedy violations when they occur." This citation does not mean, nor was it intended to imply, that the Supreme Court's decision was being interpreted by FDA as authorizing the agency to order manufacturers or distributors to initiate recalls. However, the decision does support the view that firms engaged in the production and marketing of FDA-regulated products have by the nature of their business assumed a duty to recall their products when necessary to protect the health and well-being of the public. These recall guidelines are thus founded upon this inherent responsibility of firms and are intended to provide guidance for them to carry out this responsibility. Therefore, so long as firms continue to cooperate in discharging their product recall responsibilities, there appears to be no need for FDA to possess authority under the FFDC Act in order to recall administratively. However, if experience under these regulations proves that such authority is needed, the Commissioner agrees that the agency should seek it from Congress.

Finally, under the authority claimed, Part 7 was amended to read:

§7.1 Scope

This part governs the practices and procedures applicable to regulatory enforcement actions initiated by the FDA pursuant to the FFDC Act (21 U.S.C. 301 et seq.) and other laws that it administers. This part also provides guidelines for manufacturers and distributors to follow with respect to their voluntary removal or correction of marketed violative products. This part is promulgated to clarify and explain the regulatory practices and procedures of the FDA, enhance public understanding, improve consumer protection, and assure uniform and consistent application of practices and procedures throughout the agency.

§7.3 Definitions

(a) "Agency" means the FDA.

(b) "Citation" or "cite" means a document and any attachments thereto that provide notice to a person against whom criminal prosecution is contemplated of the opportunity to present views to the agency regarding an alleged violation.

(c) "Respondent" means a person named in a notice who presents views concerning an alleged violation either in person, by designated representative, or in writing.

(d) "Responsible individual" includes those in positions of power or authority to detect, prevent, or correct violations of the Federal Food, Drug, and Cosmetic Act.

(e) (Reserved).

(f) "Product" means an article, subject to the jurisdiction of the FDA, including any food, drug, and device intended for human or animal use, any cosmetic and biologic intended for human use, and any item subject to a quarantine regulation under Part 1240 of this chapter. "Product" does not include an electronic product that emits radiation and is subject to Parts 1003 and 1004 of this chapter.

(g) "Recall" means a firm's removal or correction of a marketed product that the FDA considers to be in violation of the laws it administers and against which the agency would initiate legal action, e.g., seizure. "Recall" does not include a market withdrawal or a stock recovery.

(h) "Correction" means repair, modification, adjustment, re-labeling, destruction, or inspection (including patient monitoring) of a product without its physical removal to some other location.

(i) "Recalling firm" means the firm that initiates a recall or, in the case of an FDA-requested recall, the firm that has primary responsibility for the manufacture and marketing of the product to be recalled.

(j) "Market withdrawal" means a firm's removal or correction of a distributed product which involves a minor violation that would not be subject to legal action by the FDA of which involves no violation, e.g., normal stock rotation practices, routine equipment adjustments and repairs, etc.

(k) "Stock recovery" means a firm's removal or correction of a product that has not been marketed or that has not left the direct control of the firm, i.e., the product is located on premises owned by, or under the control of, the firm and no portion of the lot has been released for sale or use.

(l) "Recall strategy" means a planned specific course of action to be taken in conducting a specific recall, which addresses the depth of recall, need for public warnings, and extent of effectiveness checks for the recall.

(m) "Recall classification" means the numerical designation, i.e., I, II, or III, assigned by the FDA to a particular product recall to indicate the relative degree of health hazard presented by the product being recalled.

   (1) Class I is a situation in which there is a reasonable probability that the use of, or exposure to, a violative product will cause serious adverse health consequences or death.

(2) Class II is a situation in which use of, or exposure to, a violative product may cause temporary or medically reversible adverse health consequences or where the probability of serious adverse health consequences is remote.

(3) Class III is a situation in which use of, or exposure to, a violative product is not likely to cause adverse health consequences.

(n) "Consignee" means anyone who received, purchased, or used the product being recalled.

§7.40 Recall policy

(a) Recall is an effective method of removing or correcting consumer products that are in violation of laws administered by the FDA. Recall is a voluntary action that takes place because manufacturers and distributors carry out their responsibility to protect the public health and well being from products that present a risk of injury or gross deception or are otherwise defective. This section and §§7.41 through 7.59 recognize the voluntary nature of recall by providing guidelines so that responsible firms may effectively discharge their recall responsibilities. These sections also recognize that recall is an alternative to an FDA-initiated court action for removing or correcting violative, distributed products by setting forth specific recall procedures for the FDA to monitor recalls and assess the adequacy of a firm's efforts in recall.

(b) Recall may be undertaken voluntarily and at any time by manufacturers and distributors, or at the request of the FDA. A request by the FDA that a firm recall a product is reserved for urgent situations and is to be directed to the firm that has primary responsibility for the manufacture and marketing of the product that is to be recalled.

(c) Recall is generally more appropriate and affords better protection for consumers than seizure, when many lots of product have been widely distributed. Seizure, multiple seizure, or other court action is indicated when a firm refuses to undertake a recall requested by the FDA, or where the agency has reason to believe that a recall would not be effective, determines that a recall is ineffective, or discovers that a violation is continuing.

§7.41 Health hazard evaluation and recall classification

(a) An evaluation of the health hazard presented by a product being recalled or considered for recall will be conducted by an ad hoc committee of FDA scientists and will take into account, but need not be limited to, the following factors:

(1) Whether any disease or injuries have already occurred from the use of the product.

(2) Whether any existing conditions could contribute to a clinical situation that could expose humans or animals to a health hazard. Any conclusion shall be supported as completely as possible by scientific documentation and/or statements that the conclusion is the opinion of the individual(s) making the health hazard determination.

(3) Assessment of hazard to various segments of the population, e.g., children, surgical patients, pets, livestock, etc., who are expected to be exposed to the product being considered, with particular attention paid to the hazard to those individuals who may be at greatest risk.

(4) Assessment of the degree of seriousness of the health hazard to which the populations at risk would be exposed.

(5) Assessment of the likelihood of occurrence of the hazard.

(6) Assessment of the consequences (immediate or long-range) of occurrence of the hazard.

(b) On the basis of this determination, the FDA will assign the recall a classification, i.e., Class I, Class II, or Class III, to indicate the relative degree of health hazard of the product being recalled or considered for recall.

§7.42 Recall strategy

(a) *General*.

    (1) A recall strategy that takes into account the following factors will be developed by the agency for an FDA-requested recall by the recalling firm for a firm-initiated recall to suit the individual circumstances of the particular recall:

        (i) Results of health hazard evaluation.

        (ii) Ease in identifying the product.

        (iii) Degree to which the product's deficiency is obvious to the consumer or user.

        (iv) Degree to which the product remains unused in the marketplace.

        (v) Continued availability of essential products.

    (2) The FDA will review the adequacy of a proposed recall strategy developed by a recalling firm and recommend changes as appropriate. A recalling firm should conduct the recall in accordance with an approved recall strategy but need not delay initiation of a recall pending review of its recall strategy.

(b) *Elements of a recall strategy.* A recall strategy will address the following elements regarding the conduct of the recall:

    (1) *Depth of recall.* Depending on the product's degree of hazard and extent of distribution, the recall strategy will specify the level in the distribution chain to which the recall is to extend, as follows:

        (i) Consumer or user level, which may vary with product, including any intermediate wholesale or retail level; or

        (ii) Retail level, including any intermediate wholesale level; or

        (iii) Wholesale level.

    (2) *Public warning.* The purpose of a public warning is to alert the public that a product being recalled presents a serious hazard to health. It is reserved for urgent situations where other means for preventing the use of the recalled product appears inadequate. The FDA in consultation with the recalling firm will ordinarily issue such publicity. The recalling firm that decides to issue its own public warning is requested to submit its proposed public warning and plan for distribution of the warning for review and comment by the FDA. The recall strategy will specify whether a public warning is needed and whether it will issue as:

        (i) General public warning through the general news media, either national or local as appropriate, or

        (ii) Public warning through specialized news media, e.g., professional or trade press, or to specific segments of the population such as physicians, hospitals, etc.

    (3) *Effectiveness checks.* The purpose of effectiveness checks is to verify that all consignees at the recall depth specified by the strategy have received notification about the recall and have taken appropriate action. The method for contacting consignees may be accomplished by personal visits, telephone calls, letters, or a combination thereof. A guide entitled "Methods for Conducting Recall Effectiveness Checks" that describes the use of these different methods is available upon request from the Hearing Clerk (HFC-20), FDA, Room 4-65, 5600 Fishers Lane, Rockville, MD 20857, USA. The recalling firm will ordinarily be responsible for conducting effectiveness checks, but the FDA will assist in this task where necessary and appropriate. The recall strategy will specify the method(s) to be used for and the level of effectiveness checks that will be conducted, as follows:

        (i) Level A—100% of the total number of consignees to be contacted;

        (ii) Level B—Some percentage of the total number of consignees to be contacted, which percentage is to be determined on a case-by-case basis, but is greater than 10% and less than 100% of the total number of consignees;

        (iii) Level C—10% of the total number of consignees to be contacted;

        (iv) Level D—Two percent of the total number of consignees to be contacted or;

        (v) Level E—No effectiveness checks.

§7.45 FDA-requested recall

(a) The Commissioner of Food and Drugs or his designee under §5.20 of this chapter may request a firm to initiate a recall when the following determinations have been made:
  (1) That a product that has been distributed presents a risk of illness or injury or gross consumer deception.
  (2) That the firm has not initiated a recall of the product.
  (3) That an agency action is necessary to protect the public health and welfare.
(b) The Commissioner or his designee will notify the firm of this determination and of the need to begin immediately a recall of the product. Such notification will be by letter or telegram to a responsible official of the firm, but may be preceded by oral communication or by a visit from an authorized representative of the local FDA district office, with formal, written confirmation from the Commissioner or his designee afterward. The notification will specify the violation, the health hazard classification of the violative product, the recall strategy, and other appropriate instructions for conducting the recall.
(c) Upon receipt of a request to recall, the firm may be asked to provide the FDA any or all of the information listed in Section 7.46(a). The firm, upon agreeing to the recall request, may also provide other information relevant to the agency's determination of the need for the recall or how the recall should be conducted.

Additional sections address firm-initiated recalls that are voluntary or occasioned by information from the FDA that the product(s) are violative without further instruction from the agency, 21 CFR 7.46. Other sections deal with recall communications (21 CFR 7.49), public notification to the degree advisable in addition to the weekly FDA enforcement report (21 CFR 7.50), and terms and descriptions required in recall status reports 21 CFR 7.53.

A recall will be terminated when the FDA is confident product has been removed from market in accordance with the recall strategy applicable. The subject product(s) should have been removed, and proper disposition or correction made commensurate with the degree of hazard of the recalled product. The FDA's written notice to the regulatee is the real termination (21 CFR 7.55).

Finally, the advice given at 21 CFR 7.59 is the advice to follow at present as a prudent organization. Prepare and maintain a current contingency plan, in writing and communicated to others at the facility, on how to initiate and effect a proper recall that meets the regulatory requirements.

Recall is fundamentally a regulatory mechanism offering an alternative to older and more familiar statutory enforcement. But statutory increments offer more contemporary actions. Today, there are many other statutory concerns for the alleged errant manufacturer, including the explication and implication of the sentencing guidelines for organizations and responsible individuals. These of course simply demand self-surveillance and company programs to comply, or else heightened disaster may be experienced in terms of punishment. And, we are not understating the statutorily enhanced debarments and blacklisting policies that have already put numerous corporate and individual entities "out of the business."

Very recently there has been a flurry of debarment orders under Section 306(a) of the Act, and as delegated pursuant to 21 CFR 5.20. These have been directed at individuals convicted of a felony under the federal law for illegal conduct related to the development or approval of a drug product, and inclusive of the process for same, as well as for misconduct relating to the regulation of a drug product [Section 305a(a)(z)(A,B)]. Any person with an approved or pending drug product application who knowingly uses the services of the person so debarred, in any capacity, during the stated period of debarment becomes subject to civil money penalties, as does the person debarred. In addition, the FDA will not accept or review any abbreviated new drug applications (ANDAs) submitted by or with the assistance of the debarred person during his or her period of debarment.

One must understand that making a false statement to a U.S. government agency is by itself a felony federally under 18 USC 1001, as is the obstruction of an agency proceeding under 18 USC 1505. So, where an employee made a false representation in a certificate of

TABLE 4 A Subset of the 2001 to 2005 Data on Specific Drug Products[a] by Problem Area

Years	Contamination	Labeling	Potency	Other	Mfg./testing	NDA	Dissolution
2001–2005	284	197	184	141	97	96	69

[a]Oral solid, liquid; Injectable; Topical; Ophthalmic; Suppository; Inhalation; Nasal; Transdermal; API; Mouth wash; and Otic.
Source: From Ref. 6.

analysis regarding the potency of a particular lot of drug product and it was submitted to the FDA to support an ANDA submission, the employee's conduct was deemed felonious. Separately, when the FDA was conducting an audit and management or an employee agent of management destroyed samples, the FDA would need and check a previous representation that those samples met their monograph requirements. The destruction obstructed the agency's execution of their task and that conduct was deemed felonious.

Because of the severity of this remedy, the severity of the offense, the publicity given is most effective and of course is noted in the Federal Register. See, for example, FR Vol. 59, No. 250, 12/30/94, P67709; FR Vol. 60, No. 7, 1/11/95, P.2767.

RECALL STATISTICS AND COMPLIANCE ISSUES

A review of the recall results from 2001 to 2005 indicates that the number of CDER-related recalls has remained about the same (Tables 2 and 4). The occasional spikes over the years reflect specific actions such as Able Laboratories in 2005. While there is a yearly increase in the volume of drug products on the market, one can question the level of incremental improvement in the industry especially considering that the Class I levels are basically unchanged.

CDER Class I to III recall results

Year	Class I	Class II	Class III
2001	14	168	134
2002	15	235	187
2003	20	154	168
2004	16	138	134
2005	18	314	170

Source: From Ref. 3

The recall data by problem area over the last five years indicates that fundamental problems and issues still exist in the industry. Product GMP issues (contamination, labeling, potency, manufacturing, and dissolution) are >90% of recalls.

Labeling is an interesting example of the effect of compliance efforts. Over a decade ago, labeling errors, mix-ups were a major cause of recalls. The GMPs were updated with additional sections added on labeling controls (automated scanning technology and/or a 2 × 100% inspection of un-scanned cut labeling). Based on the 2005 data, recalls due specifically to labeling mix-ups are very infrequent (<10). This indicates the requirement for scanning and increased inspection has resulted in a decrease in frequency. The majority of labeling-related recalls are now due to product mix-ups and errors in the label text. However, the total number of labeling-related recalls over the last five years is still significant.

REFERENCES

1. FDA Website www.fda.gov/opacom/7alerts.html, accessed September 7, 2006.
2. FDA Website www.fda.gov/foi/warning.htm, accessed September 7, 2006.
3. Seizures by FDA Center, www.fda.gov/ora/about/enf_story/ch3/cder_charts.pdf, accessed September 7, 2006.
4. Glossary Related to Compliance for Analytical Laboratories, www.labcompliance.com/glossary/c-d-glossary.htm, accessed September 7, 2006.
5. Schering Plough Signs Consent Decree with FDA Agrees to Pay US $500 Million www.fda.gov/bbs/topics/NEWS2002/NEW00809.html, accessed September 7, 2006.

6.   The Gold Sheet, February, 2006. P. 7.

SUGGESTED READING

1.   Urban. FDA's policy on seizures, injunctions, civil fines and recalls. J Assoc Food Drug Officials 1992; 56(2):7.

Following are some FDA Guidances relevant to this chapter.

SEC. 400.900 CLASS I RECALLS OF PRESCRIPTION DRUGS (CPG 7132.01)

ISSUED: 10/1/80

Background:

A Class I recall is an emergency situation involving removal of a product from the market in which the consequences are immediate or long-range, life-threatening, and involve a direct cause-effect relationship. Class I recalls can, if necessary, require retrieval of the recalled article from consumers (users). The pattern of distribution of prescription drugs to consumers is different from that of other articles. Retrieval of drugs when in the possession of consumers must take into consideration the doctor-patient relationship.

Policy:

When there is a Class I recall of a prescription drug, retail level consignees (retail, hospital, nursing home pharmacists) will be required to review their prescription files for the appropriate time period consistent with the period of distribution of the drug, in order to identify all customers to whom the recalled drug was dispensed. The pharmacist must notify those customers' physicians' of the specific problem, and keep a record of the physician notifications. The physician will be responsible for deciding whether his patients are to be contacted.

If the pharmacist cannot distinguish in his prescription records between those customers to whom the lot(s) of recalled drug was dispensed, and those who received the same drug from a lot not under recall, or from a different manufacturer, then the pharmacist, as a precautionary measure, must notify the physicians of all customers who received the drug.

If retail level consignees (pharmacists, hospitals, dispensing physicians) cannot identify persons to whom a drug under Class I recall was dispensed, there must be a warning issued by FDA to the general public.

20 | Controlled Substances Safeguards (21 CFR 1300, et seq.)

Joseph D. Nally
Nallianco LLC, New Vernon, New Jersey, U.S.A.

There is concern with the registration and the control of the manufacture and distribution of controlled substances. There is the immediate requirement that one in such a business must annually obtain a registration for such purposes. The Drug Enforcement Administration (DEA) must register such an applicant if consistent with the public interest. In addition to the obligations of the United States under international treaties with respect to Schedule I and II substances, the statute gives standards for the public interest. These are six in number for manufacturing registration and five for distribution registration. Likewise, they are applicable to manufacture and distribution registration for Schedule III, IV, and V substances.

The additional standard for the manufacturer is that the registrant promotes technical advances in the art of manufacturing these substances and the development of new substances. The other five, applicable both to manufacturers and to distributors, are that the registrant:

1. Maintains effective control against diversion of any controlled substance or any products or substances in turn compounded from the controlled substances over which it is granted authority to manufacture and/or distribute, into other than legitimate medical, scientific, and industrial channels, and produces an adequate and uninterrupted supply of these substances under legally acceptable sales condition,
2. Complies with applicable state and local law,
3. Has a satisfactory record with regard to prior conviction under federal or state laws relating to the manufacture and distribution of controlled substances,
4. Has past experience in the manufacture of controlled substances and distribution of same, with existence in the establishment of effective controls against diversion,
5. Such miscellaneous factors as may be relevant to and consistent with the public health and safety.

The facility management recognizes that it has the primary operational function to establish and maintain effective systems in order to prevent diversion. Be confident that mutual cooperation between management and law enforcement personnel will result in fewer diversions and less accessibility for potential abusers. Guarding against drug loss involves the utilization and implementation of systems and procedures sufficient to confine drug theft to the lowest possible level.

EMPLOYEE SCREENING AS ONE OPERATIONAL SAFEGUARD

Concern with personnel security starts before an employee is hired. Pre-employment screening for the purpose of identifying potential security problems is recognized by management to be of vital importance when choosing new employees for work in and around areas where narcotics and other scheduled controlled substances, or potent drugs, are handled. The screening program includes a careful evaluation (to the fullest extent possible in light of other laws that affect employment selection) of the applicant's personal and prior employment references. Supervisors are charged with the responsibility, by management, to make similar precautionary evaluations before transferring employees to new assignments in areas where they are first exposed, or more heavily exposed, to some role in the manufacturing process, storing, and/or shipping of bulk or finished controlled substances.

The facility management continues communication and review, with the assistance of the Human Resources Department, as to the implementation of personnel programs that relate to these concerns. Absenteeism and alcoholism are only a few clues to the employee that may create insecurity. By paying attention to the health needs and problems of employees and by confidential assistance to the employees in need of diagnosis, treatment, rehabilitation, and other related procedures, management will attempt to diminish insecurity. However, all employees and union officials, where applicable, and all applicants for employment should be advised that in any case where the company has reason to suspect a violation of this policy, the company may from time to time examine the employees' lockers, tool boxes, and personal packages on company property. The company should have a policy on drug abuse and it should be clearly stated for all employees.

RESTRICTED ACCESS AS AN OPERATIONAL SAFEGUARD

As a manufacturer and as a distributor, the facility has substantial areas to be protected against unauthorized access and egress. Access restrictions consist of people as well as equipment, mechanical and electronic devices, vaults, signs, and fences.

SUPERVISORY OVERSIGHT AS AN OPERATIONAL SAFEGUARD

As to people, the security director, by whatever title known, for each facility must be on the alert for pilferage at every level of drug handling: from the receipt of raw materials through all phases of manufacturing and processing, including quality control, sampling, in-process, and reserves; through the packaging and labeling procedures, the fill, the counting, and reconciliation of estimated and actual yields; then, to the finished product storage, quarantine, and other, and to shipping. Each employee responsible must be alert to theft of dosage-form drugs, bulk raw material, and even chemical precursors. The facility contemplates careful observation of the shipping area, and when controlled substances are moved to the loading dock, it is under appropriate surveillance.

SECURITY GOALS IN TRANSPORTATION AS AN OPERATIONAL SAFEGUARD

In addition, each facility must be familiar with the transporter, selected on the basis of good reason. Elements of the same type that went into employee hiring and assignment must be used in selecting carriers. The manufacturer is concerned that controlled substances shipped are secure throughout the entire chain of handling in which it can have input. Where scheduled drugs are lost in transit, a signed statement of the facts, including a list of the controlled substances involved in the loss, is provided, along with notification to DEA and local authorities (21 CFR 1301.74).

RECORD KEEPING AS AN OPERATIONAL SAFEGUARD

The company must devise an adequate system, which it diligently seeks to test and improve, in order to keep complete and accurate records including perpetual inventories pursuant to the requirements of 21 CFR 1304.04, et seq. The effort to maintain strict product accountability is a very positive factor in carrying out the statutory command. A well-organized and properly effectuated records system is placed under strict management supervision with plant manager, director of security, and so forth, charged with proper execution.

PHYSICAL SECURITY AS AN OPERATIONAL SAFEGUARD

The DEA and the company are concerned with the provision of certain barriers and types of barriers commensurate with the quality and character of the raw materials and finished goods handled by the facility. Therefore, physical security contemplates a mixture in proper measure of mechanical and electrical apparatus directly related in the facility to maintain secure conditions for controlled substances at every phase of facility activity. Since experience by others and by the DEA indicates that losses of controlled substances usually occur when protective measures are either of such inferior quality, or simply lacking, these facilities should have incorporated physical safeguards sufficient to ensure a reasonable standard of protection for the company's controlled substances on the premises. Naturally, Schedule II compounds are afforded the maximum degree of protection in this scheme, by use of safes, vaults, and human surveillance. In addition, in other circumstances, secured rooms and wired enclosures are used, all to correspond with 21 CFR 1301.71 through 1301.74. Vaults and safes used in this facility should accord with 21 CFR 1301.72. All secured areas are backed up by effective electrical protection as required by occupancy, purpose, and schedules of materials.

Title 21 CFR, Part 1300–1399 is available at: http://www.deadiversion.usdoj.gov/21cfr/index.html. As of April 1, 2004.

- Part 1300-Definitions
- Section 1301-Registration of Manufacturers, Distributors, and Dispensers of Controlled Substances
- Part 1302-Labeling and Packaging Requirements for Controlled Substances
- Part 1303—Quotas
- Part 1304—Records and Reports of Registrants
- Part 1305—Order Forms
- Part 1306—Prescriptions
- Part 1307—Miscellaneous
- Part 1308—Schedules of Controlled Substances
- Part 1309—Registration of Manufacturers, Distributors, Importers and Exporters of List I Chemicals
- Part 1310—Records and reports of Listed Chemicals and Certain Machines
- Part 1311—[Reserved]
- Part 1312—Importation and Exportation of Controlled Substances
- Part 1313—Importation and Exportation of Precursors and Essential Chemicals
- Part 1314—[Reserved]
- Part 1315—[Reserved]
- Part 1316—Administrative Functions, Practices, and Procedures

SUGGESTED READINGS

1. See *Federal Register* January 31, 1989 Part II; Drug free workplace. Vol 54, No. 19 Government wide implementation of the Drug Free Workplace Act of 1988.
2. For General Legal Considerations: Drug Abuse in Industry and Business Symposium Enterprises No. Miami, FLA 33161. Sidney H. Willig, Author; W. Wayne Stewart M.D., Editor. L.C. Catalog Card No. 77—151994.

21 | The Inspection Procedure for Compliance in the United States: The Regulatee Is Inspected; The Rationale for Inspection (21 USC 373, 374)

Joseph D. Nally
Nallianco LLC, New Vernon, New Jersey, U.S.A.

Normally, all searches of private property, which includes both premises and papers, must be performed by authority of a search warrant issued in compliance with the warrant clause of the Fourth Amendment. Any search without a warrant is per se unreasonable, subject only to a few specifically established and well-delineated exceptions.

For example, a specific exception to warrant and probable cause requirements of the Fourth Amendment is that an "search" or an inspection conducted pursuant to valid consent is constitutionally permissible.

The federal or state agency officer, who usually presents credentials and a notice of inspection to the pharmaceutical manufacturing facility, receives express or implied consent to carry out the inspection. At a later time, if the agency and the manufacturer become adversaries, the government has the burden of proving where it acted on consent of the regulatee, and that the consent was in fact freely and voluntarily given and not retracted. This does not diminish the legality of the Food and Drug Administration's (FDA's) statutory authority to inspect or investigate a regulatee's premises and records. [*Note*: The terms Investigator and Inspector are used in this chapter. They are equivalent.]

When the employee of the FDA displays his or her credentials, their nature may vary, and that may be indicative of the FDA's concerns that are to be explored. Ordinarily, the credentials will consist of Form FDA-200A and Form FDA-200B. They authorize inspection, collection of samples, access to copy and verify records, to make seizures under 702(e)(5) of the Act, and to supervise compliance operations for the enforcement of the Act.

Inspectors or investigators who present credentials consisting of Form FDA-200D have special authority for criminal investigators and may carry firearms. On such presentation or presentation under Form FDA-200E, the regulatee should understand that he or she need to obtain advice from counsel promptly. With the recent increase in criminal-style enforcement, regulatees should know the difference in FDA employee credentials (21 CFR 5.35).

There is no doubt that the FDA staff can inspect and investigate with the aid of a warrant that they can procure in advance and for a lesser element of "probable cause" than what is required for a search warrant in police matters. The FDA can always go to the Court to enforce its claimed right of inspection under 21 USC 374. It can enlist the Federal District Court's aid under 21 USC 332, which grants such district courts jurisdiction to enjoin violations of 21 USC 331. Subsection (f) of 331 prohibits the refusal to permit entry or inspection, as authorized by 21 USC 374. Thus, if the court granted the FDA an injunction, the proprietor of the manufacturing premises, or other FDA regulatee, would be enjoined from refusing entry to FDA inspectors.

The FDA can go to the court to enforce a warrant that grants the right to inspect, but often takes the position that it has a right to inspect without a warrant and that one who refuses to permit inspection can be punished for the refusal, regardless of whether inspection subsequent to the refusal results in proof of noncompliance with the adulteration, misbranding, new drug sections, or other substantive areas of the statute. Other federal agencies and state agencies have made the same argument. Indeed, in *Daley v. Weinberger*, in which a doctor suspected of distributing a compound prepared in violation of the New Drug

Amendments persisted in refusal to allow inspection of her premises and her records, the FDA threatened, and the court found plausible, that the FDA might institute criminal proceedings against Dr. Daley under 21 USC 331(f) and 21 USC 333(a).

Given all these alternatives, it seems the situation should be clearly favorable to the FDA. However, the Supreme Court decisions have followed a pattern that seems to indicate that ultimate Court affirmance of the FDA view will depend upon many precedential factors. The most recent relevant decision was in *Marshall v. Barlow's Inc.*, 436 U.S. 307, 98 S. Ct. 1816 (1978), where the Supreme Court declared that the Fourth Amendment stands between the regulatee and the use of compliance inspections by the agency charged with enforcing the law (here, the Occupational Safety and Health Act). From the time of that decision, numerous controversies have arisen over the standards for administrative probable cause and over the procedures for issuing a warrant.

There are many cases to explore in considering the issue of refusing to allow an FDA inspection, or withdrawing consent for an inspection. The turning point of the modernization of this controversy was in 1967. In *Camarra v. The Municipal Court*, 387 U.S. 523, the Supreme Court met the question: Is an administrative inspection authority subject to the requirement of a warrant for entry and search of private premises? The Court answered: "Yes," overruling *Frank v. Maryland*, 359 U.S. 367, the leading case up to that time, which had held that inspections or searches minus a warrant were reasonable under state police power, if they were for purposes of health and sanitation, but not if they were for arms or criminal paraphernalia or narcotics. In the process, however, the Court diluted the warrant requirement in the administrative inspection case. As they put it: "Probable cause in the criminal law sense is not required. For purposes of an administrative search ... 1/4 probable cause justifying the issuance of a warrant may be based" not only on specific evidence of an existing violation, but also on a showing that "reasonable legislative or administrative standards for conducting an ... inspection are satisfied with respect to a particular establishment," 387 USC 538. Thus, "a warrant showing that a specific business has been chosen for ... 1/4 search on the basis of a general administrative plan for enforcement of the Act derived from neutral sources ... 1/4 and the desired frequency of the searches" (*Camarra v. The Municipal Court*) would protect the regulatee's Fourth Amendment rights.

Somewhat simultaneously with the Camarra case, its prohibition of warrantless inspections without consent was held to apply to business premises, but with equal restraint and qualifications by the agency inspectors and those inspected. "A person's office or place of business is quite as immune from warrantless search as is his kitchen or bedroom," if the area of the commercial enterprise is not one where the public may freely enter. *See v. Seattle*, 387 U.S. 541 (1967), therefore held that minus consent, the agency official needed a warrant to search or inspect, but again alluded to such a warrant as requiring a lesser standard of probable cause. "We do not in any way imply that business premises may not reasonably be inspected in many more situations than private homes, nor do we question such accepted regulatory techniques as licensing programs which require inspections prior to operating a business or marketing a product."

At this point, the FDA formulated an internal administrative warrant policy for inspections that were not consensual.

ADMINISTRATION INSPECTIONS WITHOUT WARRANTS FOLLOWING CAMARRA AND SEE CASES

Within the area of consent to inspect, consent refused, or consent withdrawn, the inspector and the agency face a varying group of possibilities. In *U.S. v. Crescent Kelvan*, 164 F.2d 582 (1947), consents given, admissions, records, and other documentary evidence were admissible in criminal prosecution of the defendants. In *U.S. v. Stanack*, 387 F. 2d 849 (1967), both withdrawal of consent and refusal of consent were viewed by the court as a prerogative of the defendants and necessitated a warrant.

A later case emphasizes the continuance of these circumstances in the case of the food and drug inspector. In *U.S. v. Thriftmart Inc.*, 429 F. 2d 1006, the 9th Circuit (California) upheld convictions of the managers of a supermarket food chain's warehouses that were based on evidence obtained during warrantless administrative searches of their premises by FDA inspectors. The inspections were routine and similar ones had been conducted periodically in the past. On arrival at the warehouses, the inspectors had approached the managers, filled out and presented their notices of inspection, requested permission to inspect, and in each case were told, "Go ahead" or words of similar import. The inspection notices contained a recitation of the applicable statute, which authorizes FDA inspectors to enter at reasonable times to inspect food warehouses. They did not have search warrants, nor did they advise the warehouse managers that they had a right to insist upon a search warrant.

The court of appeals affirmed the convictions, indicating:

> ... the administrative search is "neither evidence of crime" and this involves "a relatively limited invasion of the urban citizen's privacy." ... due to the public importance of the inspection process it should not be "hobbled by the blanket requirement of the safeguards necessary for a search of evidence of criminal acts." ... It is clear, therefore, that the administrative search is to be treated differently than the criminal search. ... In a criminal search, the inherent coercion of the badge and the presence of armed police make it likely that the consent to a criminal search is not voluntary. ... These circumstances are not present in the administrative inspection. ... Nothing is to be gained by demanding a warrant except that the inspectors have been put to trouble—an unlikely aim for the businessman anxious for administrative good will. ...
>
> We hold that the absence of coercive circumstances and the credibility of a consent given to an inspection justify a departure from the *Schoepflin/ Schoepflin v. United States*, 391 F2d 390 (9th Cir.), cert denied, 393 U.S. 865 (1968) rule in cases of administrative inspection. ... 1/4 In conclusion, we hold that in the context of the exclusionary rule, a warrantless inspectorial search of business premises is reasonable when entry is gained not by force or misrepresentation, but is, with knowledge of its purpose, afforded by manifestation of assent. ... Advance notice of inspection under (the statute) is not necessary. ... FDA discretion whether to proceed criminally or civilly is constitutional; the FDA is not required to prosecute every violation. ... Nor need the FDA announce at the outset whether it wishes to proceed criminally or civilly. ...

The courts have a narrow scope of review of the FDA's action. This is true in the case of any administrative agency, notwithstanding that they are or are not bound in their hearings and determinations by courtroom rules of evidence. Litigants challenging an agency's actions are not entitled to a complete rehearing on the factual and legal issues involved. They are more properly heard on new evidence not available at the time the administrative agency action was taken. Beyond that, the court must decide that the agency acted on substantial evidence, however obtained and weighted, and not in an arbitrary, capricious, and discriminatory manner.

In addition to such a narrow scope of review, the complainant faces the additional burden of overcoming the presumption of regularity afforded the acts of an administrator and other governmental personnel such as was enunciated by the court in *Pasadena Research Inc. v. U.S.*, 169F. 2d 375, *Citizens to Preserve v. Volpe*, 432 F. 2d 1307 (Tennessee) 1970. In the same way, courts will properly accord deference to an agency's interpretation of statutes and regulations that it administers, unless obviously erroneous, unreasonable, or inconsistent therewith. The court recognizes that administrative agencies have a primary task to administer broad policy mandates for the common good of our society, and they cannot be required to refine their rules to assure tailor-made equity for each of the complexities that may arise. If they issue and interpret regulations that are rational and supportable in their general application, they cannot be charged with unreasonableness because in particular instances they may grind with a rough edge.

In 1970, however, the Court in *Colonnade Catering Corp. v. U.S.*, 397 U.S. 72, dealt with the statutory authorization for warrantless inspections of federally licensed dealers in alcoholic beverages. Federal inspectors, without a warrant and without the owner's permission, had forcibly entered the owner's locked storeroom and seized bottles of liquor noncompliant with the tax and label sections of the pertinent law. The court found for the owner, excluding the liquor bottles from evidence. They held that "Congress had not expressly provided for forcible entry in the absence of a warrant and had instead given the government agents a remedy by making it a criminal offense to refuse admission to the inspectors under 26 USC 7342." The Court thus indicated that in certain kinds of business or industry, closely identifiable with instant public endangerment, such as liquor, inspection and even unannounced inspection are basic needs for an effective system of controls. They further pointed out that in the context of a regulatory inspection system of business premises that is carefully limited in time, place, and scope, the legality of the search depends not on consent but on the authority of a valid statute.

In *U.S. v. Biswell*, 406 U.S. 311, the Court two years later had an opportunity to expand on "pervasively regulated businesses" directly associated with traditional and historic close governmental control such as alcohol and firearms. A similar statute as in the Colonnade case was involved, but here the inspection, although warrantless, was carried out without force. There was a valid underlying statute that carried a right to inspect. It overlaid a traditionally pervasively regulated business associated with dangers of violence and crime to the public. The regulatee enjoyed federal favor in terms of his license or privilege to be in business.

Further, the court differentiated the circumstances that caused warrantless searches to be prohibited in *See v. Seattle*.

> In *See v. Seattle*, the mission of the inspection system was to discover and correct violations of the building code, conditions that were relatively difficult to conceal or correct in a short time. Periodic inspections sufficed and inspection warrants could be required and privacy given a measure of protection with little, if any, threat to the effectiveness of the inspection system there at issue. . . . Here, if inspection is to be effective and serve as a credible deterrent, unannounced, even frequent, inspections are essential. The prerequisite of a warrant could easily frustrate inspection . . . When a dealer chooses to engage in this pervasively regulated business and to accept a federal license, he does so with the knowledge that his business records, firearms,. . . . will be subject to effective inspections.

Following this case, many agency statutes were reexamined by the agency to determine whether they met the "warrantless inspection" prescription written by the court in the Biswell case. Thus, the Occupational Safety and Health Administration (OSHA) argued eventually that its staff could enter the workplace anytime, any portion, in implementing its own highly pervasive statute. The FDA characterized the food and drug industries as pervasively regulated. However, the Supreme Court sounded a call to caution. Liquor and arms are special businesses, and OSHA, which deals with no such special group, as pervasive as its controls might be on all its horizontal spectrum of regulatees, does not meet criteria of the Biswell case.

Many believe that the FDA, despite some few favorable lower court decisions, should look to the Barlow case rather than the Biswell case to measure its inspective authority. There are many analogies beyond the fact that most pharmaceutical plants must also consider OSHA. The law was held unconstitutional inasmuch as the Barlow case was not in such an exceptional business or industry and the statute purported to authorize inspection of business premises without warrant. The rule that warrantless searches are generally unreasonable applies to commercial premises as well as homes. *Camarra v. Municipal Court*, 387 U.S. 528; *See v. City of Seattle*, 387 U.S. 541, 87 S. Ct., 1737 (1967); note from the syllabus of the case as it appears in 98 S. Ct. on page 1818:

(e)   Requiring a warrant for OSHA inspections does not mean that, as a practical matter, warrantless search provisions in other regulatory statutes are unconstitutional, as the reasonableness of those provisions depends upon the specific enforcement needs and privacy guarantees on each statute.

The inspection rights sought to be exercised by the federal agency were statutorily set forth as in countless other federal and state statutes.

Section 8(a) of the Occupational Safety and Health Act of 1970 empowers its agents: 29 USC 657 (2)

(i)   "to enter without delay and at reasonable times any factory, plant, establishment, construction site, or other area, workplace or environment" where work is performed by an employee of an employer; and
(ii) "to inspect and investigate during regular working hours and at other reasonable times, and within reasonable limits and in a reasonable manner, any such place" of employment and all pertinent conditions, structures, machines, etc.

The regulations with respect to inspections, 29 CFR 1903, require an inspector to seek compulsory process if the regulatee demands same, or if partial or total access is refused.

Barlow's was an electrical and plumbing installation business, certainly not of Biswell character. The OSHA inspector, after entry to the customer service area of its enterprise (an area of invitation), presented his credentials to the proprietor and asked for entry to the working areas where he could conduct his inspection.

The inspector should seek and gain entry from one authorized to permit same. The proprietor, as in the Barlow case, was certainly appropriate. The time was reasonable; it was a time when the place was open for business. Inspectors for other agencies often show, in addition to credentials, an inspection form indicating that they are there to inspect pursuant to a statutory section. Whether they divulge their reasons for inspection is discretionary with them. While they are not likely to be untruthful, they may be evasive. The regulatee should always seek to ascertain the reason. Mr. Barlow did so here. In the case of OSHA, where complaints are frequently associated with employer-employee discord, the situation is particularly delicate on both sides. When Barlow asked whether there had been a complaint, the inspector replied negatively, but asked again to enter the nonpublic area of the business. Barlow said he could not unless he had a search warrant.

Regulatees often are willing to trade. The regulatee says, "If you will tell me what area of my activity is suspected of noncompliance, I will allow you access to it to show you that the report or rumor or complaint is wrong. If you won't be honest and cooperative with me, go get yourself a warrant." The inspector replies, "That's not as good or fair a trade-off as it sounds. First, I may betray my source of information, which I was supposed to protect. Second, if I warn him and he feels in danger, he'll not let me inspect anyway. Third, if he's so protective there may be other troubles here I might observe."

Does the case inhibit employee reports to OSHA? The opinion expressly says that it does not. It does say, however, "That an employee is free to report and the government free to use, any evidence of noncompliance with OSHA that the employee observes, furnishes no justification for federal agents to enter a place of business from which the public is restricted and to conduct their own warrantless search." Further, anything the inspector could have observed from the customer service area would have been admissible in evidence against the defendant.

If surprise searches have a reason for such and are contemplated, an administrative warrant can be secured, not only on specific evidence of an existing violation, but also on a showing that reasonable legislative or administrative standards for conducting an inspection are satisfied with respect to a particular establishment. The great majority of regulatees can be expected in normal course to consent to inspection without warrant. A warrant provides assurances from a neutral officer that the inspection is reasonable under the Constitution, is authorized by statute, and is pursuant to an administrative plan containing specific neutral criteria.

Among other issues raised by the government and dissenters was the mandate of OSHA to safeguard employees against hazards in work areas of businesses subject to the Occupational Health and Safety statute. These mandates have always enjoyed special status with the Court and are exceptions to the need for a warrant. Further, they say, to necessitate a warrant, the administrative warrant devoid of probable cause preconditions, is to dilute the warrant clause of the Fourth Amendment (5 to 3 decision).

In *Woods and Rhode Inc. v. Alaska*, Justice Rabinowitz writing for the Court, June 2, 1977, 565 P. 2nd 138, was in total agreement with the White opinion in the Camarra case and, had he but known, with the White opinion in the Barlow case as well.

Since the Supreme Court has emphasized in the Barlow case that an exception from the search warrant requirement may exist for industries with a compelling history of government oversight, and where no reasonable expectation of privacy could exist for a proprietor over the stock of such an enterprise, one can see that an agency may not wax confident that it falls within this narrow area of exception. The liquor and firearms industries comprise the exception, not the rule. The fact that a federal agency is granted governance over businesses involved in interstate commerce, alone, is insufficient to bring just any business or industry within such an exception.

In all of the Supreme Court inspection cases, the Court notes that the great majority of businessmen can be expected in normal course to consent to inspection without a warrant, and the FDA will not be crippled as to effectiveness by providing those owners who wish to refuse an initial requested entry with a time lapse while the inspector obtains necessary legal process.

The Court in the Barlow case also indicated that requiring a warrant should not impose serious burdens on the inspection system or the courts because consent is the usual circumstance. Some government inspectors, however, demand that the regulatee sign various documents authorizing the inspection and waiving rights, prior to commencement of the inspection.

For example, as to Controlled Substances, whether or not narcotic drugs, in 21 CFR 1316.06 through 1316.13, administrative inspections with and without warrant are described, and in conjunction with these, a printed form is provided to the regulatee with instructions that it be signed. In the author's opinion, the signature is not required to carry through a consensual inspection, and insistence by an inspector that the inspection will not be conducted without the signature is ill advised. To proceed to a magistrate for a warrant because the regulatee will only verbally consent to the inspection should be reserved for critical circumstances where criminal action is suspected on the part of the regulatee; otherwise, the judge is being abused by being troubled for naught.

In some instances, FDA inspectors, on being refused entry, have demanded that records be turned over to them for copy, notwithstanding. The Court indicated in the Barlow case that the inspection of the regulatee's documents, required to be maintained by the act and allocable regulations, may not be effected without a warrant, absent consent. Of course, the agency has subpoena power to force production of all such records.

Nothing said in defining the rights of inspectors and "inspecteds" is intended in any way to encourage the latter's taking a hostile, narrow, technical approach to the inspector's activities. Both, rather, should appreciate the mutuality of interest that applies: maintaining and delivering drugs of quality, safety and efficacy to the public. That is the mission of the inspector, and that is the only profitable way to stay in business. There is rarely an excuse for the inspector to be intrusive and the manufacturer to be obstructive.

When lawyers are asked for the legal determinants in a given situation, that is all they can supply as lawyers. Immediately following that, it becomes a policy decision for management. If they are not on good ground, then the delay they may gain may be less helpful than a sincere attempt at correction and remedy, for the FDA does have additional resources beyond Sections 703 and 704 of the Act. There is a general acceptance and recognition that the regulations are sound and that necessary compliance with them is in the best interest of the industry and the public. With proprietary or prescription drug manufacturers especially, there is frequent misunderstanding between inspectors and proprietors because management policy may differ because of personalities and the nature of trade secrets. When inspectors come to a manufacturing plant, they are looking for violations of the act and infractions of the regulations. They may come in to collect official samples, to copy documents showing interstate movement of a product (transportation record, the invoice, and the bill), to get other information made available to them, and to learn what personnel can contribute most to evidentiary needs. Of course, sometimes their visit is the prelude to an establishment inspection.

From the standpoint of the inspected, it is good to know the purpose of an inspector's visit beyond the formal notice handed to the responsible plant official. Do not hesitate to ask this. Most inspectors will be responsive to the extent that they are permitted and circumstances allow. The question of whether they must advise you that they suspect a criminal violation of the statute and are seeking evidence accordingly is one that has been raised before the Court and not elicited any degree of support up to now. They may advise you that it is a routine inspection of your facility or some special part of function of your factory. In any event, besides a sit-down discussion of their findings or recommendations on leaving, they may make valuable suggestions during the course of the inspection, which means that someone capable of understanding the explanation should accompany them and make appropriate notes. As a matter of interest, especially for those unfamiliar with inspection procedures, the date, time, and description of inspection areas should be detailed, questions asked, answers given, and so forthon, as well as identification of the inspectors—all on a report for the manufacturer's interest and assistance.

By virtue of their authority, inspectors can ask for legend drug products in their packaged preshipment state. If asked, the inspector will pay for them at a determined price. In such an instance, or if the manufacturer does not wish to make a charge, the manufacturer usually exchanges an invoice for a receipt, noting the exact facts of the distribution. Of course, here, as wherever the FDA takes a sample, duplicate equivalent packages from the same lot should be set aside for future comparative analytical needs, some portion of which should be checked by quality control immediately to assure being able to react quickly if necessary.

Inspectors who want a line sample are generally willing to await a convenient point in the processing to either merely observe or check it, or withdraw it for further action. Any such sample should be marked appropriately and receipted with the acknowledgment that it is from unreleased stock. Again, an equivalent sample taken from the same batch should be withdrawn by the manufacturer's quality control unit for immediate and later comparative analysis. What the manufacturer reserves as its own contemporary samples should be stored under appropriate and recommended conditions.

If inspectors unknowingly breach a sterile or other "untouchable" proceeding, they should be so informed and the material involved rejected with reason given in the record, as always in the case of rejected materials. Needless to say, if the inspector is about to unknowingly breach it, stop the inspector and explain.

Inspectors are given specific instructions and are quite expert at handling various types of samples. This in a sense involves their own quality control. The manufacturer, therefore, should always note, perhaps on a receipt for the sample and not necessarily over the inspector's signature, the manner in which the sample was taken and transported.

Again, to use an obvious example, the inspector knows better, but if he or she puts a few vials of one of the nitrate preparations in a glassine envelope and slips it into a breast pocket, those tablets may not do so well on final assay some hot hours later.

Experienced counsel in this field frequently advise clients that since seizures are in rem proceedings, they are better off, for many reasons, immediately destroying goods where question of adulteration and/or misbranding arises, and the value is less than some given figure. They should, of course, then record this for tax purposes also. In a sense this is associated with the FDA recall technique, for many of the same reasons.

There are aggravated situations where attorneys tell their clients to answer no questions except those that are submitted to them in writing and signed by the interrogator with a statement of authority, or to satisfy no requests for samples or copies of records except where request for same is similarly submitted in writing and signed by the interrogator with a statement of authority. This is really dealing at arm's length.

In enacting additional factory inspection legislation in the field of prescription drugs, Congress did strengthen the FDA procedures on new drug product licensing, as some regard it, and did broadly increase the subject matter of inspection to include research data, quality control information, complaint files, and so forth. It did not establish continuous factory inspection as a requirement, nor did it apply these broader inspectional prerogatives to other drugs. There is still no section of statute that makes a recall mandatory. Recalls are undertaken to retrieve from the market drug lots that fail to meet regulatory requirements. As a matter of fact, HR 6788-S.2580 (1964), which sought to expand inspection authority, as currently for prescription drugs, to foods, over-the-counter (OTC) drugs, devices, and cosmetics, by adding a subpoena section to the act, was never passed. However, device inspection has been enhanced by the Medical Devices Amendment. In some ways, device inspections for "prescription devices" exceed those for drugs.

There are occasions when the inspector feels a picture is worth a thousand words. The inspector will take photographs. These could be a floor or room condition that shows failure to segregate, or poor sanitation, personnel failures, poor security arrangements, or poor equipment, and so on. Since the FDA regards photography as adjunctive to inspection, they will only infrequently ask first. There is no denying, however, that this aspect of inspection is arguable and, therefore, capable of reasonable mutual satisfaction. In some instances, therefore, either the manufacturer takes the picture designated by the inspector exactly as the latter wants it done, and also retains a copy, or pictures are taken separately and simultaneously along with the inspector, covering the same angle and materials. There are instances where the proprietor refuses to allow photographs to be taken and refuses to provide them on constitutional grounds or for protection of trade secrets. Relying upon the Dow decision, the FDA is aggressive regarding their right to photograph in an inspection. Their faith in Dow may, however, be misplaced. The particular circumstances of the case and the need for aerial photography shown by the FDA is not really analogized to the ordinary pharmaceutical manufacturing plant.

Other problems remain in the FDA inspection area. Can the FDA take photographs without permission? Must the FDA warn the persons inspected that they are seeking evidence for criminal charges? Is it doubtful that there is any more authority for a right to photograph a premise than there is to conduct a warrantless search? Consents are not eternal. On the other hand, the Miranda case warnings are not required in an FDA inspection because it takes on a premise that remains under the control of the proprietor.

For the same reason, a mere statement of objection to FDA photography will not suffice to challenge the admissibility of the photos later. In order to accomplish that, the proprietor must order the FDA inspector to leave his camera behind and not take pictures. Then, if the inspector takes them surreptitiously rather than by procuring a warrant to inspect by such means, the photographs could be excluded.

HOW EXPERIENCED INSPECTORS USUALLY CONDUCT THE INSPECTION

The latest Compliance Program Guidance Manual for FDA staff is the Drug Manufacturing Inspections Program 7356.002 February, 2002. This guidance document provides much insight into the FDA's current thinking and approach to inspections. Some of the more important points include:

- Background section states that that the guidance is structured to provide for efficient use of resources devoted to routine surveillance coverage, recognizing that in-depth coverage is not feasible for all firms on a biennial basis.
- The inspection is defined as audit coverage of two or more systems, with mandatory coverage of the Quality System. Coverage of a system should be sufficiently detailed so that the system inspection outcome reflects the state of control.
- There are two inspections options: The first is a "full inspection," which is a broad and deep evaluation of current good manufacturing practice (CGMP) when little or no information exists on the firm or there is reason to doubt the compliance efforts (history of objectionable conditions, recidivism, warning letters, etc.). The other option is an "abbreviated inspection," which is designed to provide an efficient update evaluation of CGMP. Decisions for the type of inspection are made by the District Office.
- The guidance further lists sub-systems and compliance requirements/expectations in each of the six Quality Systems (Quality, Laboratory Controls) Production, Packaging & Labeling, Materials, Facilities & Equipment.

This guidance document is must reading for anyone preparing for an FDA inspection. It is available at: http://www.fda.gov/cder/dmpq/compliance_guide.htm

Armed with the Program Guidance Manual, the investigator/inspector will pull the firm's compliance file on the factory or establishment in question and see how it fared in prior inspections and how it stands as to recall records or even past or pending litigation with FDA. If there have been recalls, whatever follow-up was made, what reasons were given and/or found that caused the recall are worth reconsideration preceding the inspection. Certainly, if samples had been collected directly or indirectly under this factory's label, analytical results on these samples would be reviewed.

The prior inspection observations would have been prepared by the FDA representative who had previously inspected the plant. As part of that report, the section captioned "Discussion with Management" offers a considerable insight to management's attitude, capability, and willingness to be remedially responsive. The inspector will, in this inspection, want to compare the List of Inspectional Observations with what was done by management to correct them.

Following this review and in the course of the factory's normal working hours, generally between 9 a.m. and 5 p.m. on a weekday, the inspector will present himself or herself, display credentials[a], and issue a Notice of Inspection.

The Notice, of course, draws attention to the fact that inspections have a firm legal basis in the Act. The Notice is not to be confused with an administrative search warrant or summons or subpoena.

At this point, since the inspector is at the office, it is usual for the inspector to ask for any changes in the name, corporate status or officers, and the areas of responsibilities for each if they are separate.

The same kind of update may be requested as to products newly undertaken, or deleted (and why). It is usual for the proprietor to set someone in accompaniment of the inspector, and usually this person will acquaint the inspector quickly with new procedures, new policies, and new equipment that will enhance productivity and quality assurance. In fact, frequently it is a member of the regulatee's Quality Control staff who undertakes this duty. There are advantages in having one or two persons consistently undertake this assignment, since they develop "carryover" rapport and a technique of question and answer, mutually profitable to the agent and the regulatee. The disadvantage is the loss of such continuity through absence, but perhaps more importantly, the narrowing of communication of the inspector's concerns and observations.

After the initial exchange of credentials and information, an introductory meeting is usually scheduled by the firm where key company personnel meet with the FDA to go over:

- Purpose and scope of inspection (FDA)
- Identify documents required (by FDA)
- Review management structure (Firm's organization charts)
- Describe quality management system and quality policy (Firm)
- Designate personnel to accompany inspection team (Firm)
- Logistics—allocate rooms, equipment; determine schedules.

Depending on the inspection scope, the requested document list can be very extensive and almost always includes failure investigations, rejected materials, and complaints.

FAILURE INVESTIGATIONS

It is common practice for one or more investigators to start a review of failure investigations at the beginning of the inspection.. This could take a number of days, depending on how many were chosen. Failure investigations are reviewed to look for incompleteness [data collection, investigation process, root cause analysis, and conclusions and corrective/ preventive action (CAPA)]. Investigators will also look for a lack of management response (timing and/or action) or specific patterns/trends, especially for product or contamination issues. After that evaluation, they may ask to see specific areas or additional product documents. The results of this initial review of failure investigations are very important and can set the tone for the rest of the inspection. If numerous issues are found, it is more likely that the investigators will form a negative opinion and look for more evidence of quality management failures. If a good thorough job was done, they may skip review of documents that were on their list.

---

[a] Bearing the inspector's photograph, seal of the department, and enumeration of authority, as described previously.

PHYSICAL INSPECTION

When the inspection starts with a tour, it usually starts in the receival area, and the condition, security, and cover of the unloading dock may capsulize all subsequent findings. Someone who is careless at that point, who provides no shelter from the elements for articles being unloaded, is going to show carelessness elsewhere in the inspection. From that area and a study of the receival book with neat inked entries that provide an insight into the history of all articles received, and that can be used to trace the course of an ingredient or a container to the finished, packaged, shipped dosage form, the inspector's trail will usually go to storage. The key to storage observation that will meet the inspector's approval is segregation of materials in ways, standard or innovative, that actually preclude commingling of dissimilar substances, or dissimilar forms of substances, or dissimilar chronological groupings within the same substances.

Sanitary consideration in storaging, as well as security, will be a concern. How high a stack? How far off the floor a skid? Why is this article stacked on the floor without a skid? Condition of new materials, and appropriateness of containers, temperature, and humidity are noticed.

Segregation of released from unreleased materials, of goods ready for shipment from goods returned, is also seen.

Even as the receival area was observed for its room, condition, and coverage, all of the plant space will be similarly observed and analyzed by the inspector for its safety and adequacy for purpose. He or she will look for space as an indication to the presence of available segregation between operations to prevent mix-ups.

Thus, the inspector will look for the quarantine area in terms of size, location, and manner in which quarantined items are signally labeled. To prevent potential accidents through lack of information, is the quarantine situation set down in writing? How about the plan for sampling from the materials? Is there a written plan that tells the number of containers in the lot from which a sample is to be taken? The inspector will want to observe any sampling operations that may be in progress. By the way, the inspector will examine all of the implements and containers used by the samplers, for cleanliness and suitability. Unclean or reactive implements are obvious adulterants.

Then taking some one of the raw materials (a suspect one if that has already come up, or any one otherwise), the inspector will follow that to the production area.

In the production area, the inspector will be alert to product, personnel, and equipment, as well as the production environment. As to the latter, the inspector will look to facilities for sanitation, proper light, air, water, and cleanliness of general and specific production areas.

The inspector may be expected to examine all of the equipment in the production area to determine that every surface that meets with ingredients, intermediately completed or in completed dosage forms, is nonreactive.

Every major piece of equipment that is fixed to the room, along with mobile equipment that is moved from area to area as needed, should be lettered or numbered and as so identified appear in the proper place on the batch record.

Not only should it be clean for use, but its cleaning and maintenance record should be handy for inspection. The inspector, based on experience in other establishments, may not only check as to the procedures used in cleaning, maintenance, calibration, etc., but may suggest improvements in frequency and technique. The inspector will look for drains for washability, and in the drains for product. In the latter instance, the inspector will consider discrepancies and carelessness in reconciliation of production figures. He or she will look for screening, methods used for extermination, the freshness of paint and its condition—telltale spots that indicate environmental nonintegrity for the product.

Expect the inspector to ask as a primary question in the production area: "Where is the batch record for the product that you're working on?" In the light of the CGMP regulations, the inspector is entitled to see that for any drug, old or new, Rx or OTC. He will check it immediately against the present stage of the processing to assure that the recorded activities are truly contemporary. He may quickly "back check" to assure the preparation up to this processing point. In the original Willig-Tuckerman manual, the concept of "doers" and "checkers," since adopted by the federal regulations, was first established as an element of current manufacturing care. The inspector follows that guide. For example, if components are being weighed out, the inspector will look for two sets of initials on the batch record. This will show someone weighed, someone checked. Initialing omitted or preinserted is an indication of an operation that is high on formalities but low on care.

INSPECTION IN THE U.S.; REGULATEE; RATIONALE 335

The inspector will usually run a quick check of the calculations on the batch record at that point also. The batch record itself, by the way, should also have a signature and date showing that it is an accurate photocopy or other reproduction of the correct master formula record.

Of course, the inspector will continue the course of the raw materials, now the product, in terms of the formulation, the batch record, the theoretical and actual yield, and the sensory impression of the finished product, along to the packaging.

Here, many of the same observations will be sought, but special emphasis will be placed on the componential check of the packaging assembly. Are they doing the specified packaging? What kinds of assurance have they as to its identity, quantity, and quality? What is the appearance, the experience, the education of the personnel involved, and are there enough people to carry out necessary job assignments? Is the health and maintenance of the employees of such nature to contribute to manufacturing safe and unadulterated articles?

Looking carefully at the production and packaging equipment for cleanliness, appropriateness, lubrication, use according to manuals, maintenance generally, the inspection can actually gauge the character of the production management.

As to equipment, inspectors are certainly impressed by quality, newness, and size. However, they will be looking for makeshift innovations to equipment usually applied by an operator and without management's knowledge. The operator sees the flow of finished tablets pile up as they come through the stainless steel chute of the machinery, so pastes in a separation or diverter that does not come out, is not cleaned, etc., the operator operates at higher speed by adding improper lubricants.

At every stage in the handling of the equipment, the inspector will be looking also to the initials, the dates, of "doers" and "checkers."

The inspector will look for the chronologic and actual gap between completion of packaging and shipment. There must be some holding area or special warehouse at which such products will be kept quarantined until the quality control personnel have had a chance to check their results and give releases. In the case of controlled substances, there are extra considerations of which an FDA inspector will be aware, but which are more likely to be checked by state and Drug Enforcement Administration (DEA) officers. That is the question of accountability for finished goods tested by quality control and continuing security for the quarantined materials.

Similarly, with a view to continuing security in the case of controlled substances and in all cases with a review of procedures that move the finished product from the holding area to the finished goods (released for sale), shipment for sale area, the inspector will continue observation of product flow to the exit. Although the regulatory effects of the CGMPs have, for varied reasons, de-emphasized the importance attached to proof of interstate commerce, both inspectors and those inspected realize that articles segregated for filling a shipping order are held to be constructively in interstate commerce, even though they have not yet been placed in the carrier. The inspector will, therefore, not only observe the personnel and the manner in which they conduct themselves in this "exit" stage, but will also look for the same care and cover in protection and handling of articles earmarked for shipment, as looked for previously. It is not very sensible to exert much care in the manufacture of the product and then leave it to stand for an inordinate period, in an improper climate, exposed to elements and insecurity. The inspector will also be looking at paper, at distribution record keeping, and at accompanying invoices. For purposes of recall, the inspector will want to see that the distributor has recorded the customer's name, address, registry number if applicable, date, quantity of identified articles shipped, and the control or lot number thereof. The inspector will ascertain that they know and understand the 2-year record holding time requirements. The inspector will, no doubt, tie this in with a check into record keeping for products shipped out with an expiration date. The CGMPs call for records to be available for a period of time beyond that expiration date.

Having reached this point, and given sufficient time, the inspector can be expected to backtrack. Taking note of some article in the released area, he or she may go back to quality control, present them with the identification of the article including the lot number, and ask to see all the analytical data relating to the testing and release of the particular lot.

Checking the findings against the acceptance specifications established by the manufacturer or compendially for the article is an obvious opportunity to check on acceptability.

If some question exists or has been raised with the inspector previously as to verity, he or she may ask for the analyst's notebook, the raw data, and relate this to the figures provided. A look at the reviewer's signature or initials as well as those of different analysts who may have carried through various tests will provide the inspector with a reference point for further discussion, if needed. The reviewing inspector will not only look to results of testing, but will also look to methods used in testing for identity, stability, etc. While compendial tests are important for legal purposes, use by a manufacturer of other methods, including some that may be shorter or less expensive, but not as reliable, and failures in validation records will alert the inspector to different degrees of care. Similarly, for those products with established expiration dates, the inspector will want to see data indicating that adequate tests are being performed on products actually retained for the length of the expiration date's time period. It is an undesirable shortcut to base such data and findings on accelerated and otherwise contrived aging of the article, instead of on the product actually aged.

Certainly, the nature of the products, including their physical characteristics, may predetermine the kind of quality control equipment and methodology to be employed, but each inspector has a capacity to compare what one firm has and does by what number and quality of personnel, as it compares to another with a similar product output.

Again, usually close to the "exit" time in the traditional general inspection, the inspector will request a review of the manufacturer's complaint files. He or she may ask to see all those with respect to a particular product, from a particular date. The inspector is more apt to get cooperation in the latter request/complaint; files are required to be kept and maintained for all drug products under present CGMPs. However, especially for OTC drugs and cosmetics, actual access to them is not as clearly granted by the federal inspection laws. Among many larger producers, they are not kept in the manufacturing facility. The inspector is, however, clearly authorized to ascertain that the file exists and is maintained. Since the file can be opened for study by administrative subpoena, withholding free access to it may only gain time and principle for the regulatee, and that carries its own costs. Therefore, if the inspector requests, with particularity, he or she is not considered to be on a fishing expedition that may involve a look at complaints not within the FDA's primary concern, and will probably be shown that which he or she seeks.

COMPLAINTS

In examining the record keeping of the company or other distributor with respect to complaints, inspectors may remind them that the CGMPs require that the manufacturing firm shall maintain a record of all complaints, whether they are received in written form or orally communicated from some other source to the manufacturer. They are also bound to make an investigation of each complaint, and a record should be kept showing the results of that investigation. Of special interest to an inspector are those complaints that might indicate a formulation problem or some problem derivative of the product labeling and even those that might indicate unusual stability or decomposition. As a minimum, the company on receipt of a complaint should be able to show that it has examined reserve portions of the lot in question if it has been identified by the complainant, and if it has not been identified by the complainant, the company should be able to show that it has compared the complaint for the possibility of receipt of prior complaints on the same or related score. In any event, complaints of a medical nature are expected to be followed up in a great deal more depth. An appropriate evaluation of the complaint should be made by a physician on behalf of the complainant or the company, and perhaps both, and such reports should be in the file. Needless to say, it would be anticipated that in most instances an adverse action report would have been sent to the FDA and a copy of same would appear in the complaint file for the particular product.

One of the salutary changes in inspector and "inspected" relationships has been the growth of the mutual recognition of responsibility. Therefore, upon the completion of the inspection, the inspector will want to contact responsible management and officials. He or she will certainly desire to have top management "prepped" where there are important problems in the area of adulteration or misbranding that will require considerable investment in personnel, equipment, and remedial pressure to correct. Sometimes, management may represent various areas of the company structure. There may be different personnel there who head production and quality control. Certainly, the inspector wishes to be assured that responsible officials will be there to receive comments. It is at this time that the inspector will issue to them a list of the inspectional observations and discuss each point separately. The inspector will probably elaborate upon the circumstances under which the observation has been made, including references to named personnel who witnessed the activity. The inspector may take this opportunity to acknowledge corrections that were made at the time or welcome the opportunity to hear about corrections that may have been made subsequent to the observation. He or she will then ask for commentary from the management personnel present, to determine essentially in advance what their attitude will be. In short, do they intend to disagree with the findings and are they disturbed by the thought that changes will have to be made? Or, and of course this is preferable for the inspector, are they more apt to agree with the findings, even though they may offer explanations in mitigation of such findings, but yet acknowledge the need to improve and correct the situation. The inspector should always ask, and generally does, for a complete correction rather than a modification. The goal should be to educate for voluntary compliance, and at the same time to deter a repetitive violation. The inspector will, of course, carefully note the responses that are made to this presentation and incorporate specific quotations, as well as the general tenor of the response, within his or her report for each point made in the list. That essentially is a basic approach that the inspector will use for the average manufacturer of drug articles.

Very frequently, inspectors are called to investigate problems that are suspected as to various firms involved in the manufacture, distribution, and holding of prescription and over-the-counter (OTC) drugs. One such issue will occur as a follow-up to a consumer or competitor complaint. While the inspector will generally not release the name and address of the person who has complained to the FDA unless the inspector has received authorization to do so, he or she will seek to provide as detailed information as possible to the firm so it can really make a significant follow-up. Therefore, the inspector will try to determine from the consumer or competitor complaint what product or lot is involved rather than seeking to simply be concerned with every product or every lot, no matter what lot or control unit it was based upon. Certainly, the inspector will want to know whether the firm has received prior complaints on a certain product with that certain lot number. It is not surprising that in many instances the manufacturer has already been made aware of the complaint. The complaint may have followed a correspondence between the buyer and the manufacturer in which the buyer had asked for replacement expenses, or even medical expenses. In such a situation, often the inspector can proceed with work and review of what the manufacturer has already begun. In visiting the plant where a physical defect has been involved, the inspector may be expected to request a reserve sample from that lot to examine it for the defects that had been noted. The inspector will, of course, want to see the complaint file for that product, to see on a quantitative basis what involvement with similar problems has occurred. The incidence of complaints compared to the incidence of sale will provide a realistic ratio of concern when measured within a defined time period. However, this type of approach will not exclude the possibility that the public may be endangered. (For example, there is the classic case of the manufacturer who had sold millions of jars of product and noted only 40 or 50 complaints. Subsequent litigation indicated that the manufacturer had failed to sufficiently warn users.) Therefore, even where incidence is fairly small, the inspector must realize the misbranding potential, just as the manufacturer must recognize the product's liability potential.

Where there have been complaints of an adverse reaction or some other serious medical injury, the inspector will want to be able to contact all those involved in the initial prescription or dispensing of the drug, as well as those who examined and treated that person subsequently. If hospital records are available because the patient had emergency or other hospital treatment, these should be made available to the inspector. The individual who has complained has of course released for disclosure such information that might otherwise be deemed of as confidential, and therefore restrictive, nature. If the reaction has been alarming and significant in that it either quantitatively or qualitatively exceeded the kinds of reactions that appeared in that company's labeling, the firm should have notified the FDA with the promptitude required by the new drug reporting regulations found in 21 CFR 312.1 et seq.

The inspector should attempt to collect samples from the same lot of the product that has caused the complaint, if such is available in the plant warehouse. In some occasions, it may be necessary to purchase a sample from that lot from the retail establishment that provided it for the complainant. Another help along these lines will be the distribution record of the manufacturer, which may indicate the distribution of that lot and enable the inspector, either alone or with the cooperation of other inspectors, to collect a sample from that lot in some other outlet in that distribution scheme. If the matter warrants an extreme measure, the FDA may request a portion of the retention sample that was kept by the firm for its own testing purposes, or the FDA may request a portion of the sample that was perhaps provided by the complainant. Should the complaint seem both genuine and ominous, it may be anticipated that the FDA will start to look closely at similar drugs made by the same manufacturer or other manufacturers that use the active ingredient that seems responsible in the complaint.

A classic case, in which FDA inspectors found that a children's aspirin had become contaminated with diethylstilbestrol, which caused engorgement of the breasts of the children using those tablets, has led to general care on the inspector's part to determine what contaminants may have been involved in the manufacturing process of a drug that is the subject of a complaint. For example, an untoward allergic reaction from a drug that has no past history of such may be entirely due to the fact that the patient suffers from a penicillin allergy and there is penicillin contamination in the plant. The inspector therefore will press to examine the manufacturing logs to find out what products were made, perhaps on the same equipment, previously or on adjacent machines during the same processing times.

There are other situations in which the inspector may be called on to initiate an intensified inspection limited solely to determining the cause of the complaint with that product. In the event the complaint was due to mislabeling, as where the product was labeled with the name of another drug or with the wrong quantitative declaration, an in-depth review should be taken of the entire packaging and labeling operation in that plant that might be responsible for such a mix-up.

WARNING AS TO PREAPPROVAL INSPECTIONS (PAI)

In the course of the PAI, those inspected should recall that the policy of the FDA will consider criteria of importance to pharmaceutical manufacturers. If a company has attempted to subvert the integrity of the FDA's evaluation or review process through acts such as offering fraudulent submissions, bribes, or illegal gratuities, the agency must take steps to assess the integrity of the firm's marketed products as well as the data and information provided by that company in support of products submitted for approval. In such cases, the agency intends to conduct an investigation and audit to establish the extent to which the firm's illegal or unethical behavior may have affected approved or pending applications. The scope of such an investigation will be determined by the nature of the offense and will focus on the reliability of research and manufacturing information.

To approve an application, the FDA must determine that the applicant is capable of producing a safe and, in the case of some types of applications, effective product based on, among other things: (i) testing and other data provided by the applicant and (ii) the adequacy of the applicant's manufacturing processes and controls. A key element in making this determination is the reliability of data and information in the application.

When the FDA finds that the data in the application are fraudulent, the agency intends ordinarily to refuse to approve the application (in the case of a pending application) or to proceed to withdraw approval (in the case of an approved application), regardless of whether the applicant attempts to correct the falsification with a new submission in the form of an amendment or supplemental application. Thus, should the applicant wish to replace the false data with a new submission, the new submission must be in the form of a new application. The new application should identify the parts of the original application that were found to be false, and the accuracy of the application should be certified by the president, chief executive officer, or other most responsible for the firm's operation.

The FDA may also seek recalls of marketed products and request new testing of critical products. In the area of pharmaceuticals, for example, this would include products that are difficult to manufacture or that have narrow therapeutic ranges. The agency may pursue other actions under the FD&C Act or other applicable acts, including seizure, injunction, and criminal prosecution, as necessary and appropriate.

Firms engaging in fraud, material false statements, bribery, or proffering illegal gratuities will ordinarily need to take the following corrective actions to establish the reliability of data and information submitted to the FDA in support of pending applications, and to support the integrity of products on the market.

1. Cooperate fully with the FDA and other federal investigations to determine the cause and scope of any improprieties or problems related to safety, efficacy, or quality of products.
2. Identify all individuals involved in committing, or who are otherwise culpable in, the improper acts, or who have been convicted of an FD&C Act or related violation, and ensure that they are removed from any substantive authority on matters under the jurisdiction of the FDA.
3. Conduct a credible internal review in order to identify all instances of fraud, false information, or any other discrepancy in applications submitted to the agency or between conditions of approved applications and actual production. This review should involve an outside consultant who is qualified by training and experience to conduct such a review, and the results must be made available to FDA for independent verification. Such a review is intended to supplement the agency's ongoing comprehensive investigation to identify all instances of fraud and other improper conduct.
4. Commit, in writing, to an operating plan to assure the quality, safety, and integrity of its products. Such a commitment will ordinarily be in the form of a consent decree or agreement signed by the chief executive officer and submitted to the FDA. Such a plan will, as appropriate, address procedures to preclude future instances of fraud and noncompliance with regulatory requirements for approved applications as well as procedures to preclude any recurrence of other violations that may have been found (e.g., a comprehensive ethics program).

The FDA intends to inspect the firm and must be satisfied that the audit has been satisfactorily completed and that the firm's written plan has been satisfactorily implemented. Such inspections should disclose positive evidence (e.g., effective management controls, standard operating procedures, and corroborating documentation) that the firm's records are reliable and that the firm can be expected to manufacture products in compliance with CGMP and application requirements.

The firm may also be requested under existing regulatory procedures to recall products affected by false information or that otherwise lack adequate assurance of safety of quality. In addition, it may be requested to commit in writing to such retesting of any products (including, in the case of drugs, bioequivalence-bioavailability retesting) as the FDA may call for.

PRODUCT LIABILITY

With the burgeoning worldwide expansion of product liability litigation has come another force to add significance to establishment and maintenance of CGMP.

Through broadly permitted discovery procedures, today's manufacturer is subject to review by other than governmental personnel. Frequently, adversaries in litigation seeking to substantiate claims based on alleged defective products use the services of persons who are expert and conversant with CGMP in the hope of finding tangible evidence of such manufacturing or quality control features as would cause the product to be defective and dangerous to the consumer. This is especially true in the case of biological drugs, vaccines, medical devices, and their components. Thus, there exists a strong economic and potentially debilitating force to wreak havoc upon the careless or the unconscionable.

SPECIAL PROBLEMS ENCOUNTERED BY INSPECTORS

- After the inspector arrives, he or she is told to come back another time because the person in charge, or owner or operator is not at present there. Besides the obvious comment that any plant or facility in operation must be under supervisory guidance, the agent will ordinarily firmly advise that the person in charge is expected to permit the inspection, that whoever is refusing in the name of the proprietor may call counsel or owner or anyone else for advice in the matter, and that a refusal of entry will be noted whether or not that refusal is by a manager, owner, or others in charge.
- After the inspector arrives, he or she is advised to accompany a person delegated by management. This does not mean the inspector is to go on a tour. It means the inspector will go about the inspection, starting at the point deemed necessary, and that the facility person will accompany the inspector.
- After the inspector arrives, he or she is kept waiting for an inordinately long time. That is both discourteous and mutually unproductive. The regulatee should be advised, after a reasonable time, that the inspector will note it as a refusal of entry.
- After the inspector arrives, he or she is advised that there are no prescription drugs made there and therefore the inspection should be limited. While it is true that Section 704 expresses, by reciprocal observation, that certain limitations exist since the special prerogatives as to prescription drugs are set out, the area is not clearly a group of "do's" and "don'ts."

Whether the establishment does or does not handle Rx items, the inspectional procedure is fortified for both OTC and Rx articles, since the CGMPs apply equally to both. Extended as this now is for devices, and likely to extend to cosmetics in the future, one can see that inspectors have procedural guidelines for many articles, to determine whether the processes, facilities, and controls conform to the regulations. However, if a manufacturer declines to show the inspector the formula records for an OTC product, the manufacturer is doing that which is a legal entitlement under the law. The manufacturer may thus likewise refuse to allow copying of a formula for an OTC. Yet, a compromise can be reached where the inspector for CGMP purposes will be allowed to compare master formula records against corresponding batch records and finished products labeling, for errors in transcription, clarity of photocopy or other reproduction used by the plant, accuracy of calculations in transposition, and completeness of label information as stated in the CGMPs. Some means to compare actual to theoretical yields and to explore discrepancies is necessary for Rx drugs, OTCs, and new and old drugs.

- After the inspector has arrived and commenced the inspection, the regulatee refuses to allow access to particular areas or files or documents. That is viewed as a partial refusal at the least, but the reasonable inspector will neither threaten nor haggle. The inspector will go ahead with the balance of the inspection and describe and incorporate the circumstances of the denial to his or her supervisor. After that, the office will decide whether it needed that which was denied, and usually an administrative warrant will be then obtained. Sometimes, a phone call may make that unnecessary, but officials feel it is best to procure the warrant and use it when a second chance at voluntary access is refused. For many reasons, the inspector will prefer to gain access without a warrant. This can obviate later admissibility problems. But it must be voluntary, rather than coerced or misled. It is conceivable that refusal to permit entry for inspection may be prosecuted criminally, but that does not mandate inspection. In usual practice, on federal and state level, the refusal is followed by obtaining a warrant, and is added to whatever charges that arise from the inspection subsequent to the warrant.

Warrants can, of course, be challenged on diverse legal grounds, and that is true of administrative warrants also, but to a far lesser extent.

When a regulatee refuses to honor a warrant, that becomes a singular contemptuous act and makes the regulatee subject to swift punishment on that basis alone. However, inspectors should give the regulatee an opportunity to check with counsel or the home office or any other advisor prior to deeming the warrant dishonored.

GMP INSPECTION: PRACTICAL CONSIDERATIONS

Some of the legal issues associated with an FDA inspection have been addressed. In this section, some key practical points will be presented.

First, it should be accepted that the FDA inspection does provide a useful independent review of plant good manufacturing practice (GMP) activities. In an effectively managed plant, self-audit and QA audit (see Chapter 7) should have identified any significant shortfalls with respect to systems and compliance and have resulted in correction. The FDA inspection should not therefore identify any unknown deviations; if it does, management needs to relook at the way it operates and not just respond to any citations.

It also should be acknowledged that the FDA inspector is only doing a job and that a constructive working relationship should be more productive to both the FDA and the company. On arrival at the plant, the inspector should be asked to confirm his or her identity and be introduced to the key plant personnel. Any specific interests of the inspector should be identified and also the anticipated duration of the inspection. An ongoing straightforward relationship will facilitate this. These points allow management to program their time and, where appropriate, production operations, to better meet the requirements of the inspector and to minimize business disruption due to the involvement of managers in the inspection.

It is recommended that at least one individual (and a backup) be designated to lead the inspection on behalf of the company. This individual can significantly influence the tone of the inspection and the opinion of the inspector about the company's attitude to CGMP compliance. The designated individual must be knowledgeable on:

1. CGMP regulations
2. Previous FDA inspections
3. Company policies and procedures
4. Detailed operation of the facility
5. Key people within the facility
6. Quality issues over the past two years (statutory inspection frequency is every two years), but plant inspections may have been more frequent. That is usually a warning they will continue to be more frequent.
7. Any ongoing atypical activities in the facility-upgrading, expanding, decorating, new equipment

The designated person should:

1. Accompany the inspector at all times and act, where necessary, as an intermediary in the clarification of questions and answers. A "don't know" answer is preferable when there is uncertainty, provided there is follow-up to obtain an answer in a timely manner.

2. Make notes of all points raised by the inspector regardless of whether or not they are likely to appear in the inspector's written report or to result in a citation. These notes should be documented and communicated
    - for daily review with senior plant managements;
    - for identification of points requiring immediate action (this also demonstrates commitment to the inspector);
    - to highlight points requiring further elaboration or clarification with the inspector; and
    - to integrate into a composite action plan after the inspection.
3. Coordinate the assembly of requested documentation.
4. Notify the inspector of any company policies that impact on the inspection, such as the wearing of protective clothing, medical screening before entering the sterile suite, and taking of photographs.

In addition to the knowledgeable company escort and note-takers, many firms set up a specific room sometimes called a "war room," that contains relevant files and photocopying equipment. The FDA-requested documents are assembled in this area, and employees are dedicated 100% of the time to provide timely retrieval. Any document provided to the FDA should be reviewed by responsible company personnel before handing it over. This will help spot issues before the FDA finds them and provides the firm time to gather additional explanatory information where needed.

There are important considerations during the conduct of an FDA inspection or any operational audit.

Answering questions:
- Listen carefully and repeat back to the investigator/auditor to confirm understanding
- Do not read into the question and only volunteer information to clarify or correct

Words to avoid:
- Usually, hopefully, I think
- Ignored, disregarded, deleted
- Contaminated, failed, failure
- Tweaked, for information only

Investigator/auditor dislikes:
- Indirect answers, evasiveness
- Long delays in getting data
- Having to ask for items repeatedly
- Having to ask multiple people to get an answer
- Lack of action/re-action when a problem is found
- Dishonesty

Investigator/auditor likes:
- Preparedness and quick responses to requests
- Individual and upper management awareness and response (especially, when a problem is found)
- Familiarity with policies and procedures
- Summary reports for quick and easy review

An FDA inspection ends with an exit interview where the inspector presents comments and, where necessary, issues an FDA 483, identifying observed deviations from the CGMP regulations. The exit interview provides a further opportunity to clarify any misunderstandings or misinterpretations. In many instances, an inspector is prepared to take the comments from the exit interview into account when drafting the FDA 483. Any actions already implemented would be noted at this time. Since any proposed deviation may be invalid following retrospective review, the exit interviewee should be courteous, not adversarial, but not necessarily apologetic or a "breast beater." Give the inspector should be given every indication that careful review will be made of the helpful observations.

After departure of the inspector, management should review the overall impact of the inspection, not just the FDA 483 citations. Programs should be initiated to identify the basic causes of any FDA-noted deviations that stand up after internal review, and actions should be promptly implemented to correct these causes. Many, if not most, deviations are caused by inadequate attention from management and supervision, which sends mixed signals to other employees.

Most companies prepare a formal response to an FDA 483. This is an important document with possibly far-reaching results and implications; therefore, it should be done with care and foresight. This will indicate any corrective actions being taken, usually with defined time frames. Where management does not agree with a citation, they should provide comprehensive reasons for the disagreement. When the FDA inspection is a Pre-Approval Inspection (see Chapter), this approach may need to be modified. Such disagreements have frequently resulted in recommendations to not approve the applications (NDA, ANDA, and Supplement). In these circumstances, the potential adverse business impact on the decision has "encouraged" companies to accept the FDA citation and to "implement appropriate, and sometimes unjustifiable actions." Unfortunately, this can then create the position that this is now considered by the FDA to be "Current GMP." Regardless of this stressed situation, companies should continue to apply sound science and professionalism in the interpretation and compliance with the CGMPs. Every citation point should be addressed. In drafting the response, it should be remembered that it will be read by the inspector's supervisor and consequently sufficient detail should be included to allow the supervisor to clearly understand the response. A useful approach is to have the draft plant response reviewed by a function or functions outside of the plant, such as Group or Corporate. This tends to ensure greater clarity. The response will take some time to assemble if it is to include time scales for actions. However, the speed of response also sends signals to the FDA that indicate the company's attitude to quality; consequently, the response should be seen as a high-priority item. In the event of a potential delay in responding, it may be advisable to indicate this to the FDA, along with the reason.

Under normal circumstances, it should be possible with an inspection of 5 to 10 days to ascertain whether a facility is in a general state of compliance. Longer that this usually indicates it is a hunt to find deviations. With the current risk-based approach to inspections, the investigators are generally spending less time in a firm but will extend the plan if and when significant deviations are found.

In 1994, the Office of the U.S. Trade Representative and the Department of Commerce initiated discussions with the European Union (EU) on the subject of mutual recognition of inspections. Initial progress was slow, largely because of the differences in inspections between EU member states. However, discussions are continuing with the FDA requiring equivalence between the EU member states and the FDA, the right to inspect, periodic joint audits, and access to EU inspection reports. Work will also continue toward harmonization of GMPs, but not through the auspices of International Committee on Harmonization (ICH).

In addition to the exit interview and the FDA 483 (where issued), the inspector writes a comprehensive report of the inspection, the Establishment Inspection Report (EIR), which includes formal data on the company and its organization, points noted during the inspection, including those not appearing on a Form 483, and usually opinions on the attitudes of the plant personnel. Copies of an EIR are available through Freedom of Information.

Occasionally, proprietors suspect that inspections may be a cooperative enterprise between federal agencies, whether on an institutional or individual agent basis. There are interesting overlaps, for example, between OSHA and FDA inspections, and the comparatively newer inspection techniques of the former go somewhat beyond the latter. Photography and videotaping, use of employees to wear sampling devices, etc. are more frequent in areas of OSHA activities, and have been upheld by the courts.

The impropriety of having inspectors for one agency inspect and report to another have been discussed in the past, and if there is a substantial basis for suspicion of such conduct, the regulatee should consult with counsel for his or her association, or his or her own counsel, promptly.

For the reader's interest, see the suggested preparation for and management of an OSHA inspection. See also the Barlow case in the Appendix to this chapter and read in the context of such suggestions.

In the wake of successful prosecution by federal officials in a case that received much publicity and attention from the public, the Executive Branch, and Congress, and FDA officials circulated the following letter and attachments. These will certainly be influential with inspectors, although some of the positions taken may see argument in the future (see Appendix).

SUGGESTED READING

1. Program Guidance Manual for FDA Staff Drug Manufacturing Inspections Program 7356.002 February, 2002.

Following are relevant FDA Guidance documents to the subject of this chapter.

SEC. 300.100 INSPECTION OF MANUFACTURERS OF DEVICE COMPONENTS (CPG 7124.15)

Background:

*Section 510(h) of the Federal Food, Drug, and Cosmetic Act declares that all registered firms are subject to inspection pursuant to Section 704.

Some manufacturers have been confused by 21 CFR 807.65, believing that exemption from registration also exempts them from inspection. This is not true. As defined under Section 201 (h) of the Act, devices include components of devices, making manufacturers of device components subject to the provisions of Section 704. Title 21 CFR 807.65(a) exempts manufacturers of medical device components from the registration and listing provisions of Section 510 of the Act, if those components are the only items the manufacturer produces which have health care applications and they are sold only to other manufacturers. The exemption does not apply to manufacturers of components described in 21 CFR 807.20(a)(5) unless they are marketed only to registered device establishments for further processing. The exemption applies only to registration and listing.*

Policy

* Exemption from registration under 21 CFR 807 does not exempt the manufacturer of device components from inspection under Section 704 of the Act.
* All manufacturers of device components are subject to inspection under Section 704 of the Act.*

* Material between asterisks is new or revised.*
Issued: 7/29/77
Reissued: 10/1/80
Revised: 9/24/87

SUB CHAPTER 130 INSPECTIONS

SEC. 130.100 INSPECTIONAL AUTHORITY; REFUSAL TO PERMIT INSPECTION (CPG 7151.01)

Background:

The authority for duly appointed officers or employees of the Food and Drug Administration (FDA) to enter and inspect establishments under the jurisdiction of the Federal Food, Drug, and Cosmetic Act is in Section 704 of the Act (21 U.S.C. 374).

Questions concerning the right to inspect such establishments have often been raised and litigated. The courts have upheld the legality of an FDA inspection if it is conducted at a reasonable time, within reasonable limits and in a reasonable manner.* Consent is not the basis upon which a Food and Drug inspection is conducted, and permission or authorization to inspect is not required from the firm to be inspected.

The Federal Food, Drug, and Cosmetic Act provides criminal penalties for refusal to permit a lawful inspection.

Policy:

The legality of an FDA inspection, conducted at a reasonable time, and within reasonable limits, and in a reasonable manner, depends not on consent but on the validity of statutory authority. An inspection warrant is not a prerequisite to lawful inspection pursuant to such authority. Refusal to permit inspection, upon presentation of official notice by appropriately identified FDA officers or employees pursuant to 21 U.S.C. 374, exposes any person responsible for such refusal to criminal penalties under 21 U.S.C. 331(f) and 333.

* United States v. Biswell, 92 S. Ct. 1593 (1972)
United States v. Del Campo Baking Mfg. Company, 345 F. Supp. 1371 (D., Del., 1972)
United States v. Business Builders, Inc., 353 F. Supp. 1333 (N.D. Okla., 1973).
Issued: 10/1/80

SEC. 130.200 INSPECTION OF FIRMS WHEN LEGAL ACTION IS PENDING (CPG 7153.01)

Background:

Inquiries from the field have indicated that there is some confusion on whether or not to reinspect a firm while legal action is pending against that firm.

Policy:

Reinspection of a firm should be based upon public health considerations. The FDA has an obligation to determine compliance with the law even if a case is pending, and if on reinspection further violations are found, to take additional steps as necessary to bring about correction.

It must be clearly understood that cessation of a violation is not grounds for dismissal of a case. Prosecution actions particularly are based on violations that have already occurred, and nothing that takes place after the violation changes that fact.

Where a court requests, reinspection is also appropriate. A district should always be in a position to furnish the court with current information covering the defendant's operations. This does not mean, however, that the FDA should perform an inspection of each firm just prior to arraignment or trial. As already indicated, unless a court requests an inspection, reinspection at that time is based upon public health considerations in light of priorities and available manpower.

Issued: 12/3/73
Revised: 10/1/80, 8/31/89

SEC. 130.300* FDA ACCESS TO RESULTS OF QUALITY ASSURANCE PROGRAM AUDITS AND INSPECTIONS* (CPG 7151.02)

Background:

* Within all FDA-regulated industries, some firms establish quality assurance units (QAUs) to perform functions independently from the manufacturing or quality control organization. The QAU may periodically audit and critically review processes and procedures (e.g., data collection, manufacturing practices, and quality control processes) to determine whether established protocols and procedures have been followed.

In the preambles to the final regulations on Good Manufacturing Practice for Medical Devices (43 FR 31508; July 21, 1978) (21 CFR 820) and on Good Laboratory Practice for Nonclinical Laboratory Studies (43 FR 59986; December 22, 1978) (21 CFR 58), the FDA announced its policy not to review or copy a firm's records and reports that result from audits of a quality assurance program when such audits are conducted according to a firm's written quality assurance program at any regulated entity. The intent of the policy is to encourage firms to conduct quality assurance program audits and inspections that are candid and meaningful.

Policy:

*During routine inspections and investigations conducted at any regulated entity that has a written quality assurance program, the FDA will not review or copy reports and records that result from audits and inspections of the written quality assurance program, including audits conducted under 21 CFR 820.20(b) and written status reports required by 21 CFR 58.35(b)(4).

The FDA may seek written certification that such audits and inspections have been implemented, performed, and documented and that any required corrective action has been taken. District personnel should consult with the appropriate headquarters office prior to seeking written certification.

The FDA will continue to review and copy records and reports of such audits and inspections:

1. in "directed" or "for-cause" inspection and investigations of a sponsor or monitor of a clinical investigation;
2. in litigation [for example and not limited to: grand jury subpoenas, discovery, or other agency or Department of Justice law enforcement activity (including administrative regulatory actions)];
3. during inspections made by inspection warrant where access to records is authorized by statute; and
4. when executing any judicial search warrant.

The FDA will continue to have access to, review, and copy records and reports required by regulation, relating to quality control investigations of product failures and manufacturing errors.*

* Material between asterisks is new or revised.*
Issued: 03/01/83
Revised: 03/16/89
Revised: 06/03/89
Revised: 01/03/96

SEC. 130.400 USE OF MICROFICHE AND/OR MICROFILM FOR METHOD OF RECORDS RETENTION(CPG 7150.13)

Background:

The agency has received many questions concerning the use of microfiche and/or microfilm systems in lieu of the retention of original records. This Compliance Policy Guide is based on a May 11, 1979 response to a request for an Advisory Opinion on this subject (Docket Number 77A-0270).

Policy:

The Food and Drug Administration has* published several regulations that permit the maintenance of certain recordkeeping systems in lieu of the retention of original records: good manufacturing practices for medical devices (43 FR 31508, July 21, 1978); good manufacturing practices for human and veterinary drugs (43 FR 45014, September 29, 1978); and nonclinical laboratory studies (43 FR 59986, December 22, 1978). These regulations include the use of microfiche and/or microfilm. We therefore conclude that the utilization of a microfiche and/or microfilm reduction system in lieu of the retention of original pre-clinical, clinical, and related drug and medical device research records, and drug and medical device quality control and manufacturing records, is acceptable.

The preambles to these regulations, and the regulations, discuss the conditions applicable to the maintenance of reduction systems. These include the following:

1. All records must be readily available for review and copying by FDA investigators at any reasonable time.
2. All necessary equipment must be provided to facilitate viewing and copying of the records.
3. A reproduction must be a true and accurate copy of the original record. Thus, where the reproduction process results in a copy that does not reveal changes or additions to the original record, the original must be retained.

Also, the reproduced copy and any image shown on a viewing screen must note, in a suitable manner, that an alteration has been made and that the original record is available.

Material between asterisks is new or revised.
Issued: 6/19/79
Revised: 10/1/80, 8/31/89

SEC. 150.100 REQUESTS FOR PORTIONS OF INTERMEDIATE OR END PRODUCTS RESULTING FROM FDA SAMPLE ANALYSIS (CPG 7150.18)

Background:

The FDA occasionally receives requests for microbiological cultures isolated from samples analyzed by the FDA. Requests for other entities isolated, extracted, or produced by sample analysis, that is, chemical isolates, extracts, filth debris, etc., may also be received, especially regarding consumer complaint samples.

The Federal Food, Drug, and Cosmetic Act, makes no provision for the FDA to provide to requesters portions of end or intermediate products resulting from FDA sample analysis. Section 702(b) of the Act provides that, upon request, a part of an official sample of a food, drug, or cosmetic will be provided for examination or analysis to any person named on the label, the owner of the sampled product, or his attorney or agent. This section of the act applies to portions of the sampled commodity. It *does not* apply to intermediate or end products resulting from sample analysis. Portions of intermediate or end products resulting from FDA sample analyses will not be routinely provided to requesters from outside the agency, including consumers from whom samples have been collected as part of the complaint investigation.

Exceptions to this policy may be considered when the agency determines that providing portions of intermediate or end products to the requester would help resolve a serious public health matter or would benefit the public well-being. When a request appears to warrant such consideration, the request should be referred to the Office of Compliance within the appropriate center for review.

Issued: 3/23/88

ANALYTICAL METHODOLOGY USED BY FDA-DRUGS (CPG 7152.01)

Background:

There have been continuing problems concerning the appropriate analytical methodology used by FDA laboratories in support of regulatory actions. In several cases, regulatory actions have been disapproved and much analyst time wasted because the analyst did not adhere to the appropriate analytical method. Where regulatory actions are predicated upon analytical findings, the appropriate methods are generally those stated in the United States Pharmacopeia

(USP)/National Formulary (NF), an NDA, or a firm's Standard Operating Procedure, as applicable.

Policy:

Where FDA sample analysis is a basis for regulatory action, only the following procedures are considered appropriate, unless specific instructions to the contrary are given by the Center for* Drug Evaluation and Research.*

1. For official drugs (USP/NF), the official compendial analytical methods are to be used, unless the FDA has promulgated regulations under Section 501(b) (or, for antibiotics, Section 507) of the Act prescribing appropriate tests or assay methods, in which case the regulations are to be followed.
2. A non-official drug that is the subject of a new drug application is to be analyzed by the method in the NDA or ANDA.
3. A non-official drug, is not a new drug, is to be analyzed by the method used by the manufacturer as part of its standard operating procedures. If the FDA analyst has concern over the validity of the unofficial method, those concerns should be documented.
4. When analyzing a product by any of the above methods, the method must be strictly followed.
5. When a drug is not covered by the above situations, the analyst may select an appropriate method with which to analyze the product. In selecting the method, first consideration should be given to any existing AOAC method because AOAC methods have withstood the rigors of collaborative study. Any method selected must have been properly validated. If not previously validated, it must be validated when it is used. Validation data must be submitted with worksheets when regulatory action is recommended.
6. When the compendial, NDA, or firm's method is not satisfactory (e.g., due to an interfering substance, non-reproducible method, etc.), then this should be reported to the Division of Manufacturing and Product Quality (HFD-320) for further follow-up and guidance.

For surveillance samples such as those collected during a multiple drug survey, the laboratory may substitute a validated non-official method for the original analysis. However, any out-of-limit results must be confirmed by check analysis using the official or other appropriate method.

* Material between asterisks is new or revised.*
Issued: 7/1/81
Revised: 9/1/86, 3/95

APPENDIX: A JUDICIAL INTERPRETATION OF INSPECTION REQUIREMENTS

May 14, 1993
Dear Colleague:

Judge Wolin's interpretations of Good Manufacturing Practice (CGMP) issues contained in the court's ruling on 2/4/93 in *USA v. Barr Laboratories* is being forwarded for your information. This summary has been sent to Food and Drug Administration field offices and to the inspection cadres throughout the field for their use during drug GMP and pre-approval inspections.

Therefore, we felt it was important information to share with you with regard to future GMP and pre-approval inspections. You may wish to share this document with others in your organization as well.

We will continue to provide information to you that we feel may be useful. Please let us hear from you if there is a particular service or information you would like from the Food and Drug Administration. I can be reached at 301/443-6776 or fax: 301/443-5153.

Sincerely yours,
Mary Ann Danello, Ph.D.
Director, Office of Small Business,
Scientific and Trade Affairs

Enclosure

Judge Wolin's Interpretations of GMP Issues Contained in the Court's Ruling in USA v. Barr Laboratories (2-4-93)

1. United States Pharmacopeia Standards

The court ruled that USP's established standards are absolute and that firms cannot stretch the USP standards. These standards provide established criteria upon which firms release their product.

2. Failure (Out-of-Specification) Laboratory Results

Judge Wolin preferred to use the term "out-of-specification" (OOS) laboratory result rather than the term "product failure," which is more common to FDA's investigators. He ruled that an OOS result identified as a laboratory error by a failure investigation or an outlier test[b], or overcome by retesting[c] is not a product failure. The OOS results fall into three categories:

- laboratory error
- non-process-related or operator error
- process-related or manufacturing process error

A. Laboratory Errors

Laboratory errors occur when analysts make mistakes in following the method of analysis, use incorrect standards, and/or simply miscalculate the data.

Judge Wolin provided specific guidance about the matter of determining when an error can be designated a laboratory error. Laboratory errors must be determined through a failure investigation to identify the cause of the OOS. Once the nature of the OOS result has been identified, it can be classified into one of the three categories mentioned earlier. He states that the inquiry may vary with the object under investigation.

B. Laboratory Investigations

The court said that the exact cause of analyst error or mistake can be difficult to pin down and that it is unrealistic to expect that analyst error will always be determined and documented, and he ruled that the "laboratory investigation consists of more than a retest." The inability to identify an error's cause with confidence affects retesting procedures, not the investigation inquiry required for the initial OOS result.

The analyst should follow a written procedure, checking off each step as it is completed during the analytical procedure; laboratory test data must be recorded in notebooks; use of scrap paper and loose paper is to be avoided. These measures enhance the investigation process.

---

[b] The court provided explicit limitation on the use of outlier tests and these are discussed in a later segment of this document.
[c] The court ruled on the use of retesting, which is covered in a later segment of this document.

The court specifically identified procedures that must be followed when *single*, and *multiple* OOS results are investigated.

For the *single* OOS result, the investigation must include the following steps and these inquiries must be conducted before there is a retest of the sample:

1. The analyst conducting the test must report the OOS result to the supervisor.
2. The analyst and the supervisor must conduct an informal laboratory inspection, which addresses the following areas.
   - Discuss the testing procedure
   - Discuss the calculation
   - Examine the instruments
   - Review the notebooks containing the OOS result

An alternative means to invalidate an initial OOS result, provided the failure investigation proves inconclusive, is the "outlier" test. The court placed specific restrictions on the use of this test.

1. Firms cannot frequently reject results on this basis.
2. The USP standards govern its use in specific cases.
3. The test cannot be used for chemical testing results.[d]

A full-scale inquiry is required for *multiple* OOS results. This inquiry involves quality control and quality assurance personnel *in addition to laboratory workers* to identify exact process or non-process-related errors.

The court ruled that when the laboratory investigation is inconclusive (reason for the error is not identified):

1. Cannot conduct two retests and base release on average of three tests
2. Cannot use outlier test in chemical tests
3. Cannot use a resample to assume a sampling or preparation error
4. Will allow a retest of different tablets from the same sample when a retest is considered appropriate (see criteria elsewhere)

C. Formal Investigations

Judge Wolin ruled that formal investigations extending beyond the laboratory must follow the government's outline with particular attention to corrective action. He said the company must:

1. State the reason for the investigation
2. Provide summation of the process sequences that may have caused the problem
3. Outline corrective actions necessary to save the batch and prevent similar recurrence
4. List other batches and products possibly affected, the results of investigation of these batches and products, and any corrective action. Specifically:
   - examine other batches of product made by the troublemaking employee or machine
   - examine other products produced by the troublemaking process or operation.
5. Preserve the comment and signature of all production and quality control personnel who conducted the investigation and approved any reprocessed material after additional testing

---

[d]An initial content uniformity tests was OOS followed by a passing retest. The initial OOS result was claimed the result of analyst error based on a statistical evaluation of the data. The use of outlier test is inappropriate in this case.

D. Investigation Documentation

Analyst's mistakes, such as calculation errors, should be specified with particularity and supported by evidence. Investigations along with conclusions reached must be preserved with written documentation that enumerates each step of the review in the form of a "computer-generated flow sheet." This writing should be preserved in an investigation or failure report and placed into a central file.

E. Investigation Time Frames

All failure investigations must be performed within 30 business days of the problem's occurrence and recorded and written into a "failure or investigation report."

F. Product Failures

An OOS laboratory result can be overcome (disregarded) when laboratory error has been documented. However, non-process and process-related errors resulting from operators making mistakes, equipment (other than laboratory equipment) malfunctions, or a manufacturing process that is fundamentally deficient, such as an improper mixing time, represent product failures.

3. Retesting

Several opinions about retesting were issued in this decision. The number of retests performed before a firm concludes that an unexplained OOS result is invalid or that a product is unacceptable is a matter of scientific judgment. The goal of retesting is to isolate OOS results, but retesting cannot continue ad infinitum.

In the case of non-process and process-related errors, retesting is suspect. Because the initial tests are genuine, in these circumstances, additional testing alone cannot infuse the product with quality. The court acknowledges that some retesting may precede a finding of non–process or process-based errors. Once this determination is made, however, additional retesting for purposes of testing a product into compliance is not acceptable.

For example, in the case of content uniformity testing designed to detect variability in the blend or tablets, failing and nonfailing results are not inherently inconsistent and passing results on limited retesting do not rule out the possibility that the batch is not uniform. As part of the investigation, firms should consider the record of previous batches, since similar or related failures on different batches would be a cause of concern.

A very "important" ruling in this decision sets forth a procedure to govern the retesting program. The judge ruled that a firm should have a predetermined testing procedure and it should consider a point at which testing ends and the product is evaluated. If results are not satisfactory, the product is rejected.

Additionally, the company should consider all retest results in the context of the overall record of the product[e], type of test performed, and in-process test results. Failing assay results cannot be disregarded simply on the basis of acceptable content uniformity results being satisfactory.

Retesting following an OOS result is ruled appropriate only after the failure investigation is underway and the failure investigation determines in part whether retesting is appropriate. It is appropriate when analyst error is documented or the review of analyst's work is "inconclusive," but it is not appropriate for non-process or process-related errors.

The court ruled that retesting:

- must be done on the same, not a different sample
- may be done on a second aliquot from the same portion of the sample that was the source of the first aliquot
- may be done on a portion of the same larger sample previously collected for laboratory purposes

---

[e] The court ordered a recall of one batch of product on the basis of an initial content uniformity failure and no basis to invalidate the test result, and on a history of content uniformity problems with the product.

4. Resampling

Firms cannot rely on resampling[f] to release a product that has failed testing and retesting unless resampling is in accord with the USP standards (content uniformity and dissolution), or unless the failure investigation discloses evidence that the original sample is not representative or was improperly prepared.

5. Averaging Results of Analysis

Averaging can be a rational and valid approach, but as a general rule, this practice should be avoided[g] because averages hide the variability among individual test results. This phenomenon is particularly troubling if testing generates both OOS and passing individual results, which when averaged are within specification. Here, relying on the average figure without examining and explaining the individual OOS results is highly misleading and unacceptable.

Content uniformity results never should be averaged to obtain a passing value for content uniformity.

In the case of microbiological assays, an average is preferred by the USP. Also, the Judge ruled that it is good practice to include OOS results in the average, unless an outlier test (microbiological assays) suggests the OOS is an anomaly.

6. Remixing

The need to remix often is clear indication that the process is invalid and casts doubt on those batches that passed through testing without incident.

Remixing is reworking permitted under the GMP regulations. Occasional remixing is acceptable, but frequent or wholesale remixing is unacceptable.

7. Product Release

Scientific judgment can play a role when firms decide to release a batch to the public and the court said it cannot articulate specific procedures for release decision making. However, Judge Wolin said that the USP standards upon which firms release their products are absolute and cannot be stretched. For example, a limit of 90% to 110% of declared active ingredient, and test results of 89, 90, 91, or two 89s and two 92s all should be followed by more testing.

It is clear that the release evaluation depends in part on the background of the batch and product.

Secondary factors that affect the actual finished product results as well as their reliability are:

- physical properties;
- blend evaluations;
- time of mix; and
- tablet weight, thickness, and friability.

Judge Wolin ruled that context and history[h] inform many final conclusions and that one must consider past problems with the product and batch and evaluate all the data relative to the product and batch.

---

[f] The court ordered the recall of one batch of product after having concluded that a successful resample result alone cannot invalidate an initial OOS result.

[g] The court ruled that the firm must recall a batch that was released for content uniformity on the basis of averaged test results.

[h] On the basis of an initial content uniformity failure and no basis to invalidate the test result and on a history of content uniformity problems with the product, the court ordered the batch recalled.

8. BLEND TESTING

Blend testing is necessary to increase the likelihood of detecting inferior batches. Blend content uniformity testing cannot be waived in favor of total reliance on finished product testing because finished product testing is limited.

The court ruled that sample size influences ultimate blend test results and that the sample size should resemble the dosage size. Any other practice would blur differences in portions of the blend and defeat the object of the test. *The appropriate sample size for blend content uniformity in both validation and ordinary production batches is three times the active ingredient dosage size.*

Multiple individual samples taken from different areas cannot be composited. However, when variation testing is not the object of assay testing, compositing is permitted.

Firms must demonstrate through validation that their sampling technique is representative of all portions and concentrations of the blend. This means that the samples must be taken from places that might be problems, weak or hot spots in the blend.

In this case, the firm maintained that samples could be collected from the drums containing the finished blend. The court ruled that the firm must demonstrate that sampling from drums rather than the mixer is representative. He also ruled that the firm cannot composite blend samples and that they must take smaller blend content uniformity samples.

9. Validation Criteria

A. Retrospective Validation

The court ruled that batches meeting the following criteria must be included in retrospective validation studies:.

1. All batches made in the specified time period chosen for study must be included unless the batch was made from a non-process-related error.
2. Only batches made in accord with the process being evaluated can be included.

Only test results determined through an appropriate failure investigation and found to be caused by analyst or operator error can be excluded from the study. Test results that are explained but merely called into question by successful retesting must be included in the study. *The exclusion of batches and test results must be documented through failure investigation.*

The number of retrospective batches chosen for the study must be greater than the number used for prospective validation. Although the court set no exact number of batches to be chosen, guidelines have been established as follows:

- Five batches is unacceptable and also six or more may not be acceptable.
- Because a 10% batch failure is unacceptable, if one batch fails, more than 10 batches are needed for the retrospective study.
- Experts accept 20 to 30 batches.

B. Concurrent and Prospective Validation

Concurrent and prospective validation requires at least three consecutive batch runs of the process.

Mining time studies should be included in a prospective validation program and follow any problems that surface in retrospective validation batches.

Particle rise distribution specifications are widely accepted industry practice and should be included in validation studies.

10. Significance PF Application Approvals

The court ruled that a firm cannot rely on the claim that the FDA previously approved their procedures contained in an approved application. This approval cannot be used as a defense and cannot be used to shield a process that produces failures.

11. Methods Validation

Methods can be validated in a number of ways. Methods appearing in the USP are considered validated and they are considered validated if part of an approved ANDA. Also, a company can conduct a validation study on their method. System suitability data alone is insufficient for method validation.

12. Cleaning Validation

The court ruled that a firm cannot wait for contamination and other problems to reveal inadequate cleaning procedures. In order for the cleaning rules to be effective, the specific methods chosen must be shown to be effective.

The court ruled that a milling machine is a major piece of equipment and must be included in the cleaning validation program.

Firms must identify the cleaning agents used in its cleaning process. When these agents are known to cause residue, the company must check for the residue.

Provided the firm has described its cleaning methods and materials in sufficient detail and unless the cleaning material is known to cause a residue, then one run through of the cleaning procedure, in the absence of problems, is not insufficient for validation.

EFFECTIVELY MANAGING OSHA OCCUPATIONAL SAFETY AND HEALTH ADMINISTRATION INSPECTIONS

Introduction

How an employer prepares for and manages an inspection conducted by the Occupational Safety and Health Administration (OSHA) can go far to minimizing the severity of any citation that may issue in connection with that inspection. This article will discuss the basic steps in an OSHA inspection and some of the critical points that an employer should keep in mind when confronted with such an Agency investigation.

The Field Inspections Reference Manual (FIRM), which is OSHA's guidance to its compliance safety and health officers (CSHOs), describes four categories of inspections under the Occupational Safety and Health Act of 1970, 29 U.S.C. §651 et seq. (OSH Act). These are unprogrammed inspections, unprogrammed related inspections, programmed inspections, and programmed related inspections (FIRM, Chap. II, §B). An unprogrammed inspection is the type of inspection most often experienced by employers. It includes complaint, fatality/catastrophe, referral, imminent danger, and follow-up and monitoring inspections. A programmed inspection is one that has been scheduled based upon objective or neutral selection criteria where the work sites are selected according to national scheduling plans for safety and for health or pursuant to a special emphasis program such as the PetroSEP, involving the petrochemical industry, or the Ergonomics Special Emphasis Program in Regions 1 and III. The "related" inspections are inspections of employers on multi-employer work sites whose activities were not included in the underlying unprogrammed or programmed inspection.

The OSHA inspections generally consist of certain fixed stages and involve recurrent issues. Being aware of what to expect from the Agency at each of these stages in the inspection helps the employer to respond properly and, where appropriate, effectively to assert its rights when OSHA overreaches its authority.

The Complaint and/or Warrant

When OSHA arrives at the work site, employers should inquire of the CSHO as to the type of inspection. If the inspection has been catalyzed by a complaint, the employer should ask to review a copy of the complaint. In a federal OSHA jurisdiction and most state plan states (California being the exception), the employer will be given, or at minimum shown, a copy of the complaint with the name of the complaining individual redacted. Employers should carefully scrutinize the complaint. First, this will give the employer an idea of the issue(s) in which OSHA is interested. Second, in those jurisdictions that take a narrower view of the probable cause element of an OSH Act warrant, the subject matter of the complaint may provide the basis for a challenge to the scope of a warrant or a subpoena for documents.

Although the OSH Act does not mention the employer's right to demand a warrant for non-consensual inspections, that right was secured for employers in *Marshall v. Barlow's Inc.* 436 U.S. 307 (1978). Whether to ask the CSHO for a warrant before allowing an inspection is in large part a function of both the corporate culture and the court of appeals jurisdiction in which the inspected work site is located. Currently, there is a split among the circuits that have addressed themselves to the issue. For example, the Fourth, Seventh, Eighth, and Ninth Circuits have concluded that even a specific complaint, referencing a particular department or operation, is sufficient to justify a warrant for a "wall-to-wall" inspection. In contrast, the Third and Eleventh Circuits have taken a more restrictive view. As for the "corporate culture" element, most employers do not choose to make OSHA get a warrant, concluding that it is too adversarial an approach. Of course, the issue may not be up to the employer if OSHA shows up with an anticipatory warrant. In such cases, OSHA, usually based on past history with the employer, believes that the employer may turn the CSHO away without a warrant and so the Agency seeks an ex parte warrant before showing up at the employer in the first instance.

Whether or not the employer makes OSHA get a warrant or OSHA shows up with an anticipatory warrant, the employer should carefully scrutinize any warrant presented to determine its temporal as well as its geographical scope. In some instances, for example, the warrant is limited in terms of time or plant location. Clearly, if such a limitation is set forth in the warrant, OSHA should not be permitted to go beyond those limitations in conducting the inspection.

Asking the Compliance Safety and Health Officer to Wait

It is best for the employer to have preselected an employer representative to handle the inspection and interlace with the CSHO. This person should have received prior training in the rights and responsibilities of employers during OSHA inspections so as to avoid providing more data to the Agency than may be legally required and to ensure that the CSHO stays within appropriate investigative limits. If this employer representative is not immediately available but can be on-site within a limited amount of time, such as two or three hours, the employer should courteously request that the CSHO wait until the arrival of this company point person. Although the CSHO may not be happy waiting, he or she knows that it will take far longer than an hour or two to obtain a warrant entitling the Agency to mandatory access.

The Opening Conference

The OSHA inspection begins with an "opening conference." The employer should utilize the opening conference to get an idea from the CSHO as to the operations in which the Agency is interested and the scope of the inspection in terms of length and plant operations. If it appears that the inspection will be prolonged and comprehensive, the employer should try to work out with the CSHO "ground rules" for how the inspection will be conducted.

Document Requests

Generally, employers should insist on written document requests from OSHA in order to allow for analysis of possible objections, to assist in keeping track of produced documents, and to build in sufficient time to ensure proper compliance. Exceptions to this general principle are documents that OSHA requires be kept and made available for inspection in the normal course, such as OSHA 200 Log and Form 101 and commonly required written programs, e.g., Hazard Communication, Lockout/Tagout, Hearing Conservation, Respiratory Protection, Bloodborne Pathogens, Confined Space, Emergency Action, HAZWOPER, and PSM. Employers must also ensure compliance by OSHA with the Medical Access Order requirements for review of personally identifiable employee medical records.

The Walk-Around

An employer representative should "always" accompany the CSHO when he or she conducts the walk-around inspection. The CSHO will also probably ask for an hourly employee representative to participate in the walk-around. While on the walk-around, the employer representative should not get into an extended dialogue with the CSHO or try to defend when the CSHO identifies a possible violation. First, this could lead to damaging admissions. Second, the employer representative may unknowingly undermine or dilute certain defenses available to the employer. Rather, the representative should merely answer specific questions such as how something works, what something is, whether it runs for three shifts, and so on.

Employee Interviews

The employer has the right to be present at interviews by OSHA of management personnel. These interviews by the CSHO should be scheduled and management witnesses should be properly prepared as they are with virtually all other government investigation interviews or depositions. Although the CSHO will often request to interview certain management staff members immediately, they are not entitled to such instant gratification or to disrupt normal operations. Again, the employer representative should politely but firmly inform the CSHO that the individual he or she seeks to interview is not immediately available and then arrange a time and place for the interview. Whether an attorney should be present for the preparation and/or the interview is a function of the type of inspection and the potential for substantial liability. Clearly, if there is a fatality/catastrophe inspection that carries with it the potential for criminal and/or willful liability, an attorney should get involved immediately and be present for all management interviews. However, if the inspection is routine, attorney involvement in the interviews may not be required.

     With respect to interviews of hourly employees OSHA generally takes the position that it can and should interview these employees outside the presence of management. If the employer refuses OSHA that opportunity, the Agency will subpoena the home addresses and telephone numbers of employees and will pursue interviews off-site. If OSHA identifies for scheduling purposes the hourly employees it seeks to interview, the employer has the right to speak with the employee first and inform the employee that he or she may request that a management representative be present if the hourly employee so chooses.

Sampling, Videos, Trade Secrets, and Contractors' Issues

Sampling of employees and the workplace environment along with photographing and videotaping are all inspection techniques used by OSHA and upheld by the courts. Although an employee cannot be compelled to wear a sampling device and although an employer may tell the employee of his or her right to refuse to participate in sampling, it is generally advisable to respond to employee inquiries regarding the right to refuse rather than to initiate the advice in order to avoid the appearance that the employer is encouraging lack of cooperation by its employees.

     The OSH Act provides for protection of trade secrets. However, this protection does not mean that the inspection of the proprietary equipment or operation will be forestalled; rather, it is the employer's responsibility to identify to the CSHO what is a trade secret. Thereafter, the CSHO is obligated to ensure that the identified matter is treated in accordance with Agency guidelines providing for its protection.

     Contract workers are an increasing source of OSHA concern and the Agency takes an expansive approach toward their coverage. Given this clear expression of OSHA interest and the attendant liability employers may face for contractors, it is essential that the employer keep abreast of the legal issues with respect to contractors as they develop.

The Closing Conference

After the inspection is completed but before the citation issues, the CSHO will hold a closing conference to review the inspection findings in a general fashion. Generally, the CSHO will not discuss penalties and will not specifically reference the classifications of the alleged violations. Employers should primarily listen and not try to defend against the identified violations. It is usually too late at this point to change the CSHO's mind unless there is a very clear and apparent mistake of fact. For example, if the CSHO states that the employer is to be cited for not having a written Bloodborne Pathogens Program and the employer in fact has one that inadvertently was not shown to the CSHO during the inspection, then the employer should point that out during the closing conference. However, attempts by the employer to defend against the violations referred to by the CSHO in the closing conference could result in adverse admissions or other statements that could undermine available defenses. For the same reasons, it is advisable "not" to respond to the CSHO's inquiries as to how long the employer thinks it will need to abate an identified violation. The CSHO will take a response as an admission that abatement is required, although a defense may be applicable. Additionally, if the employer has not realistically estimated the time needed for abatement, it may be necessary to recant after the citation is issued and OSHA has incorporated into the citation the employer's own abatement estimate.

Given the greatly increased liability for employers under the OSH Act and under state criminal statutes for workplace illness and injury, it is crucial for employers to properly prepare for and effectively manage OSHA inspections. Employers should be aware of evolving and expanding theories of liability. They should also be careful to assert their rights during the inspection such as preparing witnesses for interviews, objecting to overbroad document requests and being very careful not to alert OSHA to documents and witnesses about which the CSHO may not be otherwise aware.

22 | FDA Pre-Approval Inspections/Investigations: The Road from Scale-Up and Post-Approval Changes to the Food and Drug Modernization Act

Joseph D. Nally
Nallianco LLC, New Vernon, New Jersey, U.S.A.

Largely as a result of the inclusion of falsified data in some approved new drug application (ANDA) submissions, the Food and Drug Administration (FDA) introduced the Pre-Approval Inspections/Investigations Compliance Program (7346.832) in 1990. This involved the FDA district offices directly in the drug-approval process. The role of the district is to "assure current good manufacturing practice (CGMP) compliance, verify the authenticity and accuracy of the data contained in the applications, and report any other data that may impact on the firm's ability to manufacture the product in compliance with the good manufacturing practices (GMPs)." The latest update of this document was March 2005 (http://www.fda.gov/cder/PAI-7346832.pdf).

A Pre-Approval Inspection (PAI) will be requested by a reviewing chemist for any of the following:

- The product has a narrow therapeutic range (a list of such products is included in the Guide)
- New chemical entities
- Generic versions of the 200 most-prescribed drugs
- When the GMP status for the dosage-form manufacturer is unacceptable or the facility has not been inspected for 2 or more years
- The initial application from a company
- The first generic application of a branded product
- The reviewing chemist identifies potential deviations or discrepancies in the submission
- Bulk pharmaceutical chemical manufacturers and support operations (contract operations for processing, labeling, packaging, or testing) where there has been no recent GMP inspection
- Supplements involving new facilities or major alterations to existing facilities—dosage form, bulk drug, or control laboratory.

District offices may also initiate their own inspections if they consider this to be necessary.

The PAI emphasis includes:

Facility CGMP compliance and the capability to produce the product. This applies to the producers of dosage forms and drug substances and may be extended to the producers of novel excipients. The PAI is sometimes used as another opportunity to evaluate a manufacturer with respect to general GMP compliance. Since facilities are routinely inspected for GMP compliance, this review should be unnecessary unless previous inspections had identified areas for improvement or the technology is new to the facility. With the limited availability of FDA resources, this would seem an activity that could be curtailed.

A significant reason for a nonapprovable recommendation has been the inability of the company to manufacture the product on a commercial scale. The scale-up work and supporting stability batches were produced on pilot-scale equipment, and the full-scale facility/equipment was not available. In such circumstances, companies can ask for a delay in the inspection.

Process validation. Early on, the FDA agreed that the validation data need not be available at the time of the PAI. However, the data will be requested and if not complete must be convincing (especially for ANDA products). As a minimum, the approved protocols must be in place and the validation must be completed before product is distributed.

The FDA inspectors may wish to evaluate the completed validation package, but this can occur after approval and commercialization of the product. In the event that the data are found to be inadequate, the product may have to be recalled or further production not allowed until the validation is repeated.

The situation is different for foreign operations. Because of the cost and resource implications associated with foreign inspections, it is usually impracticable to arrange a follow-up inspection to review validation data. Consequently, the validations are expected to be completed at the time of the PAI. This does, of course, add extra cost to the manufacturer, since in most instances these batches will have expired or be close to expiration by the time FDA approval is received. Hopefully, if the approval cycle time is reduced considerably, this may become less of a problem.

For products involving aseptic processing or sterilization, reviewing chemists will require that some validation data be included in the submission to demonstrate that these specific processes are qualified.

The Guide also indicates that the validation requirements for active pharmaceutical ingredients (APIs) need not be so extensive and the main objective is to ascertain that the raw material/product is characterized and production processes are adequately defined and perform consistently. Identification of the failure (reject, rework, and production deviation) of a significant number of batches is considered as evidence of nonconsistency.

Data accuracy and completeness. The FDA investigator will review raw data and compare this with the NDA/ANDA submission. The intent is two-fold. First is to ensure that no data has been excluded. All data should be included in the submission with appropriate comments where the data is considered to be irrelevant. The second intent is to determine whether any of the submitted data appears to be fraudulent. The FDA places a heavy emphasis on this issue, and unfortunately there have been examples of fraudulent data submission.

Biobatch. The field investigator is expected to specifically examine the GMP compliance of key batches—those used in pivotal clinical studies, for bioavailability/bioequivalence, and for stability. A good practice is to manufacture and evaluate a Biobatch as if it was a validation batch (level of process definition, identified critical process parameters and ranges, extended in-process and finished product testing). This can provide the linking data to subsequent process validation and commercialization records. By this time in the development process, that level of detail should be known.

Laboratory methodology. The FDA field analytical laboratories will validate the analytical methods included in the submission. The Guide includes comprehensive details of samples required for validation and also for other evaluations (forensic and biotest). Also refer to *Guidance for Industry Analytical Procedures and Method Validation*, Draft Guidance August, 2000 (http://www.fda.gov/cder/guidance/2396dft.htm).

The PAI provided the FDA district offices with considerably more "muscle" than ever before. Previously, any CGMP violations or interpretive violations resulted in the issuance of an FDA 483 citation. The company involved usually responded, giving details of the remedial actions being taken. Unless the violations were considered significant, requiring further action, the commercial activities of the company were not adversely impacted. For most companies the corrective actions were initiated expeditiously. The PAI provided the FDA districts with the opportunity to recommend nonapproval of an NDA/ANDA/Supplement until the appropriate corrective actions had been implemented and confirmed by reinspection. The potential loss of profit resulting from delays in approval was enough to ensure that company top management supported the corrective actions and provided any needed resources. On the negative side, an interpretive violation could also delay approval. Over the last decade, the FDA has reduced its level of control from Washington and has allowed considerable local freedom. In some (minority of) cases, this has resulted in unnecessarily aggressive interpretation of the CGMPs by some investigators being used to delay approvals.

PAIs can be scheduled at any time after the receipt and initial review of the submission for completeness. Preparation for a PAI should be a team effort frequently led by QA. Some key steps in the preparation include:

- Confirmation of GMP compliance at all sites. This could involve a review of all recent quality inspections by any source with confirmation that any required remedial actions had been completed. This should be supplemented by a further GMP audit of each facility.
- Collation of data. Raw data and supporting documentation should be collected so that they are readily available for review by the investigator. This also provides a useful opportunity to perform an internal reassessment on the data to ensure that any deviations were adequately evaluated. Special emphasis should be placed on data relating to raw material characterization, biobatches, and stability batches.
- Review of the technology transfer data from R&D to plant. This will include validation protocols (and possibly IQ and OQ data on equipment), operator and analyst training, analytical methods verification, and comparison of scale-up batch processing details with the biobatches.
- Confirmation that the development report, which describes the history of the development of the product, is comprehensive and clear.

While the preparation for a PAI, as just outlined, is essential, this is really too late to rectify any serious issues or omissions. However, it does allow time to prepare an explanation and possibly to compile a protocol for any additional work that may be required. The most effective way to assure a PAI recommendation for approval is to perform the entire development process in compliance with requirements:

- Train personnel in GMP, GLP, or GCP as appropriate and ensure that they are fully aware of the importance of compliance—good science and regulatory impact.
- Managers and supervisors to constantly monitor compliance and to initiate appropriate actions when deviations are observed. Document all deviations at the time they occur, with supporting evaluations.
- Have a development and technology transfer process that clearly defines accountabilities and responsibilities.
- Provide continuous audit of the process and data.

Since the introduction of the PAI, the level of recommendations for nonapproval has moved from the majority of nonapprovals to approvals. From an FDA perspective, there is no doubt that the PAI program interjects FDA at a critical time in the development and commercialization of a product.

The most common causes for these recommendations to withhold approval were lack of plant capability to produce the product; general GMP deviations including lack of production batch records, incomplete failure investigations, incomplete cleaning validation, and stability program deficiencies; and laboratory problems such as discarding of initial raw data after repeat testing, failure to calibrate laboratory equipment, and incomplete laboratory notebooks.

Overall, the PAI has had an impact in improving the level of GMP compliance. However, in some cases it has also increased product costs and delayed product introductions. At this time no other regulatory authority has introduced an equivalent program. Questions have been raised recently about whether the PAI program is still relevant considering the enhanced documentation and demonstration of the scientific understanding of manufacturing processes and controls (1). However, it is very likely that this program will remain in effect until the FDA is satisfied that the industry understands and has provided the scientific understanding or the product manufacturing and controls.

FDA MODERNIZATION ACT AND MANUFACTURING CHANGES TO APPROVED DRUGS

The Food and Drug Administration Modernization Act (FDAMA) was intended to be fully effective on November 26, 1999. Some small parts have been declared unconstitutional (such as the Pharmacy Compounding requirements), but they do not affect its impact on primary manufacturing and quality control. The latter, with a special note as to the manufacturing changes to approved drug applications have undergone changes somewhat favorable to manufacturers, who in the past hewed closely to the statutory requirements in the Federal Food Drug and Cosmetic (FFDC) Act and 21 CFR 314.70 of the regulations.

The Modernization Act does provide a relative approach concerning manufacturing changes unlikely to impact on quality versus those that do, and much remains for the regulators to determine as to their implementation.

Bear in mind that the FDA, a so-called executive agency, itself answered the criticism of industry in that it often required, somewhat inflexibly, an investment of time and money in the costly submission of prior approval manufacturing supplements for minor changes. The FDA Scale-Up and Post-Approval Changes (SUPAC) process was their answer, and while the spirit of the response was good for manufacturers, there has been mixed results in the industry's experience with the requirements. Regardless, the SUPAC guidance is good and has been updated in July 2005 (http://www.fda.gov/cder/guidance/SUPAC.htm).

There have been a more substantial number of change-being-effected (CBE) supplements noticed to the FDA by inclusion in an annual report, but large-scale manufacturers are hesitant because distribution is at their own risk during the pendency of review of the supplement. In general, conservative CBE supplements are related to changes that seem to make more certain the anticipated detection, and effect of the drug substance will be maintained. Less-conservative ones depend on changes that will be deemed inconsequential to maintenance of the anticipated effect of the drug substance. 21 CFR 314.70 (a), (b), (c), (d) are the regulatory guidelines for supplements that may or may not be made before FDA approval and form a reliable baseline for a determination by those who regulate and those regulated. Some changes may be for notation in the annual report 21 CFR 314.70 (d).

Therefore, when the FDA undertook the SUPAC approach to the process with the National Performance Review staff, it was made public that the FDA was proposing reforms that could save the drug and device industry a half-billion dollars annually. The FDA's Center for Drug Evaluation and Research was charged with the responsibility of developing a guidance document for drugs in tablet and capsule form (other than those for controlled release) to ease existing rules for manufacturing changes.

The SUPAC-1R Notice appeared in the Federal Register [60FR61.637 (1995)] "Immediate Release Solid Oral Dosage Forms; Scale-Up and Post Approval Changes: Chemistry, Manufacturing and Controls; In vitro dissolution testing; In vitro bioequivalence documentation." One must, however, regard this as a preliminary to the implementation of FDA's authority, apparent or assumed, as FDAMA and its regulations become the statutory guides for the end of 1999. Whether written or unwritten, as an internal strategy, the use of levels of change by the FDA is probably here to stay because the FDA has used them informally before addressing them specifically in the published SUPAC-IR Notice previously mentioned. Therefore, it remains important for the manufacturer's judgment to have the capability of documenting effectively the minor variations that can be checked by mantle of Level 1 Changes, which require that the supportive information simply be included in a subsequent annual report.

Bear in mind that the progenitor of the SUPAC-IR Notice was not merely the "reinventing government" push of the Administration. It was a product of a four-year joint effort that our prior edition has considered for its inclusion of academic and industry groups. It was actually prepared by the SUPAC Expert Working Group of the Chemistry Manufacturing Controls Coordinating Committee of the FDA's Center for Drug Evaluation and Research (CDER), with contributions from the American Association of Pharmaceutical Sciences in conjunction with the United States Pharmacopeial Convention. It incorporated suggestions generated by research from many universities and colleges, among them the University of Maryland (Baltimore) and Michigan.

The FDA apparently believes that the SUPAC-IR guidelines fit comfortably within established regulatory mechanism. See, in part, in 314.70 (a): "... an applicant shall make a change provided for ... in accordance with a guideline, notice or regulation published in the Federal Register that provides for a less burdensome notification of the change...."

Considering that the purpose of the enactment of the FDAMA of 1997 was to improve regulatory efficiency of drugs, medical devices, and food, without undermining established discretionary authority of the FDA, it was destined to be a welcome law to some and disappointing to other regulatees. The full text of the FDAMA of 1997 is available at: www.fda.gov/cder/fdama/contents.htm.

The main purpose of the Act was to:

- Reauthorize the Prescription Drug User Fee Act of 1992
- Modernize the regulation of biological products by bringing them into harmony with the regulations for drugs
- Eliminating certain batch certification and monograph requirements
- Streamline the approval process and reduce the need for environmental assessments
- Increase patients access to experimental drugs and medical devices

To date, the FDA has completed all but 17 out of greater than 300 initiatives set out by the act (2).

But as to the new section of the FFDC Act added, 506A, to which we direct ourselves here, the requirements for approval of manufacturing changes for drugs and biologics have given optimism to global manufactures. The level of change has been statutorily visualized, with aid from new regulations, to be "major" or "other." All "changes" are defined as those determined to have a substantial potential to adversely impact the identity, strength, quality, purity, or potency of the drug as those characteristics relate to the safety and effectiveness of the drug in terms of its approved labeling. Therefore "major" changes are statutorily defined at 21USC 506A[a] (see Appendix A hereto) Every established manufacturer with approximately educated and trained personnel can make such determination with reasonable accuracy.

But the Act holds promise: the SUPAC style handling of the "other" changes, the CBE supplement, seem to be simplified and time shortened; perhaps the site transfers for manufacturing and packaging and changes in equipment that are fairly common.

The FDA, it is emphasized, continues to have great force because it is the judge of what changes are "major" and which are "other"; which of course means a determination of time, expenditures, and special or usual personnel needs for the changes. Thus, the SUPAC-IR Notice and various regulations will play a part in the new statutes' constructive effect on manufactures, but it is that statute and its implementing regulations that will provide the seminal strain of law and policy.

An important role for the FDA, aside from its statutory change, will be how it uses three major areas of discretion set out in FDAMA. History bodes well. The FDA has been appreciative of the needs of manufacturers and other distributors but it is clear it will exert its own quality control mechanism in protection of consumers when any evidence of questionable product quality is surfaced.

One area that has become controversial in recent months (as we go to press) is the degree of Compounding permitted by pharmacies. The following excerpts emphasize the issue:

From Warning Letter No. 2006-NOL-04 issued on February 15,2006 to firm involved in compounding sterile products (3):

> "As you may be aware, Section 127 of the FDA Modernization Act of 1997, amended the Act by adding Section 503A, which specified certain conditions under which compounded human drugs could be exempt from certain requirements of the Act. However, in April 2002, the United States Supreme Court struck down as unconstitutional the commercial speech restrictions in Section 503A of the Act. Accordingly, all of Section 503A is now invalid.
>
> As a result, FDA utilizes its long-standing policy of exercising its enforcement discretion regarding certain types of pharmacy compounding. This policy is articulated in FDA's Compliance Policy Guide (CPG), Section 460.200, issued on June 7, 2002. The CPG contains factors FDA considers in deciding whether to exercise enforcement discretion. One factor FDA considers is whether a compounded product is a copy of a commercially available product and, if so, whether there is any documentation of a medical need for the compounded product."

---

[a]This guidance has been prepared by the *Center for Drug Evaluation and Research (CDER)* at the FDA.

In this particular case, the FDA had legitimate concerns about the safety of the products and moved to a warning letter voluntary enforcement action.

On the other hand, in recent action (August, 2006) the Federal Court in Texas ruled(4):

> *"Missouri City, TX (Aug. 30)*—US District Court Judge Robert Junell issued a written opinion in *Medical Center Pharmacy, et al. v. Gonzalez, et al.*, supporting the ten plaintiff pharmacies' assertion that the U.S. Food and Drug Administration (Rockville, MD), lacks the authority to regulate compounded drugs and inspect state-licensed retail pharmacies. The opinion followed Junell's May 25, 2006 ruling from the bench.
>
> Junell held that compounded drugs do not fit FDA's definition of new drugs and said that the cost and time involved in the new-drug approval process would prevent patients from obtaining individually tailored prescriptions. "It is in the best interest of public health to recognize an exemption for compounded drugs that are created based on a prescription written for an individual patient by a licensed practitioner," Junell wrote.
>
> The opinion also says that FDA may not inspect a pharmacy's records unless the agency can show that the pharmacy does not comply with relevant state laws and does not operate as a retail pharmacy. Junell also ruled that all plaintiffs in the case met the requirements for exemption."

There are arguments on both sides of this issue but having worked greater than three decades in pharmaceutical manufacturing and control operations, there is a definite need for oversight of compounding operations and applicable GMP standards, especially when the compounded products are readily available from pharmaceutical manufacturers.

REFERENCES

1. A Drug Quality System for the 21st Century, Breakout Session: Integrating CMC Review and Inspection. http://www.fda.gov/cder/breakout-CMCreview.htm, accessed September 11,2006.
2. Implementation of the Food and Drug Administration of the FDAM Act of 1997. http://www.fda.gov/po/modactchart/modact97num.html, accessed September 11, 2006.
3. U.S. Food and Drug Administration, *Warning Letters.* http://www.fda.gov/foi/warning_letters/g5719d.htm, accessed August 21, 2006.
4. Pharmaceutical Technology, *Federal Court Rules Compounders Are Exempt from FDA Regulation.* http://www.pharmtech.com/pharmtech/article/articleDetail.jsp?id = 370151, accessed September 8, 2006.

SUGGESTED READINGS

1. Food and Drug Administration Compliance Program Guidance Manual Program 7346.832, *Chapter 46 – New Drug Evaluation Pre-Approval Inspections / Investigations*, http://www.fda.gov/cder/gmp/PAI-7346832.pdf.
2. Food and Drug Administration Guidance for Industry, *Immediate Release Solid Oral Dosage Forms*, CDER, November 1995. http://www.fda.gov/cder/guidance/supac.htm.
3. FDA Modernization Act of 1997 CDER Related Documents. http://www.fda.gov/cder/fdama/contents.htm.
4. Berry IR. The Pharmaceutical Regulatory Process. Marcel Dekker, 2005.
5. Manufacturing Changes: "From SUPAC to FDAMA," Pinco, Jiminez, Katdare, FDCMD Law Digest, Vol. 15, No. 3, 1998.
6. "FDAMA, Its Impact on Medical Devices," Connolly, 15 FDCMOL Digest 60 (1998).
7. 15 FDCMDL Digest 64 (1998), Wu and Mazan.
8. "The Food Revisions of the Food and Drug Modernization Act", Rodriguez, 15 FDCMDL Digest 15 (1998).
9. "FDA Compliance Program Guidance Manual (7346.832)—Pre-Approval Inspections/Investigations," 1994.
10. Justice RM, Rodriguez JO, Chiasson WJ. Ten Steps to Assure a Successful Pre-NDA Approval Inspection, J Parenteral Sci Technol 1993; 47(92):89.
11. M. Wells. Foreign FDA Pre-Approval Inspections: Requirements/Preparation. J Pharm Sci Technol 1994; 48(6):30.
12. "FDA Guide to the Inspections of Oral Solid Dosage Forms. Pre-/Post Approval Issues for Development and Validation," January 1994.
13. (http://www.fda.gov/cder/guidance/SUPAC.htm)

14. "FDA Guide to Inspection of Oral Solutions and Suspensions," August 1994.
15. "FDA Guide to Inspection of Dosage Form Drug Manufacturers—CGMPs," October 1993.
16. "FDA Guide to Inspection of Topical Drug Products," July 1994.
17. "FDA Guide to Inspections of Lyophilization of Parenterals," July 1993.

Following are additional FDA Guidance documents relevant to this chapter.

COMPLIANCE POLICY GUIDES SUB CHAPTER 440 NEW DRUGS

Page updated: 06/20/2006

This CPG supersedes section 440.100, Marketed New Drugs Without Approved NDAs or ANDAs (CPG 7132c.02)

SEC. 440.100 MARKETED NEW DRUGS WITHOUT APPROVED NDAS OR ANDAS (CPG 7132C.02)

This guidance represents the (FDA's) current thinking on this topic. It does not create or confer any rights for or on any person and does not operate to bind the FDA or the public. You can use an alternative approach if it satisfies the requirements of the applicable statutes and regulations. If you want to discuss an alternative approach, contact the FDA staff responsible for implementing this guidance. If you cannot identify the appropriate FDA staff, call the appropriate number listed on the title page of this guidance.

I. Introduction

This compliance policy guide (CPG) describes how we intend to exercise our enforcement discretion with regard to drugs marketed in the United States that do not have required FDA approval for marketing. This CPG supersedes section 440.100, Marketed New Drugs Without Approved NDAs or ANDAs (CPG 7132c.02). It applies to any drug required to have FDA approval for marketing, including new drugs covered by the Over-the-Counter (OTC) Drug Review, except for licensed biologics and veterinary drugs.

The FDA's guidance documents, including this guidance, do not establish legally enforceable responsibilities. Instead, guidances describe the Agency's current thinking on a topic and should be viewed only as recommendations, unless specific regulatory or statutory requirements are cited. The use of the word "should" in Agency guidances means that something is suggested or recommended, but not required.

II. Background

A. *Reason for This Guidance*

For historical reasons, some drugs are available in the United States that lack required FDA approval for marketing. A brief, informal summary description of the various categories of these drugs and their regulatory status is provided in Appendix A as general background for this document. The manufacturers of these drugs have not received FDA approval to legally market their drugs, nor are the drugs being marketed in accordance with the OTC drug review. The new drug approval and OTC drug monograph processes play an essential role in ensuring that all drugs are both safe and effective for their intended uses. Manufacturers of drugs that lack required approval, including those that are not marketed in accordance with an OTC drug monograph, have not provided the FDA with evidence demonstrating that their products are safe and effective, and so we have an interest in taking steps to either encourage the manufacturers of these products to obtain the required evidence and comply with the approval provisions of the Federal Food, Drug, and Cosmetic Act (the Act) or remove the products from the market. We want to achieve these goals without adversely affecting public health, imposing undue burdens on consumers, or unnecessarily disrupting the market.

The goals of this guidance are to (1) clarify for FDA personnel and the regulated industry how we intend to exercise our enforcement discretion regarding unapproved drugs and (2) emphasize that illegally marketed drugs must obtain FDA approval.

B. Historical Enforcement Approach

The FDA estimates that, in the United States today, perhaps as many as several thousand drug products are marketed illegally without required FDA approval.[b] Because we do not have complete data on illegally marketed products, and because the universe of such products is constantly changing as products enter and leave the market, we first have to identify illegally marketed products before we can contemplate enforcement action. Once an illegally marketed product is identified, taking enforcement action against the product would typically involve one or more of the following: requesting voluntary compliance; providing notice of action in a Federal Register notice; issuing an untitled letter; issuing a Warning Letter; or initiating a seizure, injunction, or other proceeding. Each of these actions is time-consuming and resource intensive. Recognizing that we are unable to take action immediately against all of these illegally marketed products and that we need to make the best use of scarce Agency resources, we have had to prioritize our enforcement efforts and exercise enforcement discretion with regard to products that remain on the market.

In general, in recent years, FDA has employed a risk-based enforcement approach with respect to marketed unapproved drugs. This approach includes efforts to identify illegally marketed drugs, prioritization of those drugs according to potential public health concerns or other impacts on the public health, and subsequent regulatory follow-up. Some of the specific actions the Agency has taken have been precipitated by evidence of safety or effectiveness problems that has either come to our attention during inspections or been brought to our attention by outside sources.

III. FDA's Enforcement Policy

In the discussion that follows, we intend to clarify our approach to prioritizing our enforcement actions and exercising our enforcement discretion with regard to the universe of unapproved, illegally marketed drug products in all categories.

A. Enforcement Priorities

Consistent with our risk-based approach to the regulation of pharmaceuticals, the FDA intends to continue its current policy of giving higher priority to enforcement actions involving unapproved drug products in the following categories:

Drugs with potential safety risks. Removing potentially unsafe drugs protects the public from direct and indirect health threats.

Drugs that lack evidence of effectiveness. Removing ineffective drugs protects the public from using these products in lieu of effective treatments. Depending on the indication, some ineffective products would, of course, pose safety risks as well.

Health fraud drugs. FDA defines health fraud as "[the deceptive promotion, advertisement, distribution or sale of articles ... that are represented as being effective to diagnose, prevent, cure, treat, or mitigate disease (or other conditions), or provide a beneficial effect on health, but which have not been scientifically proven safe and effective for such purposes. Such practices may be deliberate or done without adequate knowledge or understanding of the article" (CPG Sec. 120.500). Of highest priority in this area are drugs that present a direct risk to health. Indirect health hazards exist if, as a result of reliance on the product, the consumer is likely to delay or discontinue appropriate medical treatment. Indirect health hazards will be evaluated for enforcement action based on section 120.500, Health Fraud—Factors in Considering Regulatory Action (CPG 7150.10). The FDA's health fraud CPG outlines priorities for evaluating regulatory actions against indirect health hazard products, such as whether the

---

[b]This rough estimate comprises several hundred drugs (different active ingredients) in various strengths, combinations, and dosage forms from multiple distributors and repackagers.

therapeutic claims are significant, whether there are any scientific data to support the safety and effectiveness of the product, and the degree of vulnerability of the prospective user group (CPG Sec. 120.500).

Drugs that present direct challenges to the new drug approval and OTC drug monograph systems. The drug approval and OTC drug monograph systems are designed to avoid the risks associated with potentially unsafe, ineffective, and fraudulent drugs. The drugs described in the preceding three categories present direct challenges to these systems, as do unapproved drugs that directly compete with an approved drug, such as when a company obtains approval of a new drug application (NDA) for a product that other companies are marketing without approval (see section III.C, Special Circumstances—Newly Approved Product). Also included are drugs marketed in violation of a final and effective OTC drug monograph. Targeting drugs that challenge the drug approval or OTC drug monograph systems buttresses the integrity of these systems and makes it more likely that firms will comply with the new drug approval and monograph requirements, which benefits the public health.

Unapproved new drugs that are also violative of the Act in other ways. The Agency also intends, in circumstances that it considers appropriate, to continue its policy of enforcing the pre-Approval requirements of the Act against a drug or firm that also violates another provision of the Act, even if there are other unapproved versions of the drug made by other firms on the market. For instance, if a firm that sells an unapproved new drug also violates current good manufacturing practice (CGMP) regulations, the Agency is not inclined to limit an enforcement action in that instance to the CGMP violations. Rather, the Agency may initiate a regulatory action that targets both the CGMP violation and the violation of section 505 of the Act (21 U.S.C. 355). This policy efficiently preserves scarce Agency resources by allowing the Agency to pursue all applicable charges against a drug and/or a firm and avoiding duplicative action. See *United States v. Sage Pharmaceuticals, Inc.*, 210 F.3d 475, 479-80 (5th Cir. 2000).

Drugs that are reformulated to evade an FDA enforcement action. The Agency is also aware of instances in which companies that anticipate an FDA enforcement action against a specific type or formulation of an unapproved product have made formulation changes to evade that action, but have not brought the product into compliance with the law. Companies should be aware that the Agency is not inclined to exercise its enforcement discretion with regard to such products. Factors that the Agency may consider in determining whether to bring action against the reformulated products include, but are not limited to, the timing of the change, the addition of an ingredient without adequate scientific justification [see, e.g., 21 CFR 300.50 and 330.10(a)(4)(iv)], the creation of a new combination that has not previously been marketed, and the claims made for the new product.

B. Notice of Enforcement Action and Continued Marketing of Unapproved Drugs

FDA is not required to, and generally does not intend to, give special notice that a drug product may be subject to enforcement action, unless FDA determines that notice is necessary or appropriate to protect the public health.[c] The issuance of this guidance is intended to provide notice that any product that is being marketed illegally is subject to FDA enforcement action at any time.[d] The only exception to this policy is, as set forth elsewhere, that generally products

[c]For example, in 1997, the FDA issued a Federal Register notice declaring all orally administered levothyroxine sodium products to be new drugs and requiring manufacturers to obtain approved new drug applications (62 FR 43535, August 14, 1997). Nevertheless, the FDA gave manufacturers three years (later extended to four (65 FR 24488, April 26, 2000) to obtain approved applications and allowed continued marketing without approved new drug applications because the FDA found that levothyroxine sodium products were medically necessary to treat hypothyroidism and no alternative drug provided an adequate substitute.

[d]For example, the FDA may take action at any time against a product that was originally marketed before 1938, but that has been changed since 1938 in such a way as to lose its grandfather status [21 U.S.C. 321(p)].

subject to an ongoing Drug Efficacy Study Implementation (DESI)[e] proceeding or ongoing OTC drug monograph proceeding (i.e., an OTC product that is part of the OTC drug review for which an effective final monograph is not yet in place) may remain on the market during the pendency of that proceeding[f] and any additional period specifically provided in the proceeding (such as a delay in the effective date of a final OTC drug monograph).[g] However, once the relevant DESI or OTC drug monograph proceeding is completed and any additional grace period specifically provided in the proceeding has expired, all products that are not in compliance with the conditions for marketing determined in that proceeding are subject to enforcement action at any time without further notice (see, e.g., 21 CFR 310.6).

The FDA intends to evaluate on a case-by-case basis whether justification exists to exercise enforcement discretion to allow continued marketing for some period of time after the FDA determines that a product is being marketed illegally. In deciding whether to allow such a grace period,[h] we may consider the following factors: (1) the effects on the public health of proceeding immediately to remove the illegal products from the market (including whether the product is medically necessary and, if so, the ability of legally marketed products to meet the needs of patients taking the drug); (2) the difficulty associated with conducting any required studies, preparing and submitting applications, and obtaining approval of an application; (3) the burden on affected parties of immediately removing the products from the market; (4) the Agency's available enforcement resources; and (5) any special circumstances relevant to the particular case under consideration.

C. Special Circumstances—Newly Approved Product

Sometimes, a company may obtain approval of an NDA for a product that other companies are marketing without approval.[i] We want to encourage this type of voluntary compliance with the new drug requirements because it benefits the public health by increasing the assurance that marketed drug products are safe and effective—it also reduces the resources that the FDA must expend on enforcement. Thus, because they present a direct challenge to the drug approval system, the FDA is more likely to take enforcement action against remaining unapproved drugs in this kind of situation. However, we intend to take into account the circumstances once the product is approved in determining how to exercise our enforcement discretion with regard to the unapproved products. In exercising enforcement discretion, we intend to balance the need to provide incentives for voluntary compliance against the implications of enforcement actions on the marketplace and on consumers who are accustomed to using the marketed products.

When a company obtains approval to market a product that other companies are marketing without approval, the FDA normally intends to allow a grace period of roughly 1 year from the date of approval of the product before it will initiate enforcement action (e.g., seizure or injunction) against marketed unapproved products of the same type. However, the grace period provided is expected to vary from this baseline based upon the following factors: (1) the effects on the public health of proceeding immediately to remove the illegal products

---

[e]The DESI was the process used by the FDA to evaluate for effectiveness for their labeled indications over 3,400 products that were approved only for safety between 1938 and 1962. DESI is explained more fully in the appendix to this document.

[f]OTC drugs covered by ongoing OTC drug monograph proceedings may remain on the market as provided in current enforcement policies. See, for example, CPG sections 450.200 and 450.300 and 21 CFR part 330. This document does not affect the current enforcement policies for such drugs.

[g]Sometimes, a final OTC drug monograph may have a delayed effective date or provide for a specific period of time for marketed drugs to come into compliance with the monograph. At the end of that period, drugs that are not marketed in accordance with the monograph are subject to enforcement action and the exercise of enforcement discretion in the same way as any other drug discussed in this CPG.

[h]For purposes of this guidance, the terms grace period and allow a grace period refer to an exercise of enforcement discretion by the Agency (i.e., a period of time during which the FDA, as a matter of discretion, elects not to initiate a regulatory action on the ground that an article is an unapproved new drug).

[i]These may be products that are the same as the approved product or somewhat different, such as products of different strength.

from the market (including whether the product is medically necessary and, if so, the ability of the holder of the approved application to meet the needs of patients taking the drug); (2) whether the effort to obtain approval was publicly disclosed;[j] (3) the difficulty associated with conducting any required studies, preparing and submitting applications, and obtaining approval of an application; (4) the burden on affected parties of removing the products from the market; (5) the Agency's available enforcement resources; and (6) any other special circumstances relevant to the particular case under consideration. To assist in an orderly transition to the approved product(s), in implementing a grace period, the FDA may identify interim dates by which firms should first cease manufacturing unapproved forms of the drug product, and later cease distributing the unapproved product.

The length of any grace period and the nature of any enforcement action taken by the FDA will be decided on a case-by-case basis. Companies should be aware that a Warning Letter may not be sent before initiation of enforcement action and should not expect any grace period that is granted to protect them from the need to leave the market for some period of time while obtaining approval. Companies marketing unapproved new drugs should also recognize that, while the FDA normally intends to allow a grace period of roughly 1 year from the date of approval of an unapproved product before it will initiate enforcement action (e.g., seizure or injunction) against others who are marketing that unapproved product, it is possible that a substantially shorter grace period would be provided, depending on the individual facts and circumstances.[k]

The shorter the grace period, the more likely it is that the first company to obtain an approval will have a period of de facto market exclusivity before other products obtain approval. For example, if the FDA provides a 1-year grace period before it takes action to remove unapproved competitors from the market, and it takes 2 years for a second application to be approved, the first approved product could have 1 year of market exclusivity before the onset of competition. If the FDA provides for a shorter grace period, the period of effective exclusivity could be longer. The FDA hopes that this period of market exclusivity will provide an incentive to firms to be the first to obtain approval to market a previously unapproved drug.[l]

D. Regulatory Action Guidance

District offices are encouraged to refer to CDER for review (with copies of labeling) any unapproved drugs that appear to fall within the enforcement priorities in section III.A. Charges that may be brought against unapproved drugs include, but are not limited to, violations of 21 U.S.C. 355(a) and 352(f)(1) of the Act. Other charges may also apply based on, among others, violations of 21 U.S.C. 351(a)(2)(B) (CGMP), 352(a) (misbranding), or 352(o) (failure to register or list).

---

[j]For example, at the Agency's discretion, we may provide for a shorter grace period if an applicant seeking approval of a product that other companies are marketing without approval agrees to publication, around the time it submits the approval application, of a Federal Register notice informing the public that the applicant has submitted that application. A shortened grace period may also be warranted if the fact of the application is widely known publicly because of applicant press releases or other public statements. Such a grace period may run from the time of approval or from the time the applicant has made the public aware of the submission, as the Agency deems appropriate.

[k]Firms are reminded that this CPG does not create any right to a grace period; the length of the grace period, if any, is solely at the discretion of the Agency. For instance, firms should not expect any grace period when the public health requires immediate removal of a product from the market, or when the Agency has given specific prior notice in the Federal Register or otherwise that a drug product requires FDA approval.

[l]The Agency understands that, under the Act, holders of NDAs must list patents claiming the approved drug product and that newly approved drug products may, in certain circumstances, be eligible for marketing exclusivity. Listed patents and marketing exclusivity may delay the approval of competitor products. If the FDA believes that an NDA holder is manipulating these statutory protections to inappropriately delay competition, the Agency will provide relevant information on the matter to the Federal Trade Commission (FTC). In the past, the FDA has provided information to the FTC regarding patent infringement lawsuits related to pending abbreviated new drug applications (ANDAs), citizen petitions, and scientific challenges to the approval of competitor drug products.

APPENDIX:

BRIEF HISTORY OF FDA MARKETING APPROVAL REQUIREMENTS AND CATEGORIES OF DRUGS THAT LACK REQUIRED FDA APPROVAL[m]

Key events in the history of the FDA's drug approval regulation and the categories of drugs affected by these events are described below.

A. 1938 and 1962 Legislation

The original Federal Food and Drugs Act of June 30, 1906, first brought drug regulation under federal law. That Act prohibited the sale of adulterated or misbranded drugs, but did not require that drugs be approved by the FDA. In 1938, Congress passed the Federal Food, Drug, and Cosmetic Act (the Act), which required that new drugs be approved for safety. As discussed below, the active ingredients of many drugs currently on the market were first introduced, at least in some form, before 1938. Between 1938 and 1962, if a drug obtained approval, the FDA considered drugs that were identical, related, or similar (IRS) to the approved drug to be covered by that approval, and allowed those IRS drugs to be marketed without independent approval. Many manufacturers also introduced drugs onto the market between 1938 and 1962 based on their own conclusion that the products were generally recognized as safe (GRAS) or based on an opinion from the FDA that the products were not new drugs. Between 1938 and 1962, the Agency issued many such opinions, although all were formally revoked in 1968 (see 21 CFR 310.100).

B. DESI

In 1962, Congress amended the Act to require that a new drug also be proven effective, as well as safe, to obtain FDA approval. This amendment also required the FDA to conduct a retrospective evaluation of the effectiveness of the drug products that the FDA had approved as safe between 1938 and 1962 through the new drug approval process.

The FDA contracted with the National Academy of Science/National Research Council (NAS/NRC) to make an initial evaluation of the effectiveness of over 3,400 products that were approved only for safety between 1938 and 1962. The NAS/NRC created 30 panels of six professionals each to conduct the review, which was broken down into specific drug categories. The NAS/NRC reports for these drug products were submitted to the FDA in the late 1960s and early 1970s. The Agency reviewed and re-evaluated the findings of each panel and published its findings in Federal Register notices. The FDA's administrative implementation of the NAS/NRC reports was called the DESI. DESI covered the 3,400 products specifically reviewed by the NAS/NRCs as well as the even larger number of IRS products that entered the market without FDA approval.

Because DESI products were covered by approved (pre-1962) applications, the Agency concluded that, prior to removing products not found effective from the market, it would follow procedures in the Act and regulations that apply when an approved new drug application is withdrawn:

■ All initial DESI determinations are published in the Federal Register and, if the drug is found to be less than fully effective, there is an opportunity for a hearing.
■ The Agency considers the basis of any hearing request and either grants the hearing or denies the hearing on summary judgment and publishes its final determination in the Federal Register.
■ If the FDA's final determination classifies the drug as effective for its labeled indications, as required by the Act, the FDA still requires approved applications for continued marketing

---

[m]This brief history document should be viewed as a secondary source. To determine the regulatory status of a particular drug or category of drugs, the original source documents cited should be consulted.

of the drug and all drugs IRS to it—NDA supplements for those drugs with NDAs approved for safety, or new ANDAs or NDAs, as appropriate, for IRS drugs. DESI-effective drugs that do not obtain approval of the required supplement, ANDA, or NDA are subject to enforcement action.

If FDA's final determination classifies the drug as ineffective, the drug and those IRS to it can no longer be marketed and are subject to enforcement action.

5. Products Subject to Ongoing DESI Proceedings

Some unapproved marketed products are undergoing DESI reviews in which a final determination regarding efficacy has not yet been made. In addition to the products specifically reviewed by the NAS/NRC (i.e., those products approved for safety only between 1938 and 1962), this group includes unapproved products identical, related, or similar to those products specifically reviewed (see 21 CFR 310.6). In virtually all these proceedings, the FDA has made an initial determination that the products lack substantial evidence of effectiveness, and the manufacturers have requested a hearing on that finding. It is the Agency's long-standing policy that products subject to an ongoing DESI proceeding may remain on the market during the pendency of the proceeding. See, e.g., *Upjohn Co. v. Finch*, 303 F. Supp. 241, 256-61 (W.D. Mich. 1969).[n]

6. Products Subject to Completed DESI Proceedings

Some unapproved marketed products are subject to already-completed DESI proceedings and lack required approved applications. This includes a number of products IRS to DESI products for which approval was withdrawn due to a lack of substantial evidence of effectiveness. This group also includes a number of products IRS to those DESI products for which the FDA made a final determination that the product is effective, but applications for the IRS products have not been both submitted and approved as required under the statute and long-standing enforcement policy (see 21 CFR 310.6). The FDA considers all products described in this paragraph to be marketed illegally.

C. Prescription Drug Wrap-Up

As mentioned above, many drugs came onto the market before 1962 without FDA approvals. Of these, many claimed to have been marketed prior to 1938 or to be IRS to such a drug. Drugs that did not have pre-1962 approvals and were not IRS to drugs with pre-1962 approvals were not subject to DESI. For a period of time, the FDA did not take action against these drugs and did not take action against new unapproved drugs that were IRS to these pre-1962 drugs that entered the market without approval.

Beginning in 1983, it was discovered that one drug that was IRS to a pre-1962 drug, a high potency Vitamin E intravenous injection named E-Ferol, was associated with adverse reactions in about 100 premature infants, 40 of whom died. In November of 1984, in response to this, a congressional oversight committee issued a report to the FDA expressing the committee's concern regarding the thousands of unapproved drug products in the marketplace.

---

[n]Products first marketed after a hearing notice is issued with a different formulation than those covered by the notice are not considered subject to the DESI proceeding. Rather, they need approval prior to marketing. Under long-standing Agency policies, a firm holding an NDA on a product for which a DESI hearing is pending must submit a supplement prior to reformulating that product. The changed formulation may not be marketed as a related product under the pending DESI proceeding; it is a new drug, and it must be approved for safety and efficacy before it can be legally marketed. See, for example, "Prescription Drugs Offered for Relief of Symptoms of Cough, Cold, or Allergy" (DESI 6514), 49 FR 153 (January 3, 1984) (Dimetane and Actifed); "Certain Drugs Containing Antibiotic, Corticosteroid, and Antifungal Components" (DESI 10826), 50 FR 15227 (April 17, 1985) (Mycolog). See also 21 U.S.C. 356a(c)(2)(A). Similarly, firms without NDAs cannot market new formulations of a drug without first getting approval of an NDA.

In response to the E-Ferol tragedy, CDER assessed the number of pre-1962 non-DESI marketed drug products. To address those drug products, the Agency significantly revised and expanded CPG section 440.100 to cover all marketed unapproved prescription drugs, not just DESI products. The program for addressing these marketed unapproved drugs and certain others like them became known as the Prescription Drug Wrap-Up. Most of the Prescription Drug Wrap-Up drugs first entered the market before 1938, at least in some form. For the most part, the Agency had evaluated neither the safety nor the effectiveness of the drugs in the Prescription Drug Wrap-Up.

A drug that was subject to the Prescription Drug Wrap-Up is marketed illegally, unless the manufacturer of such a drug can establish that its drug is grandfathered or otherwise not a new drug.

Under the 1938 grandfather clause [see 21 U.S.C. 321(p)(1)], a drug product that was on the market prior to passage of the 1938 Act and that contained in its labeling the same representations concerning the conditions of use as it did prior to passage of that Act was not considered a new drug and therefore was exempt from the requirement of having an approved new drug application.

Under the 1962 grandfather clause, the Act exempts a drug from the effectiveness requirements if its composition and labeling has not changed since 1962 and if, on the day before the 1962 Amendments became effective, it was (a) used or sold commercially in the United States, (b) not a new drug as defined by the Act at that time, and (c) not covered by an effective application. See Pub. L. 87–781, section 107 (reprinted following 21 U.S.C.A. 321); see also *USV Pharmaceutical Corp. v. Weinberger*, 412 U.S. 655, 662–66 (1973).

The two grandfather clauses in the Act have been construed very narrowly by the courts. The FDA believes that there are very few drugs on the market that are actually entitled to grandfather status because the drugs currently on the market likely differ from the previous versions in some respect, such as formulation, dosage or strength, dosage form, route of administration, indications, or intended patient population. If a firm claims that its product is grandfathered, it is that firm's burden to prove that assertion. See 21 CFR 314.200(e)(5); see also *United States v. An Article of Drug (Bentex Ulcerine)*, 469 F.2d 875, 878 (5th Cir. 1972); *United States v. Articles of Drug Consisting of the Following: 5,906 Boxes*, 745 F.2d 105, 113 (1st Cir 1984).

Finally, a product would not be considered a new drug if it is generally recognized as safe and effective (GRAS/GRAE) and has been used to a material extent and for a material time. See 21 U.S.C. 321(p)(1) and (2). As with the grandfather clauses, this has been construed very narrowly by the courts. See, e.g., *Weinberger v. Hynson, Westcott & Dunning, Inc.*, 412 U.S. 609 (1973); *United States v. 50 Boxes More or Less Etc.*, 909 F.2d 24, 27-28 (1st Cir. 1990); *United States v. 225 Cartons*. Fiorinal, 871 F.2d 409 (3rd Cir. 1989). See also Letter from Dennis E. Baker, Associate Commissioner for Regulatory Affairs, FDA, to Gary D. Dolch, Melvin Spigelman, and Jeffrey A. Staffa, Knoll Pharmaceutical Co. (April 26, 2001) (on file in FDA Docket No. 97 N-0314/CP2) (finding that Synthroid, a levothyroxine sodium product, was not GRAS/GRAE).

As mentioned above, the Agency believes it is not likely that any currently marketed prescription drug product is grandfathered or is otherwise not a new drug. However, the Agency recognizes that it is at least theoretically possible. No part of this guidance, including the Appendix, is a finding as to the legal status of any particular drug product. In light of the strict standards governing exceptions to the approval process, it would be prudent for firms marketing unapproved products to carefully assess whether their products meet these standards.

D. New Unapproved Drugs

Some unapproved drugs were first marketed (or changed) after 1962. These drugs are on the market illegally. Some also may have already been the subject of a formal Agency finding that they are new drugs. See, for example, 21 CFR 310.502 (discussing, among other things, controlled/timed release dosage forms).

E. Over-the-Counter Drug Review

Although OTC drugs were originally included in DESI, the FDA eventually concluded that this was not an efficient use of resources. The Agency also was faced with resource challenges because it was receiving many applications for different OTC drugs for the same indications. Therefore, in 1972, the Agency implemented a process of reviewing OTC drugs through rule-making by therapeutic classes (e.g., antacids, antiperspirants, cold remedies). This process involves convening an advisory panel for each therapeutic class to review data relating to claims and active ingredients. These panel reports are then published in the Federal Register, and after FDA review, tentative final monographs for the classes of drugs are published. The final step is the publication of a final monograph for each class, which sets forth the allowable claims, labeling, and active ingredients for OTC drugs in each class (see, e.g., 21 CFR part 333). Drugs marketed in accordance with a final monograph are considered to be generally recognized as safe and effective (GRAS/GRAE) and do not require FDA approval of a marketing application.

Final monographs have been published for the majority of OTC drugs. Tentative final monographs are in place for virtually all categories of OTC drugs. The FDA has also finalized a number of negative monographs that list therapeutic categories (e.g., OTC daytime sedatives, 21 CFR 310.519) in which no OTC drugs can be marketed without approval. Finally, the Agency has promulgated a list of active ingredients that cannot be used in OTC drugs without approved applications because there are inadequate data to establish that they are GRAS/ GRAE [e.g., phenolphthalein in stimulant laxative products, 21 CFR 310.545(a)(12)(iv)(B)].

OTC drugs covered by ongoing OTC drug monograph proceedings may remain on the market as provided in current enforcement policies (see, e.g., CPG sections 450.200 and 450.300, and 21 CFR part 330). This document does not affect the current enforcement policies for such drugs.

OTC drugs that need approval, either because their ingredients or claims are not within the scope of the OTC drug review or because they are not allowed under a final monograph or another final rule, are illegally marketed. For example, this group would include a product containing an ingredient determined to be ineffective for a particular indication or one that exceeds the dosage limit established in the monograph. Such products are new drugs that must be approved by the FDA to be legally marketed.

23 | Worldwide Good Manufacturing Practices

Joseph D. Nally
Nallianco LLC, New Vernon, New Jersey, U.S.A.

The formalization of good manufacturing practices (GMPs) commenced in the 1960s and they are now in effect in over 100 countries ranging from Afghanistan to Zimbabwe. Although many countries have developed local requirements, many also rely on the World Health Organization (WHO) GMPs for pharmaceutical products. Regional requirements have also appeared with application to several countries. Examples of these include the following.

(a) Pharmaceutical Inspection Convention (PIC)—guide to GMP for pharmaceutical products—Australia, Austria, Belgium, Canada, Denmark, Finland, France, Hungary, Ireland, Italy, Latvia, Liechtenstein, Malaysia, The Netherlands, Norway, Poland, Portugal, Romania, Singapore, Slovak Republic, Spain, Sweden, Switzerland, and the United Kingdom.
(b) Association of South-East Asia Nations (ASEAN)—GMP: general guidelines—Brunei Darussalaam, Cambodia, Indonesia, Lao PDR, Malaysia, Myanmar, Philippines, Singapore, Thailand, and Vietnam.
(c) European Economic Community (EEC)—guide to GMP for medicinal products—Austria, Belgium, Denmark, Finland, France, Germany, Greece, Ireland, Italy, Luxembourg, the Netherlands, Portugal, Spain, Sweden, and the United Kingdom.

The above guidelines are similar in design and content and model more of a quality management approach or principle when compared with product testing and control more prevalent in the U.S. GMPs. Over the years, these regulations/guides have also been supplemented by descriptive guidelines providing additional information on specific topics.

In general, GMPs have been issued as guides to the achievement of consistent product quality, with interpretation and individual variations being accepted. This was always considered by the industry to be the best approach, rather than defining specific "how to" regulations. However, within the Food and Drug Administration (FDA) areas of compliance, the variability of interpretation by individual inspectors, coupled with the authority to delay approvals of submissions because of noncompliance, has raised some questions about the viability of this approach. Obviously, "how to" regulations leave little room for interpretation and, therefore, in theory make it easier to assure compliance. However, this is a very restrictive approach and does not allow freedom to introduce alternative and more effective methods without modification to the regulations. Consequently, concept or intent regulations are still preferred.

In order to view the current good manufacturing practices (CGMPs) in the context of an international industry, we are presenting brief evaluations of some other GMPs followed in Europe, in Canada, and by WHO and highlighting major differences from the U.S. GMPs.

Bulk pharmaceuticals or APIs were dealt with in Chapter 17.

GOOD MANUFACTURING PRACTICE FOR MEDICINAL PRODUCTS IN THE EUROPEAN ECONOMIC COMMUNITY

The principles and guidelines for GMP are defined in two directives: Directive 91/356/EEC for human products and Directive 91/412/EEC for veterinary products. The guide to GMP applies to both human and veterinary medicinal products, although 2 of the 18 annexes (4 and 5) apply specifically to veterinary medicinal products. The U.S. GMPs also apply to both product types.

The preamble to the guide indicates that this replaces any national GMP requirements within the EEC; consequently, these are multinational requirements, unlike the CGMPs that only apply to products manufactured in or for the United States. It also states that alternative approaches are permitted if they are validated and provide equivalent levels of quality assurance. An equivalent situation exists with the U.S. GMP requirements. However, interpretation by individual FDA inspectors and possible opportunity to block approvals via the Pre-Approval Inspection have made this a less than ideal situation in the United States.

Each of the nine chapters of the EEC guide opens with a principle that essentially defines the intent. The guidelines that follow provide more details on the areas to be addressed. This is an admirable approach and the one that has been used in drafting internal company quality standards by one of the authors. The guide also includes 18 annexes, which address special areas such as sterile medicinal products, biological medicinal products and liquids, and creams and ointments in even more detail.

In reviewing the guide, only those topics that differ from the FDA regulations are discussed.

CHAPTER 1. QUALITY MANAGEMENT

The principle emphasizes that the achievement of quality requirements "is the responsibility of senior management and requires the participation and commitment by staff in many different departments and at all levels within the company." This clearly acknowledges the important role of senior management and that quality control (QC) alone cannot achieve the required results.

The chapter also refers to product development and requires the application of GMP and good laboratory practices (GLP) to the design and development phase. The U.S. GMPs do not specifically address design and development, although GLPs do apply to certain stages and there is an expectation that GMPs will be applied especially during the production of clinical supplies.

Self-inspection and/or quality audit is required; these are not specifically included in the U.S. GMPs. There is also the statement that materials may not be released for use before the relevant tests are performed. The term "relevant" is very subjective and could be considered to allow use at risk provided some data were available. This is similar to §211.84, which uses the term "appropriate."

CHAPTER 2. PERSONNEL

The concept of the qualified person (authorized person in PIC), with responsibility for product release, is introduced. Unlike §211.22, the qualifications and experience required for this head of QC are defined (Article 23 of Directive 75/319/EEC).

A formal qualification is required in pharmacy, medicine, veterinary medicine, chemistry, pharmaceutical chemistry, or biology. The subjects that must be included in the course are defined. The qualification is to be followed with one year of practical training, at least six months of which is to be in a pharmacy. A further two years of experience in a QC environment is required.

This chapter also differentiates between the evaluation of products imported from other EEC and non-EEC countries. In the latter case, the importing country must perform full qualitative analysis, quantitative analysis of all active ingredients, and any other tests required to confirm the product quality. Directive 75/319/EEC Article 22 does allow member states to relieve the importing qualified person of these responsibilities provided further exportation is not to occur and that arrangements have been made with the exporting country to assure that the testing was performed there. For importation from another EEC country, this testing is not required if the QC reports, signed by the qualified person, are provided.

The chapter defines the responsibilities of the head of production and the head of QC. This clearly makes the head of production responsible for production operations and compliance with procedures. This role is not specified in the CGMPs, and §211.25 only requires that personnel shall be appropriately qualified and/or experienced to perform their assigned duties. This defined role for production is a positive acknowledgment of the importance of product management in the GMP compliance.

In addition to defining the areas of responsibility for the heads of these two key functions, there is a definition of joint responsibilities, which can include approval of procedures, process validation, approval of suppliers, and monitoring of GMP compliance.

With respect to training, there is a requirement to assess effectiveness. This should be a routine requirement for any important activity.

All personnel, presumably only those more directly involved in production and support activities, are to be medically examined at the time of recruitment. Section 211.28 does refer to personnel health and medical examination, but there is no specific requirement for a medical examination on recruitment or at any other specified time.

CHAPTER 3. PREMISES AND EQUIPMENT

The maintenance and repairs to premises are to be performed so that there is no adverse impact on quality. This would seem to be obvious. However, §211.58 only requires that buildings shall be maintained in a good state of repair. For equipment, the guide and the CGMPs are essentially equivalent.

The guide requires that highly sensitizing products (e.g., penicillins) and "certain additional products such as certain antibiotics, certain hormones, certain cytotoxins, certain highly active drugs and nonmedicinal products" should be produced in different facilities or exceptionally by campaigning in the same facilities. Section 211.176 applies restrictions only to penicillin.

Sampling of starting materials is normally expected to be performed in a separate sampling area, but alternatives are allowed provided that they prevent the opportunity for cross-contamination. It is surprising that this elaboration is included since the guide overall allows alternatives. Section 211.84 does not specify a separate area, but does require the prevention of contamination.

CHAPTER 4. DOCUMENTATION

This chapter makes several references to the signing of documents—approvals, alterations, process steps (initials), process completion, and process deviations. There is also reference to electronic recording, which is considered acceptable with the usual safeguards regarding access. This would appear to be equivalent to the approach by the FDA.

There is a rather extreme requirement for the use of logbooks—for recording of equipment validation, calibration, maintenance, cleaning, and repair and also for equipment and facility usage. The only reference to logbooks in CGMPs is §211.182 with respect to equipment cleaning and maintenance. Possibly of minor significance, the guide refers to logbooks and U.S. GMPs only to logs.

CHAPTER 5. PRODUCTION

There are several references to minimizing the potential for cross-contamination, ranging from material sampling through production and from operator clothing to packaging. The U.S. GMPs address most of the same concerns. However, there is no specific reference to the manufacturing processes themselves, except to that of penicillin.

Although validation appeared to gain European acceptance slowly, a requirement now appears.

The importance of starting material quality is emphasized by the preference to buy directly from the producer rather than through an agent. Although not included in the U.S. GMPs, this is a good practice.

The preference for roll labels over cut labels is emphasized. The FDA has been more demanding in this area because of the number of recalls caused by incorrect labeling. The guide also emphasizes the need to confirm that code readers and counters are working correctly. Again there is a specific highlight, probably because of the importance of the subject.

The samples removed from packaging lines are not to be returned. The intent of this is to prevent mix-ups by the return of opened/modified units to the wrong line. There would seem to be no need for this restriction, provided adequate precautions are taken; also, is a side table adjacent to a packaging line considered to be part of the line? The CGMPs provide no such restrictions.

The guide also notes "reprocessing of rejected products should be exceptional." Although not so described in the U.S. GMPs, the FDA and industry would agree. Indeed, a validated process should only rarely produce rejects and that should be due to special cause variation, not due to normal process variation.

CHAPTER 6. QUALITY CONTROL

Storage of reference samples for products differs from that for starting materials. The guide suggests that with certain defined exceptions (solvents, gases, and water), samples of all starting materials should be retained for two years after the expiration date of the last batch of product manufactured from the material. The U.S. GMPs (§211.170) refer to retained samples of active ingredients only and for one year beyond product expiration.

The testing reports require the initials of persons performing and checking the testing and signatures for release. The earlier comments on electronic recording presumably also apply here. This is equivalent to those mentioned in §211.194.

CHAPTER 7. CONTRACT MANUFACTURE AND ANALYSIS

With the emphasis on ISO 9000 in Europe, it is not surprising that a chapter was included on arrangements.

The contract should define responsibilities with respect to purchasing of materials, testing and release of materials, process control, final testing, and product release. Additional issues include who retains samples and evaluates complaints. The drafting of the contract should involve persons with adequate knowledge, especially of GMP requirements.

Access to the contractor's premises should be agreed in the contract.

The U.S. GMPs do not include any reference to contract operations. The emphasis on technical involvement in the guide is particularly relevant since many contracts are drafted by legal departments with only limited technical input.

CHAPTER 8. COMPLAINTS AND PRODUCT RECALL

The guide provides more direction on the extrapolation of complaints to other batches. Both the guide and the U.S. GMPs require regular review of complaints data to identify potential problems and require appropriate action. The guide also provides additional guidance with respect to recalls, but these are essentially identical to the FDA expectations.

CHAPTER 9. SELF-INSPECTION

Self-inspections are to be conducted by competent persons within the company and recorded. The author considers that additionally managers and supervisors should perform frequent audits of their functions; these need not necessarily be recorded.

Surprisingly, the U.S. GMPs do not include any reference to internal inspections, which are, however, a performed and expected practice.

The 18 annexes, providing more detailed information, are entitled:

- Manufacture of sterile medicinal products;
- Manufacture of biological medicinal products for human use;
- Manufacture of radiopharmaceuticals;
- Manufacture of veterinary medicinal products other than immunologicals;
- Manufacture of immunological veterinary medicinal products;

- Manufacture of medicinal gases;
- Manufacture of herbal medicinal products;
- Sampling of starting and packaging materials;
- Manufacture of liquids, creams, and ointments;
- Manufacture of pressurized metered dose aerosol preparations for inhalation;
- Computerized systems;
- Use of ionizing radiation in the manufacture of medicinal products;
- Manufacture of investigational medicinal products;
- Manufacture of products deriver from human blood or human plasma;
- Qualification and validation;
- Certification by a qualified person and batch release;
- Parametric release;
- GMP for active pharmaceutical ingredients.

Copies of "The Rules Governing Medicinal Products in the European Community. Volume IV. Good Manufacturing Practice for Medicinal Products" may be obtained from Unipub, 4611-F Assembly Drive, Lanham, MD 20706-4391, U.S.A. (301-459-7666), and are also available at http://72.14.205.104/search?q=cache:S5mTvAMvYrwJ:www.farmaceutene.no/filer/Guide_to_good_practice.pdf.

WORLD HEALTH ORGANIZATION—GOOD MANUFACTURING PRACTICES FOR PHARMACEUTICAL PRODUCTS

The first WHO guide on GMPs was drafted in 1967. Some minor revisions were incorporated when it was published as a supplement to the International Pharmacopoeia in 1971. It was later incorporated into the WHO Certification Scheme on the Quality of Pharmaceutical Products Moving in International Commerce (WHO 28.65) in 1975. No further revisions occurred until 1992. This latest version took into account the considerable developments in GMPs since 1975 and also the ISO 9000 series and the Convention for Mutual Recognition of Inspection in Respect of the Manufacture of Pharmaceutical Products (PIC).

The WHO GMPs are presented in three parts: Part One, Quality Management in the Drug Industry: Philosophy and Essential Elements; Part Two: Good Practices in Production and Quality Control; and Part Three: Containing Two Supplementary Guidelines. It is anticipated that this part will be expended by the addition of more guidelines. The attachment of the guidelines to the GMPs makes the document easier to use, whereas the FDA guides and guidelines are totally separate.

Throughout the document, there is considerable emphasis on validation and avoidance of cross-contamination. The WHO guideline is essentially identical to the EEC guideline in content. This means that there is a high level of consistency throughout a large part of the industry.

The WHO version of GMPs is used by the pharmaceutical regulators and the pharmaceutical industry in over 100 countries worldwide, primarily in the developing world. In the European Union, the EU-GMPs, with more compliance requirements than those stated in the WHO GMPs, are in force; whereas in the United States, the FDA's version of GMPs, including requirements over and above those stated in the WHO document, are enforced. Similar forms of GMPs are used in other countries, with Australia, Canada, Japan, Singapore, and others having highly developed/sophisticated GMP requirements (1).

Since the publication in 1999 by the International Conference on Harmonization (ICH) of "GMPs for Active Pharmaceutical Ingredients," GMPs also apply in those countries and trade groupings that are signatories to ICH (the EU, Japan, and the United States) and other countries who adopt ICH guidelines (e.g., Australia, Canada, and Singapore) to the manufacture and testing of active raw materials (1).

GENERAL CONSIDERATIONS

The guide is applicable to all "large-scale operations for the production of drugs in their finished dosage forms including large-scale processes in hospitals and the preparation of clinical trials supplies." "Large-scale" is not defined, but the term is used presumably to differentiate production that is usually for multiple patient/customers from dispensing and individual patient special preparations. The specific reference to hospitals is unique. Other GMP guides/regulations do not specifically refer to hospital manufacture. It is possibly assumed that they should comply, but hospitals are rarely, if ever, inspected.

Parts One and Two (the main body of the GMPs) do not apply to the manufacture of active ingredients (bulk pharmaceutical chemicals). However, one of the two supplementary guidelines in Part Three does deal specifically with this subject. This is a welcome plus compared with the confusion over the degree of application of the CGMPs to bulk pharmaceutical chemicals. Excipients are not covered by the current guide, and this topic could become a subject for a future Part Three guideline.

As with the EEC GMPs, the WHO document is considered to provide guidance, and equivalent alternative approaches are allowed to meet individual needs provided they are validated. There is, however, an interesting footnote—"the world 'should' in the text means a strong recommendation." This implies that even more attention than usual should be given to the evaluation of alternatives in these instances.

Because of the similarity to the EEC guideline, only a limited review is presented.

PART ONE. QUALITY MANAGEMENT IN THE DRUG INDUSTRY: PHILOSOPHY AND ESSENTIAL ELEMENTS

As with ISO 9000, the important role of top management in defining and authorizing quality policy is emphasized. It also states that quality assurance (the concept and not the function) is a management tool. This appears to be indicating that quality should be used as a value-adding approach to the business, rather than being considered a regulatory requirement.

1. *Quality assurance.* In this section, the guide introduces the term "principle," with which it opens several later sections. The principle here is that quality assurance is a concept that covers everything that can impact on product quality. It includes all procedures, activities, and personnel responsibilities. It applies to product design and development (as with ISO 9001), although there is no elaboration on this aspect of product life cycle. Presumably, it is considered that although design and development can impact on future production and product quality, the GMPs do not apply—except as previously indicated to the production of clinical supplies. This is consistent with other GMP guides. However, it is stated that pharmaceutical products should be designed and developed with the application of GCP and GLP and to "take into account the requirements of GMP."

   The importance of management is denoted by the need to define managerial responsibilities in job descriptions.

   Quality audit and/or self-inspection is required to evaluate compliance. The either/or approach is novel and again is an indication of the wide-reaching role of the supervision and management in assuring compliance and de-emphasizing the role of the quality function (see also section 9).

   The overall purpose of WHO GMP compliance is to ensure that products are "fit for their intended use." This goes beyond the U.S. GMPs, which are primarily concerned that the products should be safe and effective. For example, a bottle that is difficult to open may not be fit for intended use, but the product inside could be safe and effective. It is also stated that many levels of personnel in different functions, suppliers and distributors, must participate and be committed to the achievement of the required quality, but that the responsibility lies with senior management. The FDA continues to issue a warning letter addressed to the chief executive officer of the company. One of the main reasons for this is to ensure that top management was aware of important quality deficiencies. However, one would expect that company internal communications would normally address such issues.

2. *Good manufacturing practices for pharmaceutical products (GMP).* The section states that GMPs "are directed primarily to diminishing the risks, inherent in any pharmaceutical production,

that cannot be prevented completely through the testing of final products." This statement clearly acknowledges that perfection may not be achievable. It also infers that infrequent inadvertent lapses in compliance that result in no increase in risk should not be considered a serious issue. The potential risks (not detectable by finished product testing) are identified as cross-contamination and labeling mix-ups. This appears to be a very practical approach, since deviations from specifications can usually be detected by product evaluation.

The subsection on complaints requires that action be taken to prevent reoccurrence. This is an illustration of quality being used as a tool for improvement. In more farseeing companies, one would expect the complaint handling process (reactive) to be converted into a proactive customer satisfaction process; possibly this is beyond GMP.

3. *Quality control.* This encompasses all activities that are sometimes subdivided functionally into QC (laboratory) and quality assurance (procedures and documentation). It is surprising in a modern document, with so many positive approaches, that the quality function is still termed quality control. It is becoming increasingly common for the overall function to be called quality assurance, with QC being the laboratory activity.

The required independence of QC from production is emphasized—"is fundamental." As indicated elsewhere in this book, there are several ways to assure this independence, and functional reporting of plant QC outside of plant management is only one approach.

4. *Sanitation and hygiene.* No issues.

5. *Validation.* Validation is required for processes, testing, and cleaning. There is no reference to computer systems validation. Installation and operational qualification for equipment are not mentioned, but these could be considered as subsets of process validation. The terms prospective and retrospective validation are included, but not concurrent validation. Although it is acknowledged that in most situations retrospective (currently marketed products) and prospective (new products) validation will apply, there are situations where concurrent validation is appropriate. These were described earlier in the book.

6. *Complaints.* The emphasis relates to the potential for recall.

7. *Product recalls.* Unlike the CGMPs, there is a requirement to test the recall system to ensure that it will function effectively if needed.

8. *Contract production and analysis.* ISO 9000 was probably the underlying basis for this section, which requires that the responsibilities of each party are clearly defined and agreed with respect to purchasing, testing, and releasing materials, for process controls, product testing, and also for record keeping. This is an important subject requiring special attention from the quality function since contracts are frequently handled by the legal department, which is not always aware of these important technical considerations.

There is a specific comment that the contract should permit audit of the contractor operation. The author has seen examples in which this was not included and audit was either prohibited or restricted.

There is no equivalent section in the U.S. GMPs.

9. *Self-inspection and quality audits.* This section elaborates on the general statement in section 2, thereby bringing a different interpretation to the subject. We have previously defined self-inspection as an inspection carried out by the management/supervision responsible for the particular process or activity (chapter 7). In this section, self-inspection relates to the use of an internal multifunctional team to assess compliance. Quality audit is defined as an approach to improve performance of a process using external sources. Although this can be an effective approach, the absence of continuous self-inspection by management weakens the concept that all personnel are to be involved in quality.

A subsection refers to audit of suppliers' facilities, but only if required. However, it is difficult to appreciate how GMP compliance can be assured without an audit by someone, either from the purchaser's company or from a regulatory agency that issues a report. Possibly, ISO 9000 certification could be acceptable confirmation that quality systems are in place.

10. *Personnel.* This section is essentially the same as outlined in the EEC guide. The scientific education requirements are defined for key production and QC supervisors and managers. These include an appropriate combination of chemistry or biochemistry, chemical engineering, microbiology, pharmaceutical sciences and technology, pharmacology and toxicology, physiology, or other related sciences. They must also have experience in the manufacture and quality assurance of pharmaceutical products. The individual responsibilities of the heads of production and QC and their shared responsibilities are also identified. The U.S. GMPs do not go into this amount of detail.

Training programs are to be approved by the head of production or QC, as appropriate, and effectiveness should be assessed. There is no reference to the qualifications of the personnel giving the training. This is the converse of the U.S. GMPs, which do not refer to approval of programs but do require training to be given by "qualified individuals."

Personal hygiene requirements again mirror the EEC guideline. Pre-employment health examination is required and also periodic eye examinations for those who perform visual inspections. It is surprising that color blindness evaluations are not specifically mentioned, especially for laboratory personnel.

11. *Premises.* The main differences from the U.S. GMPs relate to the identification of products (in addition to penicillin) that require self-contained or separate facilities. These include biological preparations (live organisms), some other antibiotics, hormones, cytotoxic substances, and highly active pharmaceutical products and nonpharmaceuticals. It is accepted that campaign production in the same facilities may be satisfactory provided there are validated cleaning procedures.

There is also a recommendation, usually followed in the design of newer facilities, that services should be accessible from outside of the manufacturing areas to facilitate maintenance.

12. *Equipment.* No issues.

13. *Materials.* Suppliers are to be named in the specifications and, where possible, purchases should be directly from the manufacturer. The objective here is to eliminate the possibility of an intermediary agent changing source of supply without the knowledge of the purchaser. This is established practice, but is not in the U.S. GMPs. It is also recommended that the material specification be "discussed" with the supplier. This, although a move in the right direction, is not enough. Specifications should be agreed with the supplier.

As with most GMPs, materials must be released by QC before use. However, this does not appear to preclude use before all testing is completed, provided QC gives approval.

The important subject of reference standards is also addressed in this section and allows the use of secondary (in-house) standards. These should be crosschecked against official reference standards—a good practice for laboratory operations.

14. *Documentation.* The importance of good documentation as a solid basis for GMP compliance is emphasized. This section does, however, introduce some degree of confusion with respect to signatures on documents. For example, 14.3 requires documents to be signed, whereas 14.9 indicates that electronic data-processing systems may be used for data entry. Is a signature data? In 14.28, on processing records, there are several references to requirements for signatures or initials, whereas in 14.31 on packaging records, the required entries may be by signature or electronic password. Overall, it must be assumed that a secure and validated electronic system is considered an adequate alternative to handwritten signatures and initials. This is in line with the electronic signature FDA requirements.

Logbooks are required for major pieces of equipment to record validation, calibration, maintenance, cleaning, and repair. In today's electronic age, we assume that electronic records would be acceptable, but this is not so stated.

PART TWO. GOOD PRACTICES IN PRODUCTION AND QUALITY CONTROL

This part goes into more detail on the procedures to be followed. Sterile products and bulk pharmaceutical chemicals are covered in Part Three.

15. *Good practices in production.* This section has an emphasis on the avoidance of contamination and highlights the need for effective cleaning between products, the segregation of hazardous operations, and the minimization of dust generation and potential problems of air recirculation.

A unique point (15.18) is the requirement to clean all containers before filling to eliminate any foreign contamination such as glass particles. This would seem excessive, however, as previously indicated alternative approaches are acceptable. Consequently, provided it can be demonstrated that foreign materials will not be present, the cleaning could be eliminated. It must be noted that the cleaning requirement is a "must"—a strong recommendation.

In the subsections on processing operations, it states that instruments used for analytical testing should be checked daily in addition to routine calibration. Although not specified in U.S. GMPs, this is probably a useful approach, especially for instruments located in processing environments.

There are the usual concerns about the potential for labeling mix-ups. The guide does not go as far as the current FDA approach in defining what to do, but it does recommend labeling immediately after filling and sealing and taking special care with cut labels, off-line coding, and hand packaging. The use of on-line electronic scanning is encouraged, and the need to routinely verify the operational effectiveness of such equipment is noted.

16. *Good practices in quality control.* As indicated previously, in this guide, QC is the complete quality function, encompassing QC and quality assurance functions.

Certificates of analysis from a supplier may be used as a basis for reduced testing, but there are several interesting and practical provisos: reliability of the supplier's analytical capability must be demonstrated by comparative analysis, the supplier's site must be audited, and the certificate must be an original (not a photocopy) and include a statement of the specification and the test methods used and the results and date of testing. This goes well beyond the CGMPs. Most of the points make practical sense. However, the need for original certificates is difficult to understand, unless it is to minimize the chance of fraud. The date of testing does ensure that the data are current.

The evaluation of batch failures or process deviations should not only be investigated, but should, if necessary, be extended to include other batches or products. This should be normal practice within the industry but is not specifically included in the U.S. GMPs.

The requirements for retention samples of active ingredients and products are the same as those for the U.S. GMPs. The WHO guide also has a retention requirement for excipients: with the exception of solvents, gases, and water, these are to be retained for two years.

PART THREE. SUPPORTING AND SUPPLEMENTARY GUIDELINES

17. *Sterile pharmaceutical products.* The introduction states that this is a supporting guideline and does not replace any sections of Part One or Two. Again, the emphasis is on minimizing the risk of contamination—microbial, paniculate, and pyrogen.

There are no surprises in this section, which is very similar to the EEC guideline. There is reference to the use of barrier technology and automated systems that "can produce significant advantages in ensuring the sterility of manufactured products." There is a specific point that a conveyor should not pass through a partition from a Class 100 area into an area with a lower classification unless the conveyor is continuously sterilized.

Disinfectants and detergents are to be monitored for microbial contamination, but there are no specific requirements that they should be sterile.

The bioburden impact is to be evaluated, including starting materials and product, before sterilization. Routine evaluation of the materials can be eliminated if supported by data.

The subject of aseptic processing versus terminal sterilization is briefly raised, and terminal sterilization is to be used wherever possible. Where this is not possible, "consideration should be given to complementing the filtration process with some degree of heat treatment." These sentiments are commendable, but unless some further guidance is provided or regulatory agencies rely on the individual company evaluation, this could lead to excessive amounts of evaluation and delays in product registration. Some of the issues include the following.

How much degradation is acceptable when terminal sterilization or complementary heating is used?

The amount of toxicology works to evaluate the potential impact of these degradation products.

Which heat treatments should be evaluated to complement aseptic processing (temperature and time)?

Filled containers of injections are required to be inspected individually. This is also a United States Pharmacopeia requirement. This is another area where process improvements and validation have not allowed any relaxation of evaluation. Presumably, the facts that no process can provide 100% assurance of compliance, the evaluation is nondestructive, and the product is an injection all contribute to the continuation of this test.

Links to the WHO GMPs can be found at http//www.who.int/medicines/areas/quality_safety/quality_assurance/production/en/.

CANADA GOOD MANUFACTURING PRACTICES GUIDELINES, 2002 EDITION, VERSION 2

The introduction emphasizes the guideline nature of this document. "The content of this publication should not be regarded as the only interpretation of the GMP regulations nor does it intend to cover every conceivable case. Alternate means of complying with these Regulations can be considered with the appropriate scientific justification. Different approaches may be called for as new technologies emerge."

The guideline only applies to dosage forms and there is no reference to any alternative approaches for drug substances.

The overall format of the guideline is interesting. The regulations, which are statutory requirements, are highlighted and then followed by the rationale for the regulation and an interpretation with more detail of how compliance might be assured. This appears to be a very sensible and practical approach. In the author's opinion, it is also a good approach to take for internal company policies—Policy, Rationale (Level 1) and Guidance on how to be in compliance (Level 2), followed by local site SOPs (Level 3).

The general intent is similar to other major GMPs and only unique points will be addressed.

There is a Quality Management and Quality Assurance Principle section after the introduction section. Some notable parts include as follows.

QUALITY MANAGEMENT

"The attainment of this quality objective is the responsibility of senior management and requires the participation and commitment of personnel in many different departments and at all levels within the establishment and its suppliers. To achieve the objective reliably, there must be a comprehensively designed and correctly implemented system of quality assurance that incorporates Good Manufacturing Practices and thus quality control."

"The basic concepts of quality assurance, Good Manufacturing Practices and quality control are inter-related. They are described here in order to emphasize their relationships and their fundamental importance to the production and control of drugs."

QUALITY ASSURANCE

"Quality assurance is a wide-ranging concept that covers all matters that individually or collectively influence the quality of a drug. It is the total of the organized arrangements made with the objective of ensuring that drugs are of the quality required for their intended use. Quality assurance therefore incorporates Good Manufacturing Practices, along with other factors that are outside the scope of these guidelines."

The concept of holding senior management responsible is not covered in U.S. GMP, but is certainly practiced as evidenced by the addressee in FDA Warning Letters.

PREMISES (C.02.004)

The 1989 document made no reference to critical systems. This has been addressed in the 1995 and 2002 revisions that clarify under interpretation such systems such as HVAC and DI water to be qualified when installed or changed and to be subject to periodic verification.

EQUIPMENT (C.02.005)

Some details are included in the Interpretation section, which, although obvious, are not in other GMPs.

Tanks used for the manufacture of liquids and ointments are to be equipped with fittings that can be dismantled and cleaned.

Tanks and similar fabricated equipment are equipped with covers.

Chain drives and transmission gears are enclosed or properly covered.

The use of temporary devices, such as tape, is avoided.

Validation is introduced in the 1995 proposal by requiring installation and operational qualification on all critical manufacturing equipment—"critical" is not defined. In addition, design and maintenance of equipment to assure effective operation is extended to process water systems.

PERSONNEL (C.02.006)

"For fabricators, packagers, labelers and testers, individuals in charge of the manufacturing department and quality control department:

hold a university degree equivalent in a science related to the work being carried out;
have practical experience in their responsibility area;
directly control and personally supervise on site, activities under their control; and
can delegate their duties . . ."

This is not as detailed as the EEC and WHO guidelines, but is more extensive than the CGMPs.

The regulators have identified that the cutbacks occurring in the industry could have a detrimental impact on quality and state that "The responsibilities placed on any one individual are not so extensive as to present any risk to quality." Evaluation of training effectiveness is also required.

RAW MATERIAL TESTING (C.02.009/C.02.010)

As with the other regulations, testing is expected to be complete before material is used in production. "Test methods are validated, and the results of such validation studies are documented. Full validation is not required for methods included in any standard listed in Schedule B to the Food and Drug Regulations, but the user of such a method establishes its suitability under actual conditions of use.

Note: Guidance for the validation of particular types of methods can be obtained in publications such as the ICH guidelines titled "Validation of Analytical Procedures: Methodology" or in any standard listed in Schedule B to the Food and Drug Regulations.

The Interpretation section of C02.010 on raw materials is prescriptive for vendor certification. "The testing is performed on a sample taken after receipt of the raw material on the premises of the person who formulates the raw material into dosage form, unless the vendor is certified. A raw material vendor certification program, if employed, is documented in a standard operating procedure. At a minimum, such a program includes the following:

1. A written contract outlining the specific responsibilities of each party involved. The contract specifies:
   1.1.1 the content and the format of the certificate of analysis, which exhibits actual numerical results and makes reference to the raw material specifications and validated test methods used;
   1.1.2 that the raw material vendor must inform the drug fabricator of any changes in the processing or specifications of the raw material; and
   1.1.2 that the raw material vendor must inform the drug fabricator in case of any critical deviation during the manufacturing of a particular batch of a raw material.
1.2 An audit report is issued by a qualified regulatory authority demonstrating that the API fabricator complies with the ICH Good Manufacturing Practice guide for API or with any standard or system of equivalent quality. This report should be less than 3 years old, but is valid for 4 years from the date of the inspection. If such an audit report is unavailable or is more than 4 years old, an on-site audit of the API fabricator, against the same standard or its equivalent, by a person who meets the requirements of Interpretation 1 under Section C.02.006, is acceptable.
1.3 Complete confirmatory testing is performed on the first lot of any raw material received from a new vendor. For APIs, a copy of the results of the impurity and residual solvent profile is also obtained."

MANUFACTURING CONTROL (C.02.011/C02.012)

Validation of critical production processes and changes is required in the proposed update. In this instance, critical is defined—those processes that can cause variation. The use of validated on-line label accountability systems is also proposed as an alternative to label reconciliation.

Printed packaging materials are to be "readily distinguishable" to minimize the potential for mix-up. There is no additional guidance on what is suitable.

When subcontracting takes place, the contract should define acceptable quality performance and GMP responsibility.

A self-inspection program is required. There should be written reports of the findings and the corrective actions initiated. Although inherent in most pharmaceutical operations, self-inspection is noticeably absent in the U.S. GMPs.

The HPB does on occasion ask to see copies of self-inspection reports. Industry has argued that these should be confidential internal reports in order to maximize their benefit. Otherwise, they could become a focal point for inspectors to identify and cite deficiencies that have already been identified and corrected. Previously, the FDA has indicated that it would only view these inspection reports if it found other evidence of serious compliance deficiencies. More recently, some investigators have been asking to see them as proof that internal inspections were being performed.

Contractors are permitted but only when a specific written agreement is in place that covers all applicable manufacturing, packaging, and control activities/requirements. It also states that "the fabricator, packager/labeler, distributor, or importer is responsible for assessing the contractor's continuing competence to carry out the work or tests required in accordance with the principles of GMP described in these guidelines." Assessing continuing competence is not defined, but inspection or audit is the preferred method by HPB. This detail is not specifically addressed in the U.S. GMPs, but most companies do use one or more of these approaches.

QUALITY CONTROL DEPARTMENT (C.02.013/C.02.014/C.02.015)

Requirements and responsibilities for testing, control, and product release are specified in this section. Quality management and quality assurance requirements and responsibilities are not covered here.

There is nothing novel or unusual in this section except for a point of detail: the person in charge of the laboratory must be "an experienced university graduate holding a degree in science related to the work carried out ... or reports to a person having these qualifications." Although this would seem to be self-evident, there could be some problems in a company that performs both chemical and microbiological laboratory works when two such qualified individuals may be required, depending on their qualifications and experience.

PACKAGING MATERIAL TESTING (C.02.016/017)

Testing and reduced testing requirements are the same as described earlier for raw materials.

FINISHED PRODUCT TESTING (C.02.018/019)

The rationale emphasizes that finished product testing complements the in-process controls. This is very sensible and clearly highlights the importance of in-process controls. Taken to a logical conclusion, it should then be possible to delete or reduce some finished product tests, which are more effectively performed during the production process; these could include tablet weight, thickness, hardness, fragility, and liquid volume checks.

Test methods are required to be validated and they reference the ICH method validation guideline.

This section goes on to state that an importer must perform testing on receipt of product but allows for alternate approaches. If the importer "... has evidence satisfactory to the Director to demonstrate that drugs sold to him by the vendor of that lot or batch of the drug are consistently manufactured in accordance with and consistently comply with the specifications for those drugs; and undertakes periodic complete confirmatory testing with a frequency satisfactory to the Director. ..."

RECORDS (C.02.020—C.02.024)

The requirements are specific for every fabricator, packager/labeler, distributor, and importer in that they must maintain the required records on their premises in Canada for each drug sold. Required records include master production documents; evidence that the conditions under which the drug was fabricated, packaged/labeled, tested, and stored are in compliance; evidence establishing the period of time during which the drug in the container in which it is sold will meet the specifications for that drug; and adequate evidence of the testing and full distribution records. These are far-reaching requirements for importers.

This section also allows for electronic records. "Records may be maintained in electronic format provided that backup copies are also maintained. Electronic data must be readily retrievable in a printed format. During the retention period, such records must be secured and accessible within 48 hours to the fabricator, packager/labeller, distributor, or importer. An electronic signature is an acceptable alternative to a handwritten signature. When used, such a system must be evaluated and tested for security, validity, and reliability, and records of those evaluations and tests must be maintained. The validation of electronic signature identification systems is documented."

SAMPLES (C.02.025/026)

A sample of each lot of raw material used in the production of dosage forms is to be retained for at least two years after the lot was last used. This is very different from the U.S. GMPs, which require retention samples of active ingredients only and not excipients. Also, the normal retention time is one year after the expiration of the dosage form in which it was last used—for a product with a five-year shelf life, this would relate to six years. Since the primary value of retention samples is in the evaluation of quality queries on the dosage form, the FDA retention period seems more appropriate.

The Canada regulation also requires the sample to be retained in Canada, with the manufacturer or the importer. Such a restriction is not imposed by the U.S. GMPs.

STABILITY (C.02.027/028)

Accelerated stability data are considered to be preliminary information only. These accelerated data are supported by long-term testing, and the assignment of the expiry date is based on the long-term testing.

For new chemical entities, at least three lots of each strength are sampled for the development of shelf life data.

For existing chemical entities (e.g., generic drugs), two lots of each strength are sampled. The principle of bracketing and matrixing designs may be applied if justified. The shelf life is established from the date of fabrication.

This detail is not in the U.S. GMPs, but is incorporated into the ICH guidelines for new products, where three batches are required and the length of storage is defined.

The update incorporates the ICH stability testing requirements for both new products and significant changes.

There is a statement that sterility testing of a sterile formulation may not be required if the effectiveness of the container-closure system has been demonstrated.

STERILE PRODUCTS (C.02.029)

This section provides additional details with respect to sterile products. The interpretation does indicate that terminal steam sterilization is the method of choice for sterilization when practical. This is in line with the FDA thinking, but is not in the U.S. GMPs.

The Interpretation section provides detail on the Class A, B, C, and D manufacturing areas as in EC and other regulations. This is absent from the U.S. GMPs.

ANNEXES

A. Internationally Harmonized Requirements for Batch Certification
B. Alternate Sample Retention Site Application Form
C. References

The Canada GMPs are available at http://www.hc-sc.gc.ca/dhp-mps/compli-conform/index_e.html.

REFERENCE

1. Good Manufacturing Practice. http://en.wikipedia.org/wiki/good_manufacturing_practice, accessed Sep 12, 2006.

SUGGESTED READINGS

1. Sharp J. Good Pharmaceutical Manufacturing Practice. CRC Press, 2005.
2. Griffin JP. The Textbook of Pharmaceutical Medicine. 5th ed. Blackwell Publishing, 2006.
3. Monkhouse DC. Drug Products for Clinical Trials. Marcel Dekker, 1998.
4. Anisfeld MH. International Biotechnology, Bulk Chemical and Pharmaceutical GMPs. 5th ed. CRC Press, 1999.

24 | Quality Approaches: ISO 9000, Malcolm Baldrige, and Six Sigma

Joseph D. Nally
Nallianco LLC, New Vernon, New Jersey, U.S.A.

Other quality standards and approaches are presented in this chapter to provide a broader picture of current quality philosophy and efforts that consider the business as a whole. To consistently provide a high level of product quality that is required by Good Manufacturing Practices (GMP) requires not only product quality processes but sound business processes as well.

Quality and specifically quality improvement have become recognized as business differentiators. The initial stimulation, especially in the U.S.A., was the immense adverse impact on the U.S. automotive and electronics industries from Japanese imports. The U.S.A. had a large home market with minimal external competition. This lack of competition was essentially due to three factors: the cheap-labor countries, which included Japan, were not capitalized for major items such as automobiles or televisions and had a reputation for cheap but poor-quality goods; the European countries, which were recovering from the devastating impact of two wars, were mainly high-labor-cost areas, but some luxury products were exported to the U.S.A.; and third, the distances from Europe and Japan resulted in high transportation costs. This all changed, with much support and direction from an American, W. Edwards Deming, when the Japanese realized that to expand their economy by export, they must change the international perception of poor quality and that higher and consistent quality actually costs less, due to higher output and less rework activity. This, coupled with the acceptance of a long-term view with respect to profitability, forever changed the role of quality in business. Another factor having significant impact was that products were becoming more sophisticated and consequently more complex to manufacture, and even low levels of poor quality in individual components would result in a high level of poor quality in the finished product.

The American public started to buy the more reliable Japanese imports, and eventually American industries realized that they too must adopt quality concepts if they intended to stay in business. Over the years, most of the quality improvement tools that were developed were introduced by major companies. These included quality circles, zero defects, statistical process control, quality function deployment, etc. These rarely succeeded in transforming companies since in most cases they were introduced as stand-alone techniques that were expected to resolve all the problems of the company. As each new concept came to light it was introduced with tremendous support and enthusiasm, usually by consultants—who did become successful—and the previous techniques waned. The major change in the U.S.A. came about when it was realized that quality improvement should apply to every business activity and should be integrated into the business strategy—total quality. Within the U.S.A., the government got involved and, in order to enhance attention to quality, introduced in 1987 the Malcolm Baldrige National Quality Award (MBNQA).

A successful approach used by a number of Malcolm Baldrige award winners is the Six Sigma quality process improvement methodology. This approach has been around since the 1980s when it was first championed by Motorola and is used extensively today in a variety of businesses. This approach has been successful because it is typically focused on value-adding business processes.

Within Europe there was also the realization that quality was important. Additionally, with the unification of Europe through the European Economic Community, it was considered that there should be some guiding principles that would allow for easy intercountry distribution knowing that each country was following an equivalent approach to quality. The internal application of these principles might also make it more difficult for others to import into the EEC. As a consequence, the EEC introduced the International Standards Organization Series 9000 (ISO 9000).

This chapter will discuss the ISO 9000 series standards for quality management systems, the current MBNQA program, and the Six Sigma process improvement approach/methodology.

ISO 9000 SERIES

The ISO, based in Switzerland, first published the ISO 9000 standards in 1987. The series was designed as a means to increase customer confidence in the quality of supplied materials, products, or services. This was especially important for trade between different countries with different languages and cultures. Not surprisingly, the initial emphasis was within the European Economic Community as part of the movement to a unified market.

The standard has evolved over several revisions (1).

The initial 1987 version, ISO 9000:1987, although structured like the British Standard BS 5750, included numerous documents then in use around the world. Although the Standard has gone through two more iterations, which have resulted in some radically changed language, all the core, prevention-oriented quality assurance requirements were present in the 1987 document. The language of this first version of the Standard was influenced by existing U.S. and other Defence Military Standards ("MIL SPECS"), so was more accessible to manufacturing, and was well-suited to the demands of a rigorous, stable, factory-floor manufacturing process. With its structure of twenty "elements" of requirements, the emphasis tended to be overly placed on conformance with procedures rather than the overall process of management, which was the actual intent.

The 1994 version, ISO 9000:1994 emphasized quality management and quality assurance via preventive actions, and continued to require evidence of compliance with documented procedures. Unfortunately, as with the first edition, companies tended to implement its requirements by creating shelf-loads of procedure manuals, and becoming burdened with an ISO bureaucracy. Adapting and improving processes could be particularly difficult in this kind of environment.

The 2000 version, ISO 9000:2000 that consists of ISO 9001:2000 and ISO 9004:2000, made a significant change by actually placing the concept of process management front and center in the Standard. There was no change in the essential goals of the Standard, which was all about a "documented system," not a "system of documents." The goal was always to have management system effectiveness via process performance metrics. The third edition makes this more visible and so reduced the emphasis on having documented procedures if clear evidence could be presented to show that the process was working well. Expectations of continual process improvement and tracking customer satisfaction were made explicit at this revision.

ISO 9000 has gained extensive acceptance, having been adopted by over 100 countries. Compliance is voluntary; however, some companies have made certification a condition of doing business. The registration process involves independent audit and confirmation that quality systems are in compliance with the standard. A fee is involved and periodic reinspection is required. A certificate of compliance is provided. This is the converse of the U.S. GMP situation, where compliance is assumed unless the FDA states otherwise, and no certificate of compliance is provided.

However, a major benefit of ISO 9000 has been to allow and encourage nonregulated industries to introduce a quality management system. Previously, the only widely established systems included regulations such as GMPs aimed at specific industries and award programs such as Deming and Malcolm Baldrige. None of these had universal coverage, although the World Health Organization GMPs do apply to many countries. Since the pharmaceutical industry is already subject to quality regulations, attaining ISO 9000 or 9001certification is not a goal of pharmaceutical manufacturers. However, suppliers to the pharmaceutical industry have sought it.

ISO 9000 has eight quality management principles on which the quality management system standards of the revised ISO 9000:2000 series are based. The principles are derived from the collective experience and knowledge of the international experts who participate in ISO Technical Committee ISO/TC 176, quality management, and QA, which is responsible for developing and maintaining the ISO 9000 standards (2).

The eight quality management principles are defined in ISO 9000:2000, Quality Management Systems Fundamentals and Vocabulary, and in ISO 9004:2000, Quality Management Systems Guidelines for Performance Improvements. A comparison to U.S. GMP is presented.

Principle 1: Customer Focus

Organizations depend on their customers and therefore should understand current and future customer needs, should meet customer requirements and strive to exceed customer expectations.

Customer focus is a critical business focus and is not covered in the U.S. GMPs other than the need to have products that meet the standards and specifications.

Principle 2: Leadership

Leaders establish unity of purpose and direction of the organization. They should create and maintain the internal environment in which people can become fully involved in achieving the organization's objectives.

There is no mention of Leadership in the U.S. GMPs but the concept is included in the Quality Systems Approach to Pharmaceutical GMP Regulations final guidance, September, 2006 where providing leadership is the first item in the management responsibilities section.

Principle 3: Involvement of People

People at all levels are the essence of an organization and their full involvement enables their abilities to be used for the organization's benefit.

Responsibilities of the quality unit and personnel qualifications are covered in the U.S. GMPs and the develop personnel concept is in the Quality Systems Approach guidance.

Principle 4: Process Approach

A desired result is achieved more efficiently when activities and related resources are managed as a process.

Principle 5: System Approach to Management

Identifying, understanding, and managing interrelated processes as a system contributes to the organization's effectiveness and efficiency in achieving its objective.

Principles 4 and 5 are not covered in the U.S. GMPs but the concepts are implied in the Quality Systems Approach guidance.

Principle 6: Continual Improvement

Continual improvement of the organization's overall performance should be a permanent objective of the organization.

The closest thing to continual improvement is the corrective and preventive action (CAPA) GMP practices that are typically the results of a failure investigation.

Principle 7: Factual Approach to Decision Making

Effective decisions are based on the analysis of data and information.

The basis of science and supporting data is a fundamental expectation of the U.S. GMPs and quality by design (QBD) initiatives.

Principle 8: Mutually Beneficial Supplier Relationships

An organization and its suppliers are interdependent and a mutually beneficial relationship enhances the ability of both to create value.

Control of materials and components is a U.S. GMP requirement and quality agreements and supplier contracts are expected practices but relationships are not a requirement.

As a comparison, the FDA final guidance on Quality Systems Approach provides CGMP and the concept of Modern Quality Systems key concepts:Quality (product quality)

- QBD and product development
- Risk management and risk assessment
- CAPA
- Change control
- The Quality Unit (QC/QA organization)
- Six-system inspection model (production, packaging and labeling, facilities and equipment, laboratory controls, materials, quality)

It is obvious from the comparison that the GMP guidance is product focused and the ISO is business focused. In order to run an effective, compliant pharmaceutical organization will likely require both approaches.

ISO 9000 certification does not come without criticisms. Some of which include:

1. The compliance process is costly, time consuming, and require the ongoing administration to maintain.
2. Continuous improvement can be slowed or made harder when redesigning the process.
3. Cannot prevent bad managers and management.
4. Many companies go after certification because they are forced to by the marketplace (results in minimal or no buy-in).
5. Can be easier to produce the required documents as opposed to improving the business processes.

Copies of the ISO 9000 series are available for a fee from the ISO at http://www.iso.org.

MALCOLM BALDRIGE NATIONAL QUALITY AWARD

The MBNQA is an annual award to recognize business excellence and quality achievement of U.S. companies. The Criteria for Performance Excellence is the premier quality standard for a variety of businesses.

Congress established the award program in 1987 to recognize U.S. organizations for their achievements in quality and performance and to raise awareness about the importance of quality and performance excellence as a competitive edge. Three awards may be given annually in each of these categories: manufacturing, service, small business, education, and health care. In October 2004, President Bush signed into law legislation that authorizes National Institute of Standards and Technology (NIST) to expand the MBNQA Program to include nonprofit and government organizations. The U.S. Commerce Department's NIST manages the Baldrige National Quality Program in close cooperation with the private sector (3).

The award promotes:

- Awareness of quality as an increasingly important element in competitiveness.
- Understanding the requirements for performance excellence.
- Sharing of information on successful performance strategies (best practices) and benefits derived from implementation of these strategies.

The MBNQA is an award program, unlike the other programs described in this book, which are compliance programs. Both MBNQA and ISO 9000 recognize and reward achievement by the award or certification, whereas GMP evaluations punish noncompliance. Most companies adopting the MBNQA approach do so with the aim of ultimately winning an award and the extensive visibility and prestige that results. Also, while winning an award does not guarantee success (at least one winner has gone out of business), most of the winners have shown significant business improvements such as speed to market with new products, increased market share, productivity increases, and growth in sales and profits.

The award is based on a set of core values and concepts that are designed to integrate customer satisfaction and company performance. These values and concepts are embedded beliefs and behaviors found in high-performance organizations.

Core values are:

- visionary leadership,
- customer driven excellence,
- organizational and personal learning,
- valuing employees and partners,
- agility,
- focus on the future,
- managing for innovation,
- managing by fact,
- social responsibility,
- focus on results and creating value, and
- systems perspective.

The Baldrige core values should be practiced within the Performance Excellence Framework and Criteria. Seven categories make up the award criteria:

- *Leadership*—examines how senior executives guide the organization and how the organization addresses its responsibilities to the public and practices good citizenship.
- *Strategic planning*—examines how the organization sets strategic directions and how it determines key action plans.
- *Customer and market focus*—examines how the organization determines requirements and expectations of customers and markets; builds relationships with customers; and acquires, satisfies, and retains customers.
- *Measurement, analysis, and knowledge management*—examines the management, effective use, analysis, and improvement of data and information to support key organization processes and the organization's performance management system.
- *Human resource focus*—examines how the organization enables its workforce to develop its full potential and how the workforce is aligned with the organization's objectives.
- *Process management*—examines aspects of how key production/delivery and support processes are designed, managed, and improved.
- *Business results*—examines the organization's performance and improvement in its key business areas: customer satisfaction, financial and marketplace performance, human resources, supplier and partner performance, operational performance, and governance and social responsibility. The category also examines how the organization performs relative to competitors.

Figure 1 represents a system perspective of the criteria (4).

1. *Leadership*. This is the overall driver of the process. This category examines senior executives' personal leadership and involvement in creating and sustaining customer focus, clear values and expectations, and a leadership system that promotes excellence. Also examined are legal and ethical behaviors and how the values and expectations are integrated into the company's management system, including how the company addresses governance and social responsibilities. Although going far beyond any GMP requirements, the role of senior managers as drivers of improvement is universal. This category has the second highest point values (120 out of 1000) in the Baldrige scoring.

The FDA is concerned that senior management are not sufficiently involved in quality compliance and that they may, in some instances, not even be aware of serious quality issues within their companies. This resulted in an increase in the number of Warning Letters issued with respect to quality deficiencies. These Warning Letters are addressed to the

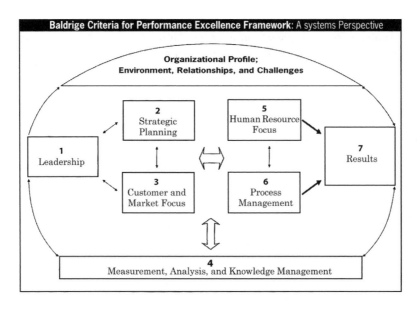

2006 Criteria for Performance Excellence

FIGURE 1   Baldrige criteria for performance excellence framework: a system's perspective.

company CEO—the rationale being that this will guarantee the availability of any resources needed to correct the deficiencies.

2. *Strategic Planning.* Strategic planning "examines how the company sets strategic directions, and how it determines key plan requirements. Also examined is how the plan requirements are translated into an effective performance management system." Strategic planning is not an element in GMPs. However, FDA investigators will sometimes take into account a company's plan for improvement when deciding what action to take on finding a compliance deviation.

3. *Customer and Market Focus.* Customer and market focus examines the areas of customer and market knowledge, and customer relation and satisfaction. Again, this is well beyond the scope of the GMPs that only address customer dissatisfaction related to product complaints.

4. *Measurement Analysis and Knowledge Management.* This section examines measurement, analysis, and review of organizational performance as well as information and knowledge management. Organizational performance is not directly covered in the GMPs but technical (product) information and knowledge is a compliance expectation. Within the pharmaceutical industry, especially within plant operations, significant amounts of data are generated. More use could be made of this data to drive improvements.

The FDA does expect to see data analyzed to identify trends and where necessary result in appropriate remedial activity. The annual report is one example where all product quality data is reviewed and reported. An excellent use of the annual review would be to incorporate the data, along with process changes that occur, into a "living" product quality document. This document would contain the original validation protocol, data, and approval; all significant process changes and the decisions regarding the need for revalidation; the annual quality reports and actions taken; all revalidation processes and data; customer feedback data; and production performance.

5. *Human Resource Focus*. This section involves the evaluation of work systems, employee learning, motivation, well being, and satisfaction. It covers how the work force is enabled to develop and utilize its full potential, aligned with the company's performance objectives. Also examined are the company's efforts to build and maintain an environment conducive to performance excellence, full participation, and personal and organizational growth. This coincides with a current trend for flatter organizational structures, delegation of responsibility, and use of self-managed work teams. The FDA does pay attention to training and levels of individual competency but the reliance on the QC unit for ultimate control is counter to some of the Baldrige philosophy. The pharmaceutical industry is still grappling with this issue— how to gain the maximum benefits from delegation without adversely impacting on product quality or safety.

The consistent achievement of product quality standards and compliance with regulatory requirements will be enhanced by the awareness by all employees of the importance of these issues. Only when every employee is fully committed to the achievement of consistent quality and held accountable and rewarded will real progress be made. The introduction of self-directed work teams as a means of involving all employees in quality has had some success. The team takes responsibility for all or most elements of the work activity. This can include hiring, dismissal and disciplinary action, production scheduling, work hours, and even product quality. A QC function can only examine a limited sample of production and consequently the possibility of detecting low-frequency defects is small. However, the production team is exposed to a much larger sample and consequently is in a better position to detect and correct problems. Policing has never been an effective way to achieve compliance, although a high level of enforcement activity coupled with significant penalties can have some major impact.

6. *Process Management*. This section examines the value creation processes, support processes, and operational planning systems in a company. Some key aspects of process management include customer-focused design, product and service delivery processes, support services, and supply management involving all work units, including research and development. This category examines how key processes are designed, effectively managed, and improved to achieve higher performance.

Unlike the GMPs, this goes into both product design, support, and product distribution service. Whereas the latter may not be too important to product quality and efficacy, the design stage is critical. A poorly designed product could be difficult to produce to a consistent quality. The inherent process variability could result in failures of individual units to fully comply with the defined specifications.

Product delivery service is obviously crucial for business success. Failure to deliver the right product on time with the correct paperwork (invoices, etc.) will soon result in loss of customers. Delivery service also includes the process for dealing with trade customer queries.

There are some similarities in this section and the FDA Quality Systems Approach guidance relative to process or system requirements.

7. *Results*. This category examines the company's performance and improvement in key business areas of products and services. Also examined are performance levels relative to competitors. The importance of this category is evidenced by the fact it is worth 450 out of the 1000-point scoring total.

Evaluation and demonstration of product quality data is a GMP requirement but not in the context of business performance. Evaluation of measures is usually directed toward compliance issues such as rejections, reworks, and complaints as indicators of inadequately validated production processes and process deviations as the examples of system noncompliance. In fact, FDA investigators frequently visit the reject area early in an inspection in order to identify failures. They can then go back into the respective batch records to evaluate the cause and to establish whether adequate remediation activity had been initiated. An important role for QA/QC is to provide senior management with data to demonstrate positive business improvements resulting from quality-related activities.

Benchmarking competitive processes is part of demonstrating performance. This is not a GMP requirement but is a good business and quality practice. This is essentially a new arena for pharmaceutical QA/QC professionals. Previously, they have focused on the narrow areas of technology and regulations—areas where they were acknowledged as professionally capable and responsible. These new areas require the establishment of new sources of data (benchmarking) with clear definition of the metrics used and the ability to communicate and influence management in these business-related areas. QA/QC management must earn respect in these new areas so that they will be heard and so that appropriate improvement actions will be initiated.

Companies not intending to apply for the award could find the MBNQA process to be a valuable tool in helping to improve business performance based on quality improvement and customer satisfaction. Useful benchmarking data and best practices information can be obtained via the extensive communications resulting from each award cycle or by direct interaction with other companies. Internal self-assessment can also provide useful data, since minimum scores required to be considered for an award are known.

Copies of the MBNQA criteria, along with application forms and instructions, can be obtained from MBNQA, National Institute of Standards and Technology, Route 270 and Quince Orchard Road, Administration Building, Room A537, Gaithersburg, MD 20899-0001, U.S.A.

The Criteria for Performance Excellence is available at http://baldrige.nist.gov/Business_Criteria.htm.

SIX SIGMA APPROACHES

There is a note in the current Malcolm Baldrige Criteria (2006) under the Value Creation Processes section of Process Management that states "To achieve better process performance and reduce variability, you might implement approaches such as a Lean Enterprise System, Six Sigma Methodology, use of ISO9000:2000 standards, or other process improvement tools."

Six Sigma is a popular process improvement methodology that can be limited to specific process improvement or actually change a company culture (for the better) if rolled out, embraced, and institutionalized. Considering the current FDA approach to quality systems, a proven process improvement methodology is a good companion to have when the goal is sustainable compliance.

Six Sigma is a process improvement methodology typically focused on business process. It is a rigorous and disciplined methodology that utilizes data and statistical analysis to measure process performance or variation. When a process is operating at a Six Sigma level, the variation is very small and the resulting products and services are 99.9997% defect free (or 3.4 defects per million units). In addition to being a statistical measure of variation, Six Sigma is also a business philosophy that starts with understanding customer needs and analyzing, measuring, and improving key business and quality processes. It was originally developed by Motorola in the 1980s and helped them win the Baldrige award. It has gained popularity recently in part because it was adopted by General Electric's CEO Jack Welsch. General Electric was seeking business excellence and improved profitability and needed Six-Sigma process capability in a competitive environment. Not all processes require Six-Sigma capability but any process can benefit from the approach.(5)

The most common Six Sigma methodology involves five steps: Define, Measure, Analyze, Improve, and Control (DMAIC). There are many variations of process improvement methodology in use today. In fact, variations in methodology are beneficial when confronted with different levels of complexity or level of improvement desired. However, the fundamentals presented in the DMAIC model are an excellent starting point for any project. This methodology is heavily dependent on the use of tools to help define and measure the processes and project. Following is a table describing the activities in each step and possible tools to use.

Project Step and Activities	Possible Tools
1. Define the project or problem	
■ Determine the scope and purpose ■ Develop a high-level process map ■ Determine customer requirements (VOC-voice of customer) ■ Determine the target or future state ■ Establish the benefits ■ Complete project charter	■ Surveys, Pareto analysis ■ Process mapping ■ SOP mapping ■ SIPOC (suppliers, inputs, outputs, customer) ■ FMEA (failure mode effect analysis)
2. Measure the current situation	
■ Gather data (input, output and in-process attributes, and variables data) ■ Plot defect data over time and analyze ■ Visualize and summarize data ■ Calculate process sigma or performance output ■ Complete detailed process map (current situation) ■ Challenge metrics/measures versus VOC	■ Pareto, histogram, run charts ■ Control charts, surveys, SOPs ■ Process maps (process, flow, value stream, swim lanes etc) ■ Summary statistics (mean, SD, range, mode, Cp, Cpk)
3. Analyze the gaps, identify root causes, and confirm data	
■ Verify the gap between the current state and target ■ Challenge/confirm the data ■ Determine root causes of defects or problems ■ Identify special causes (if any) ■ Identify potential improvement opportunities	■ Audit results ■ Hypothesis testing, DOE ■ Histograms, run & control charts ■ Cause and effect diagram/analysis ■ Brainstorming, five "whys" ■ Correlations, regressions ■ Inter-relationship diagrams
4. Improve the process	
■ Identify and select possible solutions to root causes ■ Develop implementation plan ■ Pilot plans ■ Implement and measure results ■ Evaluate improvement/benefits	■ Value stream mapping ■ Brainstorming, prioritization ■ QFD, DOE, Gantt charts ■ Project planning tools, FMEA ■ Control charts, hypothesis testing ■ Confidence intervals
5. Control the process	
■ Freeze the improved process and standardize ■ Update SOPs and train personnel ■ Monitor the performance ■ Complete after action review and lessons learned ■ Recommend future plans/improvements	■ Communication plan ■ Control charts ■ RACI (responsible, accountable, consult, inform) ■ Project closure report

The FDA guidance on Quality Systems Approach to Pharmaceutical Good Manufacturing Practice Regulations (see Chapter 14) is intended to help manufacturers that are implementing modern quality systems and risk management approaches to meet the requirements of the Agency's CGMP regulations.

Six Sigma or similar process improvement methodologies can be used to determine the health and ruggedness of a quality system to meet FDA (one of the customers) requirements. Typically 483 observations and/or adverse audit findings highlight and accelerate the need to improve the system. In a proactive company, one would not wait for significant failures before reacting.

Although it is not mandatory that a company use modern quality systems and risk management approaches, there does not appear to be many viable alternatives to running a sustainable business.

REFERENCES

1. ISO 9000 Revisions, http://en.wikipedia.org/wiki/ISO_9000, accessed Sept 12, 2006.
2. ISO 9000 and ISO 14000—in brief, http://www.iso.org/iso/en/iso9000-14000/understand/inbrief.html.
3. Frequently Asked Questions about the Malcolm Baldrige National Quality Award, http://www.nist.gov/public_affairs/factsheet/baldfaqs.htm.
4. Criteria for Performance Excellence, http://baldrige.nist.gov/Business_Criteria.htm, accessed Aug 12, 2006.
5. The Six Sigma Memory Jogger II. Goal QPC, 1–7.

SUGGESTED READINGS

1. Cantner R. The ISO 9000 Answer Book. John Wiley & Sons, 2002.
2. Brown MG. Baldrige Award Winning Quality. 13th ed. ASQ, 2004.
3. U.S. Congress. The MBNQA Program: An Oversight Review, 1995.
4. George ML. Lean Six Sigma, McGraw Hill, 2002.
5. Eckes G. Six Sigma Team Dynamics, John Wiley & Sons.
6. Kieffer R, Nally J. Implementing Total Quality in the Pharmaceutical Industry. Pharm Technol 1991; 15(9):130.
7. Michel BJ. A critical appraisal of the ISO certification. Chem Process 1994; 34.
8. Zuckerman A. One size doesn't fit all. Industry Week, 37, Jan 8, 1995; 37.
9. Romano SJ, Landsman C, Mason PJ. ISO 9000 and the pharmaceutical industry: an overview. Pharm Eng 1994; 14(3): 16.
10. Wayman QR. ISO 9000: a guide to effective design reviews. Qual Dig 1994; 45.
11. Zuckerman A. The sleeper issue of the 90's. Industry Week, 99, Aug 15, 1994.
12. Greising D. Quality: how to make it pay. Business Week, 54, Aug 8, 1994.
13. Morrow M. ISO 9000 is not dead. Qual Dig 1994; 23.
14. Herrington M. Why not a do-it-yourself Malcolm Baldrige Award?. Across the Board 1994; 34.
15. Leach KE. Lessons from the Baldrige. Industry Week, 42, Sept 19, 1994.
16. Russo CWR. How to select a Registrar. Qual Dig 1995; 15(6).
17. Zuckerman A. The future of ISO 9000. Qual Dig 1995; 15(5): 23.
18. Halperin JA. International harmonization in the pharmaceutical sector. FDCML Digest 1996; 13(3).
19. The Six Sigma Memory Jogger II, Goal QPC.
20. Keki Bhote, The Power of Ultimate Six Sigma, Amacom, 2003.

A | Center for Drug Evaluation and Research: List of Guidance Documents

FDA's guidance documents do not establish legally enforceable responsibilities. Instead, the documents describe the agency's current thinking on a topic and should be viewed only as recommendations, unless specific regulatory or statutory requirements are cited. The use of the word "should" in agency guidances means that something is suggested or recommended, but not required.

Information on developing and issuing guidance can be found at http://www.fda.gov/cder/mapp/4000-2_10_05.pdf.

The following guidance list can be found at http://www.fda.gov/cder/guidance/complist_200609.pdf.

Advertising   Issued Date

Aerosol Steroid Product Safety Information in Prescription Drug Advertising and Promotional Labeling (I)	1/12/1998
Consumer-Directed Broadcast Advertisements (I)	8/9/1999
Industry-Supported Scientific and Educational Activities (I)	12/3/1997

Advertising Draft   Issued Date

"Help-Seeking" and Other Disease Awareness Communications by or on Behalf of Drug and Device Firms (I)	2/10/2004
Accelerated Approval Products -- Submission of Promotional Materials (I)	3/26/1999
Brief Summary: Disclosing Risk Information in Consumer-Directed Print Advertisements(I)	2/10/2004
Product Name, Placement, Size, and Prominence in Advertising and Promotional Labeling (I)	3/12/1999
Promoting Medical Products in a Changing Healthcare Environment; Medical Product Promotion by Healthcare Organizations or Pharmacy Benefits Management Companies (PBMs) (I)	1/5/1998

Biopharmaceutics   Issued Date

Bioanalytical Method Validation (I)	5/23/2001
Bioavailability and Bioequivalence Studies for Orally Administered Drug Products - General Considerations (Revised) (I)	3/19/2003
Cholestyramine Powder In Vitro Bioequivalence (I)	7/15/1993
Clozapine Tablets: In Vivo Bioequivalence and In Vitro Dissolution Testing (I)	6/20/2005
Corticosteroids, Dermatologic (topical) In Vivo (I)	6/2/1995

Dissolution Testing of Immediate Release Solid Oral Dosage Forms (I)	8/25/1997
Extended Release Oral Dosage Forms: Development, Evaluation, and Application of In Vitro/In Vivo Correlations (I)	9/26/1997
Food-Effect Bioavailability and Fed Bioequivalence Studies (I)	1/31/2003
Metaproterenol Sulfate and Albuterol Metered Dose Inhalers In Vitro (I)	6/27/1989
Statistical Approaches to Establishing Bioequivalence (I)	2/2/2001
Waiver of In Vivo Bioavailability and Bioequivalence Studies for Imediate Release Solid Oral Dosage Forms Based on a Biopharmaceutics Classification System (I)	8/31/2000

Biopharmaceutics Draft   Issued Date

Antifungal (topical) (I)	2/24/1990
Antifungal (vaginal)	2/24/1990
Bioavailability and Bioequivalence Studies for Nasal Aerosols and Nasal Sprays for Local Action - 2nd Draft (I)	4/3/2003

Chemistry   Issued Date

Botanical Drug Products (I)	6/9/2004
Changes to an Approved Application for Specified Biotechnology and Specified Synthetic Biological Products (I)	7/24/1997
Changes to an Approved NDA or ANDA (Revised) (I)	4/8/2004
Changes to an Approved NDA or ANDA: Questions and Answers (I)	1/22/2001
Changes to an Approved New Drug Application or Abbreviated New Drug Application; Specifications - Use of Enforcement Discretion for Compendial Changes (I)	11/22/2004
Container Closure Systems for Packaging Human Drugs and Biologics (I)	7/7/1999
Demonstration of Comparability of Human Biological Products Including Therapeutic Biotechnology Derived Products (I)	3/26/1996
Development of New Stereoisomeric Drugs (I)	5/1/1992
Drug Master Files (I)	9/1/1989
Drug Master Files for Bulk Antibiotic Drug Sustances (I)	11/29/1999
Environmental Assessment of Human Drug and Biologics Applications (I)	7/27/1998

Format and Content for the CMC Section of an Annual Report (I)	9/1/1994
Format and Content of the Microbiology Section of an Application* (I)	2/1/1987
IND Meetings for Human Drugs and Biologics; Chemistry, Manufacturing, and Controls Information (I)	5/25/2001
INDs for Phase 2 and 3 Studies; Chemistry, Manufacturing, and Controls Information (I)	5/20/2003
Monoclonal Antibodies Used as Reagents in Drug Manufacturing (I)	3/29/2001
Nasal Spray and Inhalation Solution, Suspension, and Spray Drug Products -- Chemistry, Manufacturing, and Controls Documentation (I)	7/5/2002
NDAs: Impurities in Drug Substances (I)	2/25/2000
PAC-ALTS: Postapproval Changes - Analytical Testing Laboratory Sites (I)	4/28/1998
Submission Documentation for Sterilization Process Validation Applications for Human and Veterinary Drug Products (I)	11/1/1994
Submitting Documentation for the Manufacturing of and Controls for Drug Products* (I)	2/1/1987
Submitting Samples and Analytical Data for Methods Validation* (I)	2/1/1987
Submitting Supporting Documentation in Drug Applications for the Manufacture of Drug Products (I)	2/1/1987
Submitting Supporting Documentation in Drug Applications for the Manufacture of Drug Substances* (I)	2/1/1987
SUPAC-IR Immediate-Release Solid Oral Dosage Forms: Scale-Up and Post-Approval Changes: Chemistry, Manufacturing and Controls, In Vitro Dissolution Testing, and In Vivo Bioequivalence Documentation (I)	11/30/1995
SUPAC-IR Questions and Answers (I)	2/18/1997
SUPAC-IR/MR: Immediate Release and Modified Release Solid Oral Dosage Forms, Manufacturing Equipment Addendum (I)	2/26/1999
SUPAC-MR: Modified Release Solid Oral Dosage Forms: Scale-Up and Postapproval Changes: Chemistry, Manufacturing, and Controls, In Vitro Dissolution Testing, and In Vivo Bioequivalence Documentation (I)	10/6/1997
SUPAC-SS - Nonsterile Semisolid Dosage Forms; Scale-Up and Postapproval Changes: Chemistry, Manufacturing, and Controls; In Vitro Release Testing and In Vivo Bioequivalence Documentation (I)	6/13/1997
The Sourcing and Processing of Gelatin to Reduce the Potential Risk Posed by Bovine Spongiform (I)	12/20/2000
Validation of Chromatographic Methods -- Reviewer's Guidance (I)	11/1/1994

Chemistry Draft   Issued Date

Analytical Procedures and Methods Validation: Chemistry, Manufacturing, and Controls Documentation (I)	8/30/2000
Comparability Protocols - Chemistry, Manufacturing, and Controls Information (I)	2/25/2003
Drugs, Biologics, and Medical Devices Derived From Bioengineered Plants for Use in Humans and Animals (I)	9/12/2002
Interpreting Sameness of Monoclonal Antibody Products Under the Orphan Drug Regulations (I)	7/26/1999
Liposome Drug Products: Chemistry, Manufacturing, and Controls; Human Pharmacokinetics and Bioavailability; and Labeling Documentation (I)	8/21/2002
Metered Dose Inhalers (MDI) and Dry Powder Inhalers (DPI) Drug Products; Chemistry, Manufacturing, and Controls Documentation (I)	11/19/1998
Submitting Supporting Chemistry Documentation in Radiopharmaceutical Drug Applications*	11/1/1991
SUPAC-SS: Nonsterile Semisolid Dosage Forms Manufacturing Equipment Addendum (I)	1/5/1999

Clinical Antimicrobial   Issued Date

Antiretroviral Drugs Using Plasma Human Immunodeficiency Virus Ribonucleic Acid Measurements - Clinical Considerations for Accelerated and Traditional Approval (I)	11/1/2002
Antiviral Product Development - Conducting and Submitting Virology Studies to the Agency	6/5/2006
Clinical Development and Labeling of Anti-Infective Drug Products (I)	10/26/1992
Clinical Evaluation of Anti-Infective Drugs (Systemic) (I)	9/1/1977

Clinical Antimicrobial Draft   Issued Date

Acute Bacterial Exacerbation of Chronic Bronchitis; Developing Antimicrobial Drugs for Treatment (I)	7/22/1998
Acute Bacterial Meningitis; Developing Antimicrobial Drugs for Treatment (I)	7/22/1998
Acute Bacterial Sinusitis; Developing Antimicrobial Drugs for Treatment (I)	7/22/1998
Acute or Chronic Bacterial Prostatitis; Developing Antimicrobial Drugs for Treatment (I)	7/22/1998
Acute Otitis Media; Developing Antimicrobial Drugs for Treatment (I)	7/22/1998
Bacterial Vaginosis; Developing Antimicrobial Drugs for Treatment (I)	7/22/1998
Catheter-Related Bloodstream Infections - Developing Antimicrobial Drugs for Treatment (I)	10/18/1999

Community Acquired Pneumonia; Developing Antimicrobial Drugs for Treatment (I)	7/22/1998
Complicated Urinary Tract Infections and Pylonephritis; Developing Antimicrobial Drugs for Treatment (I)	7/22/1998
Developing Antimicrobial Drugs -General Considerations for Clinical Trials (I)	7/22/1998
Developing Drugs to Treat Inhalational Anthrax (Post-Exposure) (I)	3/18/2002
Empiric Therapy of Febrile Neutropenia; Developing Antimicrobial Drugs for Treatment (I)	7/22/1998
Evaluating Clinical Studies of Antimicrobials in the Division of Anti-Infective Drug Products (I)	2/17/1997
Lyme Disease; Developing Antimicrobial Drugs for Treatment (I)	7/22/1998
Nosocomial Pneumonia; Developing Antimicrobial Drugs for Treatment (I)	7/22/1998
Role of HIV Drug Resistance Testing in Antiretroviral Drug Development (I)	11/29/2004
Secondary Bacterial Infections of Acute Bronchitis; Developing Antimicrobial Drugs for Treatment (I)	7/22/1998
Streptococcal Pharyngitis and Tonsillitis; Developing Antimicrobial Drugs for Treatment (I)	7/22/1998
Uncomplicated and Complicated Skin and Skin Structure Infections; Developing Antimicrobial Drugs for Treatment (I)	7/22/1998
Uncomplicated Gonorrhea -- Cervical, Urethral, Rectal, and/or Pharyngeal; Developing Antimicrobial Drugs for Treatment (I)	7/22/1998
Uncomplicated Urinary Tract Infections; Developing Antimicrobial Drugs for Treatment (I)	7/22/1998
Vaccinia Virus -- Developing Drugs to Mitigate Complications From Smallpox Vaccination (I)	3/9/2004
Vuvlovaginal Candidiasis; Developing Antimicrobial Drugs for Treatment (I)	7/22/1998

Clinical Medical   Issued Date

Acceptance of Foreign Clinical Studies (I)	3/13/2001
Antianxiety Drugs -- Clinical Evaluation (I)	9/1/1977
Antidepressant Drugs -- Clinical Evaluation (I)	9/1/1977
Antiepileptic Drugs (adults and children) -- Clinical Evaluation (I)	1/1/1981
Anti-Inflammatory and Antirheumatic Drugs (adults and children) -- Clinical Evaluation (I)	4/1/1988

Available Therapy (I)	7/23/2004
Calcium DTPA and Zinc DTPA Drug Products -- Submitting a New Drug Application (I)	8/13/2004
Cancer Drug and Biological Products - Clinical Data in Marketing Applications (I)	10/5/2001
Chronic Cutaneous Ulcer and Burn Wounds - Developing Products for Treatment (I)	6/2/2006
Clinical and Statistical Sections of an Application -- Format and Content* (I)	7/1/1988
Clinical Development Programs for Drugs, Devices, and Biological Products for the Treatment of Rheumatoid Arthritis (RA) (I)	2/17/1999
Collection of Race and Ethnicity Data in Clinical Trials for FDA Regulated Products (I)	9/19/2005
Content and Format of Investigational New Drug Applications (INDs) for Phase 1 Studies of Drugs, Including Well-Characterized, Therapeutic, Biotechnology-Derived Products (I)	11/20/1995
Developing Medical Imaging Drug and Biological Products, Part 1: Conducting Safety Assessments (I)	6/22/2004
Developing Medical Imaging Drug and Biological Products, Part 2: Clinical Indications (I)	6/22/2004
Developing Medical Imaging Drug and Biological Products, Part 3: Design, Analysis, and Interpretation of Clinical Studies (I)	6/22/2004
Development and Use of Risk Minimization Action Plans (I)	3/29/2005
Development of Vaginal Contraceptive Drugs (NDA) (I)	4/19/1995
Establishing Pregnancy Exposure Registries (I)	9/23/2002
Evaluating the Risks of Drug Exposure in Human Pregnancies	4/28/2005
Exocrine Pancreatic Insufficiency Drug Products-Submitting New Drug Applications (I)	4/14/2006
FDA Approval of New Cancer Treatment Uses for Marketed Drug and Biological Products (I)	2/2/1999
FDA Requirements for Approval of Drugs to Treat Non-Small Cell Lung Cancer (I)	1/29/1991
Formatting, Assembling and Submitting New Drug and Antiobiotic Applications* (I)	2/1/1987
General Anesthetics -- Clinical Evaluation (I)	5/1/1982
General Considerations for the Clinical Evaluation of Drugs (I)	12/1/1978

General Considerations for the Clinical Evaluation of Drugs in Infants and Children (I)   9/1/1977

Good Pharmacovigilance Practices and Pharmacoepidemiologic Assessment (I)   3/29/2005

Hypnotic Drugs -- Clinical Evaluation (I)   9/1/1977

IND Exemptions for Studies of Lawfully Marketed Drug or Biological Products for the Treatment of Cancer (Revised) (I)   1/15/2004

Integration of Dose-Counting Mechanisms Into Metered-Dose Inhaler Drug Products (I)   3/13/2003

Internal Radioactive Contamination - Development of Decorporation Agents (I)   3/2/2006

Levothyroxine Sodium Tablets -- In Vivo Pharmacokinetic and Bioavailability Studies and In Vitro Dissolution Testing (I)   3/8/2001

Local Anesthetics -- Clinical Evaluation (I)   5/1/1982

MDI and DPI Drug Products -- Clinical Development and Programs (I)   9/19/1994

Oncologic Drugs Advisory Committee Discussion on FDA Requirements for Approval of New Drugs for Treatment of Colon and Rectal Cancer (I)   4/19/1988

Oncologic Drugs Advisory Committee Discussion on FDA Requirements for Approval of New Drugs for Treatment of Ovarian Cancer (I)   4/13/1988

Pediatric Use Supplements -- Content and Format (I)   5/24/1996

Postmarketing Adverse Experience Reporting for Human Drugs and Licensed Biological Products; Clarification of What to Report (I)   8/27/1997

Postmarketing Reporting of Adverse Drug Experiences (I)   3/1/1992

Premarketing Risk Assessment (I)   3/29/2005

Preparation of Investigational New Drug Products (Human and Animal) (I)   11/1/1992

Providing Clinical Evidence of Effectiveness for Human Drug and Biological Products (I)   5/15/1998

Prussian Blue for Treatment of Internal Contamination With Thallium or Radioactive Cesium (I)   2/4/2003

Psychoactive Drugs in Infants and Children -- Clinical Evaluation (I)   7/1/1979

Study and Evaluation of Gender Differences in the Clinical Evaluation of Drugs (I)   7/22/1993

Study of Drugs Likely to be Used in the Elderly (I)   11/1/1989

Submission of Abbreviated Reports and Synopses in Support of Marketing Applications (I)	9/13/1999
Summary for New Drug and Antibiotic Applications -- Format and Content* (I)	2/1/1987

Clinical Medical Draft   **Issued Date**

Abuse Liability Assessment (I)	7/1/1990
Acne Vulgaris: Developing Drugs for Treatment (I)	9/19/2005
Allergic Rhinitis: Clinical Development Programs for Drug Products (I)	6/21/2000
Anti-Anginal Drugs -- Clinical Evaluation (I)	1/1/1989
Anti-Arrhythmic Drugs -- Clinical Evaluation	7/1/1985
Antihypertensive Drugs -- Clinical Evaluation	5/1/1988
Clinical Development Programs for Drugs, Devices, and Biological Products Intended for the Treatment of Osteoarthritis (OA) (I)	7/15/1999
Clinical Endpoints for the Approval of Cancer Drugs and Biologics	4/4/2005
Clinical Evaluation of Drugs for the Treatment of Congestive Heart Failure (I)	12/1/1987
Clinical Lactation Studies - Study Design, Data Analysis and Recommendations for Labeling	2/8/2005
Clinical Trial Sponsors on the Establishment and Operation of Clinical Trial Data Monitoring Committees (I)	11/20/2001
Combination Products Timeliness of Premarket Reviews (I)	5/4/2004
Computerized Systems Used in Clinical Trials (I)	10/4/2004
Development and Evaluation of Drugs for the Treatment of Psychoactive Substance Use Disorders	2/12/1992
Development of Parathyroid Hormone for the Prevention and Treatment of Osteoporosis (I)	6/14/2000
Drugs, Biologics, and Medical Devices Derived from Bioengineered Plants for Use in Humans and Animals	9/12/2002
Estrogen and Estrogen/ Progestin Drug Products to Treat Vasomotor Symptoms and Vulvar and Vaginal Atrophy Symptoms - Recommendations for Clinical Evaluation (Revised) (I)	1/31/2003
Evaluation of the Effects of Orally Inhaled and Intranasal Corticosteroids on Growth in Children (I)	11/6/2001
Exercise-Induced Bronchospasm (EIB) - Development of Drugs to Prevent EIB (I)	2/20/2002

Female Sexual Dysfunction: Clinical Development of Drug Products for Treatment (I)	5/19/2000
Gingivitis: Development and Evaluation of Drugs for Treatment or Prevention (I)	6/28/2005
Guidance for Institutional Review Boards, Clinical Investigators, and Sponsors: Exception from Informed Consent Requirements for Emergency Research (I)	3/30/2000
Inhalation Drug Products Packaged in Semipermeable Container Closure Systems (I)	7/26/2002
Lipid-Altering Agents in Adults and Children -- Clinical Evaluation (I)	9/1/1990
OTC Treatment of Herpes Labialis with Antiviral Agents (I)	3/8/2000
Patient-Reported Outcome Measures: Use in Medical Product Development to Support Labeling Claims (I)	2/3/2006
Pediatric Oncology Studies in Response to a Written Request (I)	6/21/2000
Preclinical and Clinical Evaluation of Agents Used in the Prevention or Treatment of Postmenopausal Osteoporosis (I)	4/1/1994
Preparation of IND Applications for New Drugs Intended for the Treatment of HIV-Infected Individuals	9/1/1991
Recommendations for Complying with the Pediatric Rule (I)	12/4/2000
Systemic Lupus Erythematosus - Developing Drugs for Treatment (I)	3/29/2005
The Use of Clinical Holds Following Clinical Investigator Misconduct (I)	9/2/2004
Weight-Control Drugs -- Clinical Evaluation (I)	9/24/1996

Clinical Pharmacology	Issued Date
Drug Metabolism/Drug Interaction Studies in the Drug Development Process: Studies In Vitro (I)	4/7/1997
Exposure-Response Relationships - Study Design, Data Analysis, and Regulatory Applications (I)	5/6/2003
Format and Content of the Human Pharmacokinetics and Bioavailability Section of an Application (I)	2/1/1987
In Vivo Metabolism/Drug Interaction Studies - Study Design, Data Analysis, and Recommendations for Dosing and Labeling (I)	11/24/1999
Pharmacogenomic Data Submissions	3/23/2005
Pharmacokinetics in Patients With Impaired Hepatic Function; Study Design, Data Analysis, and Impact on Dosing and Labeling (I)	5/30/2003

Pharmacokinetics in Patients with Impaired Renal Function: Study Design, Data Analysis, and   5/15/1998
Impact on Dosing and Labeling (I)

Population Pharmacokinetics (I)   2/10/1999

Clinical Pharmacology Draft   Issued Date

Drug Interaction Studies--Study Design, Data Analysis, and Implications for Dosing and Labeling   9/12/2006
(I)

General Considerations for Pediatric Pharmacokinetic Studies for Drugs and Biological Products   11/30/1998
(I)

Pharmacokinetics in Pregnancy - Study Design, Data Analysis, and Impact on Dosing and   11/1/2004
Labeling (I)

Combination Products (Drug/Device/Biologic)   Issued Date

Application User Fees for Combination Products   4/21/2005

Combination Products (Drug/Device/Biologic) Draft   Issued Date

Combination Products Timeliness of Premarket Reviews; Dispute Resolution (I)   5/4/2004

Current Good Manufacturing Practices for Combination Products (I)   10/4/2004

Compliance   Issued Date

A Review of FDA's Implementation of the Drug Export Amendments of 1986 (I)   5/1/1990

Bar Code Label Requirements - Questions and Answers (I)   4/27/2006

Compressed Medical Gases (I)   12/1/1989

Computerized Systems Used in Clinical Trials (I)   5/10/1999

Expiration Dating and Stability Testing of Solid Oral Dosage Form Drugs Containing Iron (I)   6/27/1997

General Principles of Process Validation (I)   5/1/1987

Good Laboratory Practice Regulations -- Questions and Answers (I)   6/1/1981

Guidance for Hospitals, Nursing Homes, and Other Health Care Facilities (I)   4/6/2001

Guideline for Validation of Limulus Amebocyte Lysate Test as an End-Product Endotoxin Test for   12/1/1987
Human and Animal Parenteral Drugs, Biological Products, and Medical Devices (I)

Marketed Unapproved Drugs;Compliance Policy Guide (I)   6/9/2006

Monitoring of Clinical Investigations (I)	1/1/1988
Nuclear Pharmacy Guideline Criteria for Determining When to Register as a Drug Establishment (I)	5/1/1984
Pharmacy Compounding -- Compliance Policy Guide (I)	6/7/2002
Possible Dioxin/PCB Contamination of Drug and Biological Products (I)	8/23/1999
Prescription Drug Marketing Act Regulations for Donation of Prescription Drug Samples to Free Clinics (I)	3/14/2006
Sterile Drug Products Produced by Aseptic Processing (I)	5/1/1987
Street Drug Alternatives (I)	4/3/2000

Compliance Draft | | Issued Date

Current Good Manufacturing Practice for Positron Emission Tomography Drug Products (I)	9/20/2005
Current Good Manufacturing Practices for Medical Gases (3rd Revision) (I)	5/6/2003
Expiration Dating of Unit-Dose Repackaged Drugs: Compliance Policy Guide	5/31/2005
Good Manufacturing Practice for Positron Emission Tomrgraphy Drug Products (I)	4/1/2002
Guidance for IRBs, Clinical Investigators, and Sponsors: Exception from Informed Consent Requirements for Emergency Research (I)	5/12/2000
Investigating Out of Specification (OOS) Test Results for Pharmaceutical Production (I)	9/30/1998
Manufacture, Processing or Holding of Active Pharmaceutical Ingredients (I)	4/17/1998
Repackaging of Solid Oral Dosage Form Drug Products	2/1/1992

Current Good Manufacturing Practices | | Issued Date

Formal Dispute Resolution: Scientific and Technical Issues Related to Pharmaceutical Current Good Manufacturing Practices (I)	1/12/2006
Part 11, Electronic Records, Electronic Signatures - Scope and Application	9/5/2003
Process Analytical Technology -- A Framework for Innovative Pharmaceutical Manufacturing and Quality Assurance (I)	10/4/2004
Sterile Drug Products Produced by Aseptic Processing (I)	10/4/2004

Current Good Manufacturing Practices (CGMPs)   **Issued Date**

Quality Systems Approach to Pharmaceutical Current Good Manufacturing Practice Regulations (I)   10/2/2006

Current Good Manufacturing Practices Draft   **Issued Date**

Comparability Protocols -- Protein Drug Products and Biological Products -- Chemistry,   9/5/2003
Manufacturing, and Controls Information (I)

INDs - Approaches to Complying with Current Good Manufacturing Practice During Phase 1 (I)   1/17/2006

Powder Blends and Finished Dosage Units--Stratified In-Process Dosage Unit Sampling and   11/7/2003
Assessment (I)

Quality Systems Approach to Pharmaceutical Current Good Manufacturing Practice Regulations (I)   10/4/2004

Drug Safety Draft   **Issued Date**

FDA's "Drug Watch" for Emerging Drug Safety Information   5/10/2005

Electronic Submissions   **Issued Date**

Providing Electronic Submissions in Electronic Format - ANDAs (I)   6/27/2002

Providing Regulatory Submissions in Electronic Format -- Content of Labeling (I)   4/21/2005

Providing Regulatory Submissions in Electronic Format -- Human Pharmaceutical Product   10/19/2005
Applications and Related Submissions (I)

Regulatory Submissions in Electronic Format; General Considerations (I)   1/28/1999

Regulatory Submissions in Electronic Format; NDAs (I)   1/28/1999

SPL Standard for Content of Labeling Technical Qs & As (I)   12/8/2005

Electronic Submissions Draft   **Issued Date**

Providing Regulatory Submissions in Electronic Format -- Annual Reports for New Drug   8/28/2003
Applications and Abbreviated New Drug Applications (I)

Providing Regulatory Submissions in Electronic Format - Postmarketing Expedited Safety Reports   5/4/2001
(I)

Providing Regulatory Submissions in Electronic Format -- Postmarketing Periodic Adverse Drug   6/24/2003
Experience Reports (I)

Providing Regulatory Submissions in Electronic Format, Prescription Drug Advertising and   1/31/2001
Promotional Labeling (I)

Providing Regulatory Submissions in Electronic Format--General Considerations (I)   10/22/2003

Generic Drug   Issued Date

180-Day Exclusivity When Multiple Abbreviated New Drug Applications Are Submitted on the Same Day (I)	8/1/2003
Alternate Source of Active Pharmaceutical Ingredients in Pending ANDAs (I)	12/12/2000
ANDAs: Impurities in Drug Substances (I)	12/3/1999
Court Decisions, ANDA Approvals, and 180-Day Exclusivity Under the Hatch-Waxman Amendments to the Federal Food, Drug, and Cosmetic Act (I)	3/30/2000
Handling and Retention of Bioavailability and Bioequivalence Testing Samples (I)	5/26/2004
Letter announcing that the OGD will now accept the ICH long-term storage conditions as well as the stability studies conducted in the past (I)	8/18/1995
Letter describing efforts by the CDER & the ORA to clarify the responsibilities of CDER chemistry review scientists and ORA field investigators in the new & abbreviated drug approval process in order to reduce duplication or redundancy in the process (I)	10/14/1994
Letter on incomplete Abbreviated Applications, Convictions Under GDEA, Multiple Supplements, Annual Reports for Bulk Antibiotics, Batch Size for Transdermal Drugs, Bioequivalence Protocols, Research, Deviations from OGD Policy (I)	4/8/1994
Letter on the provision of new information pertaining to new bioequivalence guidelines and refuse-to-file letters (I)	7/1/1992
Letter on the provision of new procedures and policies affecting the generic drug review process (I)	3/15/1989
Letter on the request for cooperation of regulated industry to improve the efficiency and effectiveness of the generic drug review process, by assuring the completeness and accuracy of required information and data submissions (I)	11/8/1991
Letter on the response to 12/20/84 letter from the Pharmaceutical Manufacturers Association about the Drug Price Competition and Patent Term Restoration Act (I)	3/26/1985
Letter to all ANDA and AADA applicants about the Generic Drug Enforcement Act of 1992 (GDEA), and the Office of Generic Drugs intention to refuse-to-file incomplete submissions as required by the new law (I)	1/15/1993
Letter to regulated industry notifying interested parties about important detailed information regarding labeling, scale-up, packaging, minor/major amendment criteria, and bioequivalence requirements (I)	8/4/1993
Major, Minor, and Telephone Amendments to Abbreviated New Drug Applications (I)	12/21/2001
Potassium Chloride Modified-Release Tablets and Capsules: In Vivo Bioequivalence and In Vitro Dissolution Testing (I)	10/26/2005
Revising ANDA Labeling Following Revision of the RLD Labeling (I)	4/25/2000
Variations in Drug Products that May Be Included in a Single ANDA (I)	1/27/1999

Generic Drug Draft   Issued Date

ANDAs: Impurities in Drug Products; Chemistry, Manufacturing and Controls Information (I)	8/29/2005
ANDAs: Impurities in Drug Substances; Chemistry, Manufacturing and Controls Information (I)	1/31/2005

ANDAs: Pharmaceutical Solid Polymorphism; Chemistry, Manufacturing and Controls Information (I)	12/20/2004
Listed Drugs, 30-Month Stays, and Approval of Abbreviated New Drug Applications and 505 (b)(2) Applications Under Hatch-Waxman, as Amended by the Medicare Prescription Drug Improvement, and Modernization Act of 2003, Questions and Answers (I)	11/4/2004

Good Review Practices

Issued Date

Conducting a Clinical Safety Review of a New Product Application and Preparing a Report on the Review (I)	2/18/2005
Good Review Management Principles for Prescription Drug User Fee Act Products (I)	3/31/2005
Pharmacology/Toxicology Review Format (I)	5/10/2001

ICH - Efficacy

Issued Date

E10 - Choice of Control Group and Related Issues in Clinical Trials (I)	5/14/2001
E11 - Clinical Investigation of Medicinal Products in the Pediatric Population (I)	12/15/2000
E14 - Clinical Evaluation of QT/QTc Interval Prolongation and Proarrhythmic Potential for Non-Antiarrhythmic Drugs (I)	10/20/2005
E1A - The Extent of Population Exposure to Assess Clinical Safety: for Drugs Intended for Long-Term Treatment of Non-Life-Threatening Conditions (I)	3/1/1995
E2A - Clinical Safety Data Management: Definitions and Standards for Expedited Reporting (I)	3/1/1995
E2B - Data Elements for Transmission of Individual Case Safety Reports (I)	1/15/1998
E2B(M) - Data Elements for Transmission of Individual Case Safety Reports (Revised) (I)	4/3/2002
E2B(M): Data Elements for Transmission of Individual Case Safety Reports -- Questions and Answers (Revision 2) (I)	3/9/2005
E2C - Clinical Safety Data Management: Periodic Safety Update Reports for Marketed Drugs (I)	5/19/1997
E2C Addendum - Clinical Safety Data Management: Periodic Safety Update Reports for Marketed Drugs (I)	2/5/2004
E2E - Pharmacovigilance Planning (I)	4/1/2005
E3 - Structure and Content of Clinical Study Reports (I)	7/17/1996
E4 - Dose-Response Information to Support Drug Registration (I)	11/9/1994
E5 - Ethnic Factors in the Acceptability of Foreign Clinical Data (I)	6/10/1998
E5 - Ethnic Factors in the Acceptability of Foreign Clinical Data, Questions and Answers (I)	9/27/2006

E6 - Good Clinical Practice: Consolidated Guideline (I)	5/9/1997
E7 - Studies in Support of Special Populations: Geriatrics (I)	8/2/1994
E8 - General Considerations for Clinical Trials (I)	12/24/1997
E9 - Statistical Principles for Clinical Trials (I)	9/16/1998

ICH - Joint Safety/Efficacy (Multidisciplinary)   Issued Date

Companion Document for M2: eCTD Specification Questions & Answers and Change Requests (I)	8/1/2006
M2 - Electronic Common Technical Document Specification (eCTD) (I)	4/2/2003
M3 - Nonclinical Safety Studies for the Conduct of Human Clinical Trials for Pharmaceuticals (I)	11/25/1997
M4 - Common Technical Document for the Registration of Pharmaceuticals for Human Use - Granularity Annex (I)	10/17/2005
M4 - Organization of the Common Technical Document (CTD) (I)	10/16/2001
M4 - The CTD -- Efficacy Questions and Answers (Revised) (I)	12/22/2004
M4 - The CTD -- General Questions and Answers (Revised) (I)	12/22/2004
M4 - The CTD - Quality Questions and Answers/Location Issues (I)	6/9/2004
M4 - The CTD -- Safety Questions and Answers (I)	2/4/2003

ICH - Quality   Issued Date

Q1A(R2) - Stability Testing of New Drug Substances and Products (I)	11/21/2003
Q1B - Photostability Testing of New Drug Substances and Products (I)	5/16/1997
Q1C - Stability Testing for New Dosage Forms (I)	5/9/1997
Q1D - Bracketing and Matrixing Designs for Stability Testing of New Drug Substances and Products (I)	1/16/2003
Q1E - Evaluation of Stability Data (I)	6/8/2004
Q2A - Text on Validation of Analytical Procedures (I)	3/1/1995
Q2B - Validation of Analytical Procedures: Methodology (I)	5/9/1997

Q3A(R) - Impurities in New Drug Substances (I)   2/11/2003

Q3B(R) - Impurities in New Drug Products (I)   7/31/2006

Q3C - Impurities: Residual Solvents (I)   12/24/1997

Q3C - Tables and Lists (Revised) Recommendations for Methylpyrrolidone and Tetrahydrofuran (I)   11/13/2003

Q5A - Viral Safety Evaluation of Biotechnology Products Derived From Cell Lines of Human or   9/24/1998
Animal Origin (I)

Q5B - Quality of Biotechnology Products: Analysis of the Expression Construct in Cells Used for   2/23/1996
Production of r-DNA Derived Protein Products (I)

Q5C - Quality of Biotechnological Products: Stability Testing of Biotechnology/Biological Products 7/10/1996
(I)

Q5D - Quality of Biotechnological/Biological Products: Derivation and Characterization of Cell   9/21/1998
Substrates Used for Production of Biotechnological/Biological Products (I)

Q5E - Comparability of Biotechnological/Biological Products Subject to Changes in Their   6/30/2005
Manufacturing Process (I)

Q6A - Specifications: Test Procedures and Acceptance Criteria for New Drug Substances and   12/29/2000
New Drug Products: Chemical Substances (I)

Q6B - Test Procedures and Acceptance Criteria for Biotechnological/Biological Products (I)   8/18/1999

Q7A - Good Manufacturing Practice for Active Pharmaceutical Ingredients (I)   9/25/2001

Q8 - Pharmaceutical Development (I)   5/22/2006

Q9 - Quality Risk Management (I)   6/2/2006

ICH - Safety   **Issued Date**

S1A - The Need for Long-Term Rodent Carcinogenicity Studies of Pharmaceuticals (I)   3/1/1996

S1B - Testing for Carcinogenicity in Pharmaceuticals (I)   2/23/1998

S1C - Dose Selection for Carcinogenicity Studies of Pharmaceuticals (I)   3/1/1995

S1C(R) - Dose Selection for Carcinogenicity Studies of Pharmaceuticals: Addendum on a Limit   12/4/1997
Dose and Related Notes (I)

S2A - Specific Aspects of Regulatory Genotoxicity Tests for Pharmaceuticals (I)   4/24/1996

S2B - Genotoxicity: Standard Battery Testing (I)   11/21/1997

S3A - Toxicokinetics: The Assessment of Systemic Exposure in Toxicity Studies (I)   3/1/1995

S3B - Pharmacokinetics: Repeated Dose Tissue Distribution Studies (I)	3/1/1995
S4A - Duration of Chronic Toxicity Testing in Animals (Rodent and Nonrodent Toxicity Testing) (I)	6/25/1999
S5A - Detection of Toxicity to Reproduction for Medicinal Products (I)	9/22/1994
S5B - Detection of Toxicity to Reproduction for Medicinal Products: Addendum on Toxicity to Male Fertility (I)	4/5/1996
S6 - Preclinical Safety Evaluation of Biotechnology-Derived Pharmaceuticals (I)	11/18/1997
S7A - Safety Pharmacology Studies for Human Pharmaceuticals (I)	7/13/2001
S7B - Nonclinical Evaluation of the Potential for Delayed Ventricular Repolarization (QT Interval Prolongation) by Human Pharmaceuticals (I)	10/20/2005
S8 - Immunotoxicity Studies for Human Pharmaceuticals (I)	4/13/2006

ICH Draft - Efficacy   Issued Date

E12A Principles for Clinical Evaluation of New Antihypertensive Drugs (I)	8/9/2000
E2B(R) - Clinical Safety Data Management: Data Elements for Transmission of Individual Case Safety Reports (I)	10/3/2005
E2D - Postapproval Safety Data Management: Definitions and Standards for Expedited Reporting (I)	9/15/2003

ICH Draft - Joint Safety/Efficacy (Multidisciplinary)   Issued Date

M5 - Data Elements and Standards for Drug Dictionaries (I)	9/6/2005
Submitting Marketing Applications According to the ICH/CTD Format; General Considerations (I)	9/5/2001

ICH Draft - Quality   Issued Date

Q4B: Regulatory Acceptance of Analytical Procedures and/or Acceptance Criteria (RAAPAC) (I)	8/8/2006
Q4B: RAAPAC - Annex 1: Residue on Ignition/Sulphated Ash General Chapter Analytical Procedures and/or Acceptance Criteria (APAC) (I)	8/8/2006

INDs   Issued Date

Content and Format of INDs for Phase 1 Studies of Drugs Including Well-Characterized, Therapeutic, Biotechnology-Derived Products (I)	10/4/2000

Industry Letters   Issued Date

A Revision in Sample Collection Under the Compliance Program Pertaining to Pre-Approval Inspections	7/15/1996
Certification Requirements for Debarred Individuals in Drug Applications	6/1/1990

Continuation of a series of letters communicating interim and informal generic drug policy and guidance. Availability of Policy and Procedure Guides, and further operational changes to the generic drug review program (I)   3/2/1998

Fifth of a series of letters providing informal notice about the Act, discussing the statutory mechanism by which ANDA applicants may make modifications in approved drugs where clinical data is required (I)   4/10/1987

Fourth of a series of letters providing informal notice to all affected parties about policy developments and interpretations regarding the Act. Three year exclusivity provisions of Title I (I)   10/31/1986

Implementation of the Drug Price Competition and Patent Term Restoration Act. Preliminary Guidance (I)   10/11/1984

Implementation Plan USP injection nomenclature (I)   10/2/1995

Instructions for Filing Supplements Under the Provisions of SUPAC-IR   4/11/1996

Seventh of a series of letters about the Act providing guidance on the "180-day exclusivity" provision of section 505(j)(4)(B)(iv) of the FD&C (I)   7/29/1988

Sixth of a series of informal notice letters about the Act discussing 3- and 5-year exclusivity provisions of sections 505(c)(3)(D) and 505(j)(4)(D) of the FD&C Act (I)   4/28/1988

Streamlining Initiatives   12/24/1996

Supplement to 10/11/84 letter about policies, procedures and implementation of the Act (Q & A format) (I)   11/16/1984

Third of a series of letters regarding the implementation of the Act (I)   5/1/1985

Year 2000 Letter from Dr. Janet Woodcock (I)   10/19/1998

Labeling   **Issued Date**

Adverse Reactions Section of Labeling for Human Prescription Drug and Biological Products; Content and Format (I)   1/24/2006

Barbiturate, Single Entity-Class Labeling   3/1/1981

Clinical Studies Section of Labeling for Human Prescription Drug and Biological Products; Content and Format (I)   1/24/2006

Content and Format for Geriatric Labeling (I)   10/5/2001

Hypoglycemic Oral Agents - Federal Register   4/1/1984

Labeling Over-the-Counter Human Drug Products; Updating Labeling In Reference Listed Drugs and Abbreviated New Drug Applications (I)   10/18/2002

Local Anesthetics - Class Labeling   9/1/1982

Labeling Draft   Issued Date

Labeling for Combined Oral Contraceptives (I)	3/5/2004
Labeling for Human Prescripstion Drug and Biological Products; Implementing the New Content and Format Requirements (I)	1/24/2006
Noncontraceptive Estrogen Drug Products for the Treatment of Vasomotor Symptoms and Vulvar and Vaginal Atrophy Symptoms — Recommended Prescribing Information for Health Care Providers and Patient Labeling (I)	11/16/2005
OTC Topical Drug Products for the Treatment of Vaginal Yeast Infections (Vulvovaginal Candidiasis) (I)	7/16/1998
Public Availability of Labeling Changes in "Changes Being Effected" Supplements (I)	9/20/2006
Referencing Discontinued Labeling for Listed Drugs in Abbreviated New Drug Applications (I)	10/26/2000
Warnings and Precautions, Contraindications, and Boxed Warning Sections of Labeling for Human Prescription Drug and Biological Products; Content and Format (I)	1/24/2006

OTC   Issued Date

Enforcement Policy on Marketing OTC Combination Products (CPG 71320.16) (I)	5/1/1984
General Guidelines for OTC Combination Products (I)	11/28/1978
Labeling OTC Human Drug Products -- Updating Labeling in ANDAs (I)	2/22/2001
Labeling OTC Human Drug Products Using a Column Format (I)	12/19/2000
Upgrading Category III Antiperspirants to Category I (43 FR 46728 - 46731) (I)	10/10/1978

OTC Draft   Issued Date

Labeling OTC Human Drug Products - Submitting Requests for Exemptions and Deferrals (I)	12/19/2000
Labeling Over-the-Counter Human Drug Products; Questions and Answers	1/13/2005
OTC Actual Use Studies	7/22/1994
OTC Nicotine Substitutes	3/1/1994
Small Business Entities on Labeling Over-the-Counter Human Drug Products (I)	12/9/2004
Time and Extent Applications (I)	2/10/2004

Pharmacology/Toxicology   **Issued Date**

Carcinogenicity Study Protocol Submissions (I)	5/23/2002
Estimating the Maximum Safe Starting Dose in Initial Clinical Trials for Therapeutics in Adult Healthy Volunteers (I)	7/22/2005
Exploratory IND Studies (I)	1/17/2006
Format and Content of the Nonclinical Pharmacology/ Toxicology Section of an Application (I)	2/1/1987
Immunotoxicology Evaluation of Investigational New Drugs (I)	11/1/2002
Nonclinical Pharmacology/Toxicology Department of Topical Drugs Intended to Prevent the Transmission of Sexually Transmitted Diseases (STD) and/or the Development of Drugs Intended to Act as Vaginal Contraceptives (I)	10/16/1996
Nonclinical Safety Evaluation of Drug or Biologic Combinations (I)	3/15/2006
Nonclinical Safety Evaluation of Pediatric Drug Products (I)	2/15/2006
Nonclinical Studies for the Safety Evaluation of Pharmaceutical Excipients	5/19/2005
Photosafety Testing (I)	5/7/2003
Recommended Approaches to Integration of Genetic Toxicology Study Results (I)	1/4/2006
Reference Guide for the Nonclinical Toxicity Studies of Antiviral Drugs Indicated for the Treatment of N/A Non-Life Threatening Disease: Evaluation of Drug Toxicity Prior to Phase I Clinical Studies (I)	2/1/1989
Single Dose Acute Toxicity Testing for Pharmaceuticals - Revised (I)	8/26/1996

Pharmacology/Toxicology Draft   **Issued Date**

Integration of Study Results to Access Concerns About Human Reproductive and Developmental Toxicities	11/13/2001
Nonclinical Evaluation of Late Radiation Toxicity of Therapeutic Radiopharmaceuticals (I)	6/20/2005
Safety Testing of Drug Metabolites	6/6/2005
Statistical Aspects of the Design, Analysis, and Interpretation of Chronic Rodent Carcinogenicity Studies of Pharmaceuticals (I)	5/8/2001

Procedural   **Issued Date**

180-Day Generic Drug Exclusivity Under the Hatch-Waxman Amendments to the Federal Food, Drug, and Cosmetic Act (I)	7/14/1998
Continuous Marketing Applications: Pilot 1--Reviewable Units for Fast Track Products Under the Prescription Drug User Fee Act of 1992 (I)	10/6/2003

Continuous Marketing Applications: Pilot 2--Scientific Feedback and Interactions During Development of Fast Track Products Under the Prescription Drug User Fee Act of 1992 (I)	10/6/2003
Court Decisions, ANDA Approvals, and 180-Day Exclusivity Under the Hatch-Waxman Amendments to the Federal Food, Drug, and Cosmetic Act	3/27/2000
Disclosure of Materials Provided to Advisory Committees in Connection with Open Advisory Committee Meetings Convened by the Center for Drug Evaluation and Research Beginning on January 1, 2000 (I)	11/30/1999
Drug Products Containing Ensulizole, Hypromellose, Meradimate, Octinoxate, and Octisalate - Labeling Enforcement Policy (I)	6/3/2003
Enforcement Policy During Implementation of Section 503A of the Federal Food, Drug, and Cosmetic Act (I)	11/23/1998
Fast Track Drug Development Programs: Designation, Development, and Application Review (I)	11/18/1998
FDA Export Certificate (I)	7/12/2004
Financial Disclosure by Clinical Investigators (I)	3/28/2001
Formal Dispute Resolution: Appeals Above the Division Level (I)	3/7/2000
Formal Meetings With Sponsors and Applicants For PDUFA Products (I)	3/7/2000
Implementation of Section 120 of the Food and Drug Administration Modernization Act of 1997- Elimination of Certain Labeling Requirements (I)	11/2/1998
Implementation of Section 126 of the FDA Modernization Act of 1997 - Elimination of Certain Labeling Requirements, (I)	7/21/1998
Information Program on Clinical Trials for Serious or Life-Threatening Diseases and Conditions (I)	3/18/2002
Levothyroxine Sodium Products - Enforcement of August 14, 2001, Compliance Date and Submission of New Applications (I)	7/13/2001
National Uniformity for Nonprescription Drugs Ingredient Labeling for OTC Drugs (I)	4/9/1998
Potassium Iodide (KI) in Radiation Emergencies - Questions and Answers (I)	12/23/2002
Potassium Iodide as a Thyroid Blocking Agent in Radiation Emergencies (I)	12/11/2001
Potassium Iodide Tablets Shelf Life Extension for Federal Agencies and State and Local Governments (I)	3/8/2004
Qualifying for Pediatric Exclusivity Under Section 505A of the Federal Food, Drug, and Cosmetic Act - Revised (I)	10/1/1999
Refusal to File (I)	7/12/1993
Repeal of Section 507 of the Federal Food, Drug, and Cosmetic Act (I)	6/15/1998
Reports on the Status of Postmarketing Studies - Implementation of Section 130 of the Food and Drug Administration Modernization Act of 1997 (I)	2/16/2006

Special Protocol Assessment (I)	5/17/2002
Standards for the Prompt Review of Efficacy Supplements, Including Priority Efficacy Supplements (I)	5/15/1998
Submitting and Reviewing Complete Responses to Clinical Holds (Revised) (I)	10/26/2000
The Leveraging Handbook; an Agency Resource for Effective Collaborations - Guidance for FDA Staff (I)	6/19/2003
Useful Written Consumer Medication Information (CMI) (I)	7/18/2006
Women and Minorities Guidance Requirements	7/20/1998

Procedural Draft   Issued Date

Applications Covered by Section 505(b)(2) (I)	12/8/1999
Centralized IRB Review Proceedings in Multicenter Clinical Trials	3/23/2005
Clinical Trial Sponsors On the Establishment and Operation of Clinical Trial Data Monitoring Committees (I)	11/15/2001
Content and Format of New Drug Applications and Abbreviated New Drug Applications for Certain Positron Emission Tomography Drug Products (I)	3/10/2000
Disclosing Information Provided to Advisory Committees in Connection with Open Advisory Committee Meetings Related to the Testing or Approval of New Drugs and Convened by CDER, Beginning January 1, 2000 (I)	12/22/1999
Disclosure of Conflicts of Interest for Special Government Employees Participating in FDA Product Specific Advisory Committees	2/14/2002
Emergency Use Authroization of Medical Products (I)	7/5/2005
Fixed Dose Combination and Co-Packaged Drug Products for Treatment of HIV (I)	5/19/2004
Forms for Registration of Producers of Drugs and Listing of Drugs in Commercial Distribution (I)	5/15/2001
Good Review Management Principles for PDUFA Products (I)	7/28/2003
How to Comply with the Pediatric Research Equity Act (I)	9/7/2005
Independent Consultants for Biotechnology Clinical Trial Protocols (I)	5/7/2003
Information Program on Clinical Trials for Serious or Life-Threatening Diseases and Conditions (I)	1/27/2004
Pharmacogenomic Data Submissions (I)	11/4/2003
Postmarketing Safety Reporting for Human Drug and Biological Products Including Vaccines (I)	3/12/2001

Submitting Debarment Certification Statements (I)   10/2/1998

The Use of Clinical Holds Following Clinical Investigator Misconduct (I)   8/27/2002

Small Entity Compliance Guides   Issued Date

Sterility Requirements for Aqueous-Based Drug Products for Oral Inhalation (I)   11/7/2001

User Fee   Issued Date

Applicability of User Fees to (1) Applications Withdrawn Before Filing, or (2) Applications the   7/12/1993
Agency Has Refused to File and That Are Resubmitted or Filed Over Protest (Attachment F)

Application, Product, and Establishment Fees: Common Issues and Their Resolution (Revised)   12/16/1994
(Attachment D) (I)

Classifying Resubmissions in Response to Action Letters (I)   5/14/1998

Fees-Exceed-the-Costs Waivers Under the Prescription Drug User Fee Act (I)   8/25/1999

Information Request and Discipline Review Letters Under the Prescription Drug User Fee Act (I)   11/21/2001

Submitting Separate Marketing Applications and Clinical Data for Purposes of Assessing User   1/3/2005
Fees (I)

User Fee Draft   Issued Date

Document for Waivers of and Reductions in User Fees (Attachment G) (I)   7/16/1993

User Fee Waivers for Fixed Dose Combination Products and Co-Packaged Human   4/18/2005
Immunodeficiency Virus Drugs for the President's Emergency Plan for Acquired
Immunodeficiency Syndrome Relief (I)

B | ICH Guidelines

Home page for Quality Guideline Index and Links is available at http://www.nihs.go.jp/dig/ich/qindex-e.html.

Guideline	Date	Address
Q1A (R2) Stability Testing of New Drug Substances and Products	2/6/2003	http://www.nihs.go.jp/dig/ich/quality/q1a/q1astep4.pdf
Q1B Stability Testing: Photostability Testing of New Drug Substances and Products	11/6/1996	http://www.nihs.go.jp/dig/ich/quality/q1b/q1bstep4-e.pdf
Q1C Stability Testing of New Dosage Forms	11/6/1996	http://www.nihs.go.jp/dig/ich/quality/q1c/q1cstep4.pdf
Q1D Bracketing and Matrixing Designs for Stability Testing of Drug Substances and Drug Products	2/7/2002	http://www.nihs.go.jp/dig/ich/quality/q1d/q1dstep4.pdf
Q1E Evaluation of Stability Data	2/6/2003	http://www.nihs.go.jp/dig/ich/quality/q1e/q1e_gl_E.pdf
Q2A Text on Validation of Analytical Procedures	10/27/1994	http://www.nihs.go.jp/dig/ich/quality/q2a/q2astep4.pdf
Q2B Validation of Analytical Procedures: Methodology	11/6/1996	http://www.nihs.go.jp/dig/ich/quality/q2b/q2bstep4.pdf
Q3A(R) Impurities in New Drug Substances	2/7/2002	http://www.nihs.go.jp/dig/ich/quality/q3a/q31216e.pdf
Q3B Impurities in New Drug Products	10/7/1999	http://www.nihs.go.jp/dig/ich/quality/q3br/q3brstep2.pdf
Q3B(R) Questions and Answers on Attachment 2 of Q3B(R)	7/18/03	http://www.nihs.go.jp/dig/ich/quality/q3br/030718q3br-e.pdf
Q3C Impurities: Guideline for Residual Solvents	7/17/1997	http://www.nihs.go.jp/dig/ich/quality/q3c/q3c-e.pdf
Q3C(M) Impurities: Residual Solvents (Maintenance) PDE for NMP	9/12/2002	http://www.nihs.go.jp/dig/ich/quality/q3c(m)/q3cm0301-nmp.pdf
Q3C(M) Impurities: Residual Solvents (Maintenance) PDE for Tetrahydrofuran	9/12/2002	http://www.nihs.go.jp/dig/ich/quality/q3c(m)/qcm0301.thf.pdf
Q5A Viral Safety Evaluation of Biotechnological Products Derived from Cell Linings of Human or Animal Origin	3/5/1997	http://www.nihs.go.jp/dig/ich/quality/q5a/q5astep4.pdf
Q5B Quality of Biotechnological Products: Analysis of the Expression Construct in Cells Used for Production of R-DNA Derived Protein Products	11/30/1995	http://www.nihs.go.jp/dig/ich/quality/q5b/q5b-e.pdf

(Continued)

(*Continued*)

Guideline	Date	Address
Q5C Quality of Biotechnological Products: Stability testing of Biotechnological/Biological Products	11/30/1995	http://www.nihs.go.jp/dig/ich/quality/q5c/q5cstep4.pdf
Q5D Derivation and Characterization of Cell Substrates Used for Production of Biotechnological/Biological Products	7/16/97	http://www.nihs.go.jp/dig/ich/quality/q5d/q5dstep4-e.pdf
Q5E Comparability of Biotechnological/Biological Products Subject to Changes in Their Manufacturing Process	11/18/2004	http://www.nihs.go.jp/dig/ich/quality/q5e/q5e_041118_e.pdf
Q6A Specifications: Test Procedures and Acceptance Criteria for New Drug Substances and New Drug Products; Chemical Substances	10/6/1999	http://www.nihs.go.jp/dig/ich/quality/q6a/q6a-e.pdf
Q6B Specifications: Test procedures and Acceptance Criteria for Biotechnological/ Biological Products	3/10/93	http://www.nihs.go.jp/dig/ich/quality/q6b/q16b-e.pdf
Q7A Good manufacturing Practice Guide for Active Pharmaceutical Ingredients	11/10/2000	http://www.nihs.go.jp/dig/ich/quality/q7a/q7astep4e.pdf
Q8 Pharmaceutical Development	11/18/2004	http://www.nihs.go.jp/dig/ich/quality/q8/q8-041118_e.pdf
Q9 Quality Risk Management	3/22/2005	http://www.nihs.go.jp/dig/ich/quality/q9/q9-050322_e.pdf

C | Food and Drug Administration (FDA)/Office Regulatory Affairs (ORA) Compliance Policy Guides

The Food and Drug Administration (FDA) Compliance Policy Guides (CPGs) manual provides a convenient and organized system for statements of FDA compliance policy, including those statements which that contain regulatory action guidance information. The CPG Manual is the repository for all agency compliance policy that has been agreed to by the center(s) and the Associate Commissioner for Regulatory Affairs. Examples of sources from which CPGs are prepared include: (i) statements or correspondence by headquarters offices or centers reflecting new policy or changes in compliance policy including Office of the Commissioner memoranda, center memoranda and other informational issuances, agency correspondence with trade groups and regulated industries, and advisory opinions; ii) precedent court decisions; iii) multi-center agreements regarding jurisdiction over FDA regulated products; iv) preambles to proposed or final regulations or other Federal Register documents; and v) individual regulatory actions.

Compliance Policy Guides (CPGs) explain the FDA policy on regulatory issues related to the FDA laws or regulations. These include Current Good Manufacturing Practice (CGMP) regulations and application commitments. They advise the field inspection and compliance staffs as to the the Agency's standards and procedures to be applied when determining industry compliance. CPGs may derive from a request for an advisory opinion, from a petition from outside the the Agency, or from a perceived need for a policy clarification by FDA personnel. These are available at: http://www.fda.gov/ora/compliance_ref/cpg/default.htm.

The following are the general and drug drug-related chapters:

FDA/ORA COMPLIANCE POLICY GUIDES

Chapter 1:

Subchapter 100—General
Subchapter 110—Exports/Imports
Subchapter 120—Fraud
Subchapter 130—Inspections
Subchapter 140—Labeling
Subchapter 150—Laboratory/Analytic
Subchapter 160—Regulatory
Subchapter 170—Specific Problems Non-Food, Drug, or Cosmetic Related

Chapter 4: Human Drugs (Updated: 06-20-2006)

400-General
410-Bulk Drugs
20-Compendial/Test Requirements
425-Computerized Drug Processing
430-Labeling and Repackaging
435-Medical Gases
440-448-New Drugs
450-457-OTC
460-Pharmacy Issues
470-Post Approval Issues
480-Stability/Expiration
490-Validation

Subchapter 400—General
    400.100 Drugs, Human-Failure to Register CPG 7132.07
    400.200 Consistent Application of CGMP Determinations CPG 7132.12
    400.210 Radiofrequency Identification Feasibility Studies and Pilot Programs
    400.325 Candy "Pills" Representation as Drug CPG 7132.04
    400.335 Fructose-Containing Drugs CPG 7123b.02
    400.400 Conditions Under Which Homeopathic Drugs May be Marketed CPG 7132.15
    400.500 Identical or Similar Product Names CPG 7132b.14
    400.600 Drugs-Declaration of Quantity of Active CPG 7132.03
    400.700 Drug Product Entries in Periodic Publications CPG 7132b.17
    400.800 Collection and Charitable Distribution of Drugs CPG 7132.08
    400.900 Class I Recalls of Prescription Drugs CPG 7132.01

Subchapter 410—Bulk Drugs
    410.100 Finished Dosage Form Drug Products in Bulk Containers-Applications of Current Good Manufacturing
      Practice Regulations CPG 7132a.06

Subchapter 420—Compendial/Test Requirements
    420.100 Adulteration of Drugs Under Sections 501(b) and 501(c) of the Act. Direct Reference Seizure
      Authority for Adulterated Drugs Under Section 501(b) CPG 7132a.03
    420.200 Compendium Revisions and Deletions CPG 7132.02 420.300 Changes in Compendial Specifications
      and NDA CPG 7132c.04
    420.400 Performance of Tests for Compendial Requirements on Compendial Products CPG 7132.05
    420.500 Interference with Compendial Tests CPG 7132a.01

Subchapter 425—Computerized Drug Processing
    425.100 Computerized Drug Processing; CGMP Applicability to Hardware and Software CPG 7132a.17
    425.200 Computerized Drug Processing; Vendor Responsibility CPG 7132a.12
    425.300 Computerized Drug Processing; Source Code for Process Control Application Programs CPG
      7132a.15
    425.400 Computerized Drug Processing; Input/Output Checking CPG 7132a.07
    425.500 Computerized Drug Processing; Identification of "Persons" on Batch Production and Control Records
      CPG 7132a.08

Subchapter 430—Labeling and Repackaging
    430.100 Unit Dose Labeling for Solid and Liquid Oral Dosage Forms CPG 7132b.10
    430.200 Repacking of Drug Products-Testing/Examination under CGMPs CPG 7132.13
    430.300 Labeling Shipping Containers of Drugs CPG 7132b.13
    430.400 Urinary Preparations-Misbranding-Lack of Rx Legend and Claims CPG 7132b.04

Subchapter 435—Medical Gases
    435.100 Compressed Medical Gases-Warning Letters for Specific Violations Covering Liquid and Gaseous
      Oxygen CPG 7132a.16

Subchapter 440—448 New Drugs
    440.100 Marketed New Drugs Without Approved NDAs CPG 7132c.02 06/20/2006 This CPG supersedes
      section 440.100, Marketed New Drugs Without Approved NDAs or ANDAs (CPG 7132c.02)
    442.100 New Drugs-Export CPG 7132c.01

Chapter 6: Veterinary Medicine

Subchapter 600 Veterinary Drugs

Index

USP *See* United States Pharmacopeia
USV Pharmaceutical Corp. v. Weinberger, 330

validation, process *See* process validation
vendor qualification, 78–80

warehousing procedures
    definitions, 145
    special storage conditions, 146
warning letters, 268
Washington Legal Foundation v. Henney, 136
Weinberger v. Hynson,Westcott, and Dunning,
        Inc., 270, 330
WHO GMPs
    general, 337–338
    production and quality control, 341
    quality management, 338–341
    supporting and supplementary
        guidelines, 342
Woods and Rhode Inc. v. Alaska, 287
Worldwide GMPs
    Canada
        equipment, 343
        finished product testing, 345
        manufacturing control, 344
        packaging material testing, 345
        personnel, 343

[Worldwide GMPs, Canada *contd.*]
    premises, 343
    quality assurance, 343
    quality control department, 345
    quality management, 342–343
    raw material testing, 343–344
    records, 345
    samples, 345
    stability, 345
    sterile products, 345
Europe
    complaints and product
        recall, 336
    contract manufacture and
        analysis, 336
    documentation, 335
    personnel, 334–345
    premises and equipment, 335
    production, 335–336
    quality management, 334
    self-inspection, 336–337
WHO
    general, 337–338
    production and quality
        control, 341
    quality management, 338–341
    supporting and supplementary
        guidelines, 342